Algorithmic Graph Theory
and Perfect Graphs

Second Edition

ANNALS OF DISCRETE MATHEMATICS   5

Series Editor: Peter L. HAMMER
Rutgers University, Piscataway, NJ, U.S.A

Algorithmic Graph Theory and Perfect Graphs

Second Edition

Martin Charles Golumbic

Caesarea Rothschild Institute
University of Haifa
Haifa, Israel

2004

ELSEVIER

Amsterdam – Boston – Heidelberg – London – New York – Oxford
Paris – San Diego – San Francisco – Singapore – Sydney – Tokyo

ELSEVIER B.V.
Sara Burgerhartstraat 25
P.O. Box 211, 1000 AE
Amsterdam, The Netherlands

ELSEVIER Inc.
525 B Street, Suite 1900
San Diego, CA 92101-4495
USA

ELSEVIER Ltd
The Boulevard, Langford Lane
Kidlington, Oxford OX5 1GB
UK

ELSEVIER Ltd
84 Theobalds Road
London WC1X 8RR
UK

First edition 1980 (Academic Press, ISBN 0-12-289260-7)
Second edition 2004

Library of Congress Cataloging in Publication Data
A catalog record is available from the Library of Congress.

British Library Cataloguing in Publication Data
A catalogue record is available from the British Library.

ISBN 10: 0-444-51530-5
ISBN 13: 9780444515308

Transferred to digital printing in 2007

Dedicated to my parents

לכבוד הורי היקרים
אברם בן יצחק גולומביק
חיענע בת מנדיל הכהן

Contents

Appendix

Foreword 2004:
the *Annals* edition

The publication of this new edition of *Algorithmic Graph Theory and Perfect Graphs* marks twenty three years since its first appearance. My original motivation for writing the book was to collect and unify the topic to act as a spring board for researchers, and especially graduate students, to pursue new directions of investigation. The ensuing years have been an amazingly fruitful period of research in this area. To my great satisfaction, the number of relevant journal articles in the literature has grown tenfold. I can hardly express my admiration to all these authors for creating a success story for algorithmic graph theory far beyond my own imagination.

The world of perfect graphs has grown to include over 200 special graph classes. The Venn diagrams that I used to show some of the inclusions between classes in the First Generation, for example Figure 9.9 (on page 212), have yielded to Hasse diagrams for the Second Generation, like the one from Golumbic and Trenk [2004] reprinted in Figure 13.3 at the end of this edition.

Perhaps the most important new development in the theory of perfect graphs is the recent proof of the Strong Perfect Graph Conjecture by Chudnovsky, Robertson, Seymour and Thomas, announced in May 2002. News of this was immediately passed on to Claude Berge, who sadly passed away on June 30, 2002.

On the algorithmic side, many of the problems which were open in 1980 have subsequently been settled, and algorithms on new classes of perfect graphs have been studied. For example, tolerance graphs generalize both interval graphs and permutation graphs, and coloring tolerance graphs in polynomial time is important in solving scheduling problems where a measure of flexibility or tolerance is allowed for sharing or relinquishing resources when total exclusivity prevents a solution.

At the end of this new edition, I have added a short chapter called

Epilogue 2004 in which I survey a few of my favorite results and research directions from the Second Generation. Its intension is to whet the appetite.

Six books have appeared recently which cover advanced research in this area. They have thankfully relieved me from a pressing need to write my own encyclopedia sequel. They are the following, and are a must for any graph theory library.

- A. Brandstädt, V.B. Le and J.P. Spinrad, *"Graph Classes: A Survey"*, SIAM, Philadelphia [1999], is an extensive and invaluable compendium of the current status of complexity and mathematical results on hundreds on families of graphs. It is comprehensive with respect to definitions and theorems, citing over 1100 references.
- P.C. Fishburn, *"Interval Orders and Interval Graphs: A Study of Partially Ordered Sets"*, John Wiley & Sons, New York [1985], gives a comprehensive look at the research on this class of ordered sets.
- M.C. Golumbic and A.N. Trenk, *"Tolerance Graphs"*, Cambridge University Press [2004], is the youngest addition to the perfect graph bookshelf. It contains the first thorough study of tolerance graphs and tolerance orders, and includes proofs of the major results which have not appeared before in books.
- N.V.R. Mahadev and U.N. Peled, *"Threshold Graphs and Related Topics"*, North-Holland [1995], is a thorough and extensive treatment of all research done in the past years on threshold graphs (chapter 10 of my book), threshold dimension and orders, and a dozen new concepts which have emerged.
- T.A. McKee and F.R. McMorris, *"Topics in Intersection Graph Theory"*, SIAM, Philadelphia [1999], is a focused monograph on structural properties, presenting definitions, major theorems with proofs and many applications.
- W.T. Trotter, *"Combinatorics and Partially Ordered Sets"*, Johns Hopkins, Baltimore [1992], is the book to which I referred at the bottom of page 136. It covers new directions of investigation and goes far beyond just dimension problems on ordered sets.

Algorithmic Graph Theory and Perfect Graphs has now become the classic introduction to the field. It continues to convey the message that intersection graph models are a necessary and important tool for solving real-world problems. Solutions to the algorithmic problems on these special graph classes are continually integrated into systems for a large variety of application areas, from VLSI circuit design to scheduling, from resource allocation to physical mapping of DNA, from temporal reasoning in artificial intelligence to pavement deterioration analysis. On the mathematical side, perfect graph classes have provided rich soil for deep theoretical results. In short, it remains a stepping stone from which the reader may embark on one of many fascinating research trails.

Martin Charles Golumbic
Haifa, Israel

Foreword

Research in graph theory and its applications has increased considerably in recent years. Typically, the elaboration of new theoretical structures has motivated a search for new algorithms compatible with those structures. Rather than the arduous and systematic study of every new concept definable with a graph, the main task for the mathematician is to eliminate the often arbitrary and cumbersome definitions, keeping only the "deep" mathematical problems.

Of course, the deep problems may well be elusive; indeed, there have been many definitions (from Dieudonné, among others) of what a deep problem is. In graph theory, it should relate to a variety of other combinatorial structures and must therefore be connected with many difficult practical problems. Among these will be problems that classical algebra is not able to solve completely or that the computer scientist would not attack by himself.

This book, by Martin Golumbic, is intended as an introduction to graph theory through just these practical problems, nearly all of them related to the structure of permutation graphs, interval graphs, circle graphs, threshold graphs, perfect graphs, and others.

The reader will not find motivations drawn from number theory, as is usual for most of the extremal graph problems, or from such refinements of old riddles as the four-color problem and the Hamiltonian tour. Instead, Golumbic has selected practical problems that occur in operations research, scheduling, econometrics, and even genetics or ecology.

The author's point of view has also enjoyed increasing favor in the area of complexity analysis. Each time a new structure appears, the author immediately devotes some effort to a description of efficient algorithms, if any are known to exist, and to a determination of whether a proposed algorithm is able to solve the problem within a reasonable amount of time.

Certainly a wealth of literature on graph theory has developed by now. Yet it is clear that this book brings a new point of view and deserves a special place in the literature.

CLAUDE BERGE

Preface

The notion of a "perfect" graph was introduced by Claude Berge at the birth of the 1960s. Since that time many classes of graphs, interesting in their own right, have been shown to be perfect. Research, in the meantime, has proceeded along two lines. The first line of investigation has included the proof of the perfect graph theorem (Theorem 3.3), attempts at proving the strong perfect graph conjecture, studies of critically imperfect graphs, and other aspects of perfect graphs. The second line of approach has been to discover mathematical and algorithmic properties of special classes of perfect graphs: comparability graphs, triangulated graphs, and interval graphs, to name just a few. Many of these graphs arise quite naturally in real-world applications. For example, uses include optimization of computer storage, analysis of genetic structure, synchronization of parallel processes, and certain scheduling problems.

Recently it appeared to me that the time was ripe to assemble and organize the many results on perfect graphs that are scattered throughout the literature, some of which are difficult to locate. A serious attempt has been made to coordinate the mélange of some 200 papers referenced here in a manner that would make the subject more accessible to those interested in algorithmic and algebraic graph theory. I have tried to include most of the important results that are currently known. In addition, a few new results and new proofs of old results appear throughout the text. In particular, Chapter 9, on superperfect graphs, contains results due to Alan J. Hoffman, Ellis Johnson, Larry J. Stockmeyer, and myself that are appearing in print for the first time.

The emphasis of any book naturally reflects the bias of the author. As a mathematician and computer scientist, I am doubly biased. First, I have tried to present a rigorous and coherent theory. Proofs are constructive and are streamlined as much as possible. The notation has been chosen to facilitate these matters. Second, I have directed much attention to the algorithmic aspects of every problem.

Algorithms are expressed in a manner that will make their adaptation to a particular programming language relatively easy. The complexity of every algorithm is analyzed so that some measure of its efficiency can be determined.

These two approaches enhance one another very well. By exploiting the mathematical properties satisfied a priori by a structure, one is often able to reduce the time or space complexity required to solve a problem. Conversely, the algorithmic approach often leads to startling theoretical results. To illustrate this point, consider the fact that certain NP-complete problems become tractable when restricted to certain classes of perfect graphs, whereas the algorithm for recognizing comparability graphs gives rise to a matroid associated with the graph.

A glance at the table of contents will provide a rough outline of the topics to be discussed. The first two chapters are introductory in the sense that they provide the foundations, respectively, of the graph theoretic notions and the algorithmic design and analysis techniques that will be used in the remaining chapters. The reader may wish to read these two chapters quickly and refer to them as needed. The chapters are structured in such a way that the book will be suitable as a textbook in a course on algorithmic combinatorics, graph theory, or perfect graphs. In addition, the book will be very useful for applied mathematicians and computer scientists at the research level. Many applications of the theoretical and computational aspects of the subject are described throughout the text. At the end of each chapter there are numerous exercises to test the reader's understanding and to introduce further results. An extensive bibliography follows each chapter, and, when possible, the *Mathematical Reviews* number is included for further reference.

The topics covered in this book have been chosen to fill a vacuum in the literature, and their interrelation and importance will become evident as early as Section 1.3. Since the intersection of this volume with the traditional material covered by most graph theory books has been designed to be small, it is highly recommended that the serious student augment his studies with one of these excellent textbooks. A one-year course with two concurrent texts is suggested.

MARTIN CHARLES GOLUMBIC

Acknowledgments

I would like to express my gratitude to the many friends and colleagues who have assisted me in this project. Special thanks are due to Claude Berge for the kind words that introduce this volume. I am happy to acknowledge the help received from Mark Buckingham, particularly in Chapters 3 and 11. He is the coauthor of Sections 3.3–3.5. The suggestions and critical comments of my "trio" of students, Clyde Kruskal, Larry Rudolph, and Elia Weixelbaum, led to numerous improvements in the exposition. Over the past three years I have been fortunate to receive support from the Courant Institute of Mathematical Sciences, the National Science Foundation, the Weizmann Institute of Science, and l'Université de Paris VI.

I would also like to express my appreciation to Alan J. Hoffman for many interesting discussions and for his help with the material in Chapter 9. My thanks go to Uri Peled, Fred S. Roberts, Allan Gottlieb, W. T. Trotter, Peter L. Hammer, and László Lovász for their comments, as well as to Lisa Sabbia Walsh, Daniel Gruen, and Joseph Miller for their assistance. I am also indebted to my teacher, Samuel Eilenberg, for the guidance, insight, and kindness shown me during my days at Columbia University.

But the greatest and most crucial help has come from my wife Lynn. Although not a mathematician, she managed to unconfound much of this mathematician's gibberish. She also "axed" some of my worst (best) jokes, much to my dismay. More importantly, she has been the rock on which I have always relied for encouragement and inspiration, during our travels and at home, in the course of the research and writing of this book. As it is written in Proverbs:

פיה פתחה בחכמה, ותורת־חסד על־לשונה.
רבות בנות עשו חיל, ואת עלית על־כלנה.

List of Symbols

6	$\alpha(G)$	The *stability number* of G.
7	$\chi(G)$	The *chromatic number* of G.
113	$t(G)$	The number of *transitive orientations* of G.
126	$r(G)$	The *rank* of the Γ^*-matroid of G.
220	$\theta(G)$	The *threshold dimension* of G.
203	$\chi(G;w)$	The *interval chromatic number* of a weighted graph $(G;w)$.
206	$\omega(G;w)$	The maximum *weighted clique number* of $(G;w)$.
9	K_n	The *complete graph* on n vertices.
9	C_n	The *chordless cycle* on n vertices.
9	P_n	The *chordless path graph* on n vertices.
9	$K_{m,n}$	The *complete bipartite graph* on $m+n$ vertices partitioned into an m-stable set and an n-stable set.
9	$K_{1,n}$	The *star graph* on $n+1$ vertices.
9	mK_n	m disjoint copies of K_n.
47	$G_1 \times G_2$	The *Cartesian product* of graphs G_1 and G_2.
77	$G \cdot H$	The *normal product* of graphs G and H.
109	$H_0[H_1,\ldots,H_n]$	The *composition* of graphs.
95	$\mathcal{G}$	The class of undirected graphs satisfying the property that every odd cycle of length greater than or equal to 5 has at least two chords.
105	Γ	The *forcing* relation on edges.
106	Γ^*	The reflexive, transitive closure of Γ.
106	$\mathcal{I}(G)$	The collection of *implication classes* of G.
106	$\hat{\mathcal{I}}(G)$	The collection of *color classes* of G.
135	$\mathcal{L}(P)$	The collection of *linear extensions* of a partial order P.
135	$\dim(P)$	The *dimension* of a partial order P.
157	$G[\pi]$	The *permutation graph* of π.
235	$H[\pi]$	The *stack sorting graph* of π.
157	π^{-1}	The *inverse* of the permutation π.
158	π^ρ	The *reversal* of the permutation π.
228	$ш$	The *shuffle product*.
236	$\mathcal{H}$	The class of *stack sorting graphs*.
23	$O(f(m))$	Computational complexity *on the order of* $f(m)$.
26	P	The class of *deterministic polynomial-time* problems.
27	NP	The class of *nondeterministic polynomial-time* problems.
27	$\Pi_1 \leqslant \Pi_2$	Problem Π_1 is *polynomially transformable* to problem Π_2.
32	Λ	The *null* or *undefined* symbol in an algorithm.
176	$T \equiv T'$	The *PQ*-trees T and T' are *equivalent*.
177	$\Pi(\mathcal{I})$	The collection of all permutations π of X such that the members of each subset $I \in \mathcal{I}$ occur consecutively in π where $\mathcal{I} \subseteq \mathcal{P}(X)$.
53	$G \circ \mathbf{h}$	The graph G *multiplied* by the vector $\mathbf{h}$.
62	$\mathbb{R}^n$	The *n-dimensional vector space* over the *real numbers*.
62	$P(\mathbf{A})$	The *polyhedron* of matrix $\mathbf{A}$.
62	$P_I(\mathbf{A})$	The *integral polyhedron* of matrix $\mathbf{A}$.
59	$\mathbf{1}$	The vector of all ones.
62	$\mathbf{0}$	The vector of all zeros.
60	$\mathbf{J}$	The matrix of all ones.
60	$\mathbf{I}$	The identity matrix.
256	$G(\mathbf{M})$	The graph of matrix $\mathbf{M}$.
256	$B(\mathbf{M})$	The bipartite graph of matrix $\mathbf{M}$.

Corrections and Errata to:
Algorithmic Graph Theory and Perfect Graphs, the original 1980 edition

We apologize to Prof. George Lueker for misspelling his family name throughout the text. Hence all occurrences "Leuker" should be "Lueker".

Page 18: The graph in Figure 1.17 is a circular-arc graph.

Page 48: Exercise 21 is false.

Page 49: Garey and Johnson [1978]: add "MR80g:68056"

Page 78: Bland, et al. [1979]: add "MR80g:05034"
Chvátal, et al. [1979]: add "MR81b:05044"
de Werra [1978]: add "MR81a:05052"
Greenwell [1978]: add "MR80d:05044"

Page 79: Olaru [1977]: add "MR58#5411"

Page 80: Parthasarathy and Ravindra [1979]: add "MR80m:05045"
Pretzel [1979]: add "MR80d:06003"
Tucker [1979]: add "MR81c:05041"
Wagon [1978]: add "MR80i:05078"

Page 85: Figure 4.3: The edge (b, e) is missing.

Page 102: Exercise 24: The claim in the first sentence is false. For example, it can use as many as 7 colors on the graph G_1, in Figure 4.1. A different technique can be used to obtain a linear time coloring algorithm for triangulated graphs, which is due to Martin Farber.

Line 21: change "$Adj(w)$" to "$Adj(u)$"

Gavril [1978]: add "MR81g:05094"

Page 104: Wagon [1978]: add "MR80i:05078"

Page 138: The second footnote can be updated since M. Yannakakis has now proved that the complexity of determining if a poset has dimension 3 is NP-complete.

Page 145: Pretzel [1979]: add "MR80d:06003"

Page 146: Gysin [1977]: add "MR58#5393"

Page 147: Rabinovitch [1978b]: add "MR58#5424"

Page 156: Burkard and Hammer [1977]: change to the following:
[1980] A note on Hamiltonian split graphs, *J. Combin. Theory B* **28**, 245–248. MR81e:05095.

A necessary condition for the existence of a Hamiltonian cycle in split graphs is proved.

Erdos and Gallai [1960]: change "272" to "274"

Foldes and Hammer [1978]: add "MR80c:05111"

Hammer, Ibaraki, and Simeone [1978]: change to the following:
[1978] Degree sequences of threshold graphs, *Proc. 9th Southeastern Conf. on Combinatorics, Graph Theory and Computing, Congressus Numeratium* **21**, Utilitas Math., Winnipeg, Man., 329–355. MR80j:05088.

Page 163: There should be edges between 3–4 and 6–7 (corrected in this edition).

Page 179: Figure 8.7: The second tree on the right should have its rightmost leaf "F" rather than "E". The leaves should read from left to right as follows: B C E A D F

Page 190: line 6: change "will appear in Tucker [1979]" to: "appears in Tucker [1980]"

Page 197: line 26: change "Griggs and West [1979]" to: "Griggs and West [1980]"
Abbott and Katchalski [1979]: add "MR80b:05038"

Page 198: Booth and Lueker [1976]: add "MR55#6932"

Page 199: Griggs [1979]: add "MR81h:05083b"
Griggs and West: change to the following:
[1980] Extremal values of the interval number of a graph, *SIAM J. Algebraic Discrete Methods* 1, 1–7. MR81h:05083a.

Page 201: Roberts [1979a]: add "MR81e:05120"
Roberts [1979b]: add "MR81e:05071"
Trotter and Harary [1979]: add "MR81c:05055"

Page 202: Tucker [1979]: change to the following:
[1980] An efficient test for circular-arc graphs, *SIAM J. Comput.* 9, 1–24. MR81a:68074.

Page 203: line 17: add the following:
Vertices x of weight $w(x) = 0$ are the mapped into the empty interval.

Page 206: $\omega(T; w)$ should be $\omega(G; w)$

Page 212: Figure 9.9:
  (1) The nonsuperperfect, interval graph with the chordless 5-cycle should have two chords connecting the top two vertices to the bottom vertex. It will then be the same as the "bull's head" graph on page 16, (corrected in this edition).
  (2) The noncomparability, nontriangulated comparability graph on 7 vertices has too many edges. The two vertical edges should be removed, (corrected in this edition).
  (3) The nonsuperperfect, interval graph which has 5 triangles, is, in fact, superperfect; it should be moved into the superperfect, non-compatability, interval area of the figure. See also Section 13.9 of the Epilogue to this edition.

Page 234: Golumbic [1978a]: add "MR81e:68080"
Hammer, Ibaraki, and Simeone [1978]: change to the following:
[1978] Degree sequences of threshold graphs, *Proc. 9th Southeastern Conf. on Combinatorics, Graph Theory and Computing, Congressus Numeratium* **21**, Utilitas Math., Winnipeg, Man., 329–355. MR80j:05088.

Page 253: Gavril [1973]: change "minimum independent" to "maximum independent"

Page 267: Golumbic [1979]: add "MR81c:05077"
           Golumbic and Goss [1978]: add "MR80d:05037"
           Ohtsuki, Cheung, and Fujisawa [1976]: add "MR58#5379"

Page 280: Lueker, G. S.: change "25" to "24"
           Put name into alphabetical order.

CHAPTER 1

Graph Theoretic Foundations

1. Basic Definitions and Notations

Functions and Relations

Let X and Y be sets. A *function* (or *mapping*) f from X to Y, denoted

$$f: X \rightarrow Y,$$

is a rule which associates to each element x of X a corresponding element y of Y. It is usual to call y the *image* of x under f and denote it by $y = f(x)$. We call f an *injective* or *one-to-one* function if no pair of distinct members of X has the same image under f, that is,

$$x \neq x' \Rightarrow f(x) \neq f(x') \qquad (x, x' \in X),$$

or equivalently,

$$f(x) = f(x') \Rightarrow x = x' \qquad (x, x' \in X).$$

The function f is called *surjective* or *onto* if each y in Y is the image of some x in X, that is,

$$(\forall y \in Y)(\exists x \in X) \qquad \text{such that} \quad y = f(x).$$

A function which is both injective and surjective is called a *bijection*. A *permutation* is simply a bijection from a set to itself.

Following the usual notation of mathematics, $x \in X$ indicates that x is a member of the set X and $A \subseteq X$ means that A is a (not necessarily proper) subset of X. The *cardinality* or *size* of X is denoted by $|X|$. For subsets A and B of X, the notation $A \cap B$ and $A \cup B$ are the usual set intersection and set union operations. When A and B are disjoint subsets, we often write their union with a plus sign. That is,

$$C = A + B \qquad \text{indicates} \quad A \cap B = \varnothing \quad \text{and} \quad C = A \cup B,$$

where $\varnothing$ is the empty set. Throughout this book we will deal exclusively with finite sets. A collection $\{X_i\}_{i \in I}$ of subsets of a set X is said to *cover* X if their union equals X. The collection is called a *partition* of X if the subsets are pairwise disjoint and the collection covers X.

Let $\mathcal{P}(X)$ denote the *power set* of a set X, i.e., the collection of all subsets of X. It is well known that $|\mathcal{P}(X)| = 2^{|X|}$. A *binary relation* on X is defined to be a function

$$R: X \rightarrow \mathcal{P}(X)$$

from X to the power set of X. For each $x \in X$, the image of x under R is a subset $R(x) \subseteq X$ called the set of *relatives* of x. It is customary to represent the relation R as a collection of *ordered* pairs $\mathcal{R} \subseteq X \times X$, where

$$(x, x') \in \mathcal{R} \qquad \text{if and only if} \quad x' \in R(x).$$

In this case we say that x' is *related* to x. Notice that this does *not* necessarily imply that x is related to x'. (Perhaps one should read "will inherit from" instead of "is related to," as in the case of a poor nephew with ten children and his rich widowed childless aunt.)

A binary relation R on X may satisfy one or more of the following properties:

symmetric property

$$x' \in R(x) \Rightarrow x \in R(x') \qquad (x, x' \in X),$$

antisymmetric property

$$x' \in R(x) \Rightarrow x \notin R(x') \qquad (x, x' \in X),$$

reflexive property

$$x \in R(x) \qquad (x \in X),$$

irreflexive property

$$x \notin R(x) \qquad (x \in X),$$

transitive property

$$z \in R(y), \; y \in R(x) \Rightarrow z \in R(x) \qquad (x, y, z \in X).$$

Such a relation is said to be an *equivalence* if it is reflexive, symmetric, and transitive. A binary relation is called a *strict partial order* if it is irreflexive and transitive. It is a simple exercise to show that a strict partial order will also be antisymmetric.

Graphs

Let us formally define the notion of a graph. A *graph** G consists of a finite set V and an irreflexive binary relation on V. We call V the set of *vertices*. The binary relation may be represented either as a collection E of *ordered* pairs or as a function from V to its power set,

$$\text{Adj}: V \to \mathscr{P}(V).$$

Both of these representations will be used interchangeably. We call $\text{Adj}(v)$ the *adjacency set* of vertex v, and we call the ordered pair $(v, w) \in E$ an *edge*. Clearly

$$(v, w) \in E \qquad \text{if and only if} \qquad w \in \text{Adj}(v).$$

In this case we say that w is *adjacent* to v and v and w are *endpoints* of the edge (v, w). The assumption of irreflexivity implies that

$$(v, v) \notin E \qquad (v \in V),$$

or equivalently,

$$v \notin \text{Adj}(v) \qquad (v \in V).$$

We further denote

$$N(v) = \{v\} + \text{Adj}(v),$$

which is called the *neighborhood* of v.

In this book we will usually drop the parentheses and the comma when denoting an edge. Thus

$$xy \in E \qquad \text{and} \qquad (x, y) \in E$$

will have the same meaning. This convention, we believe, improves the clarity of exposition.

We have defined a graph as a set and a certain relation on that set. It is often convenient to draw a "picture" of the graph. This may be done in many ways. Usually one draws a circle for each vertex and connects vertex x and vertex y with a directed arrow whenever xy is an edge. If both xy and yx are edges, then sometimes a single line joins x and y without arrows. Figure 1.1 shows three of the many possible drawings that one could use to represent the same graph. In each case the adjacency structure remains unchanged. Occasionally, very intelligent persons will become extremely angry because one does not like the other's pictures. When this happens it is best to remember that our figures are meant simply as a tool to help understand the underlying mathematical structure or as an aid in constructing a mathematical model for some application.

---

* Some authors use the term *directed graph* or *digraph*.

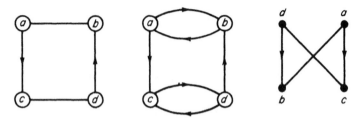

Figure 1.1.   Three pictures of the same graph.

Two graphs $G = (V, E)$ and $G' = (V', E')$ are called *isomorphic*, denoted $G \cong G'$, if there is a bijection $f: V \to V'$ satisfying, for all $x, y \in V$,

$$(x, y) \in E \Leftrightarrow (f(x), f(y)) \in E'.$$

Two edges are *adjacent* if they share a common endpoint; otherwise they are *nonadjacent*.

Let $G = (V, E)$ be a graph with vertex set V and edge set E. The graph $G^{-1} = (V, E^{-1})$ is said to be the *reversal* of G, where

$$E^{-1} = \{(x, y) \mid (y, x) \in E\},$$

that is,

$$xy \in E^{-1} \Leftrightarrow yx \in E \qquad (x, y \in V).$$

We define *symmetric closure* of G to be the graph $\hat{G} = (V, \hat{E})$, where

$$\hat{E} = E \cup E^{-1}.$$

A graph $G = (V, E)$ is called *undirected* if its adjacency relation is symmetric, i.e., if

$$E = E^{-1},$$

or equivalently,

$$E = \hat{E}.$$

We occasionally denote an undirected edge by $\widehat{ab} = \{ab\} \cup \{ba\}$. A graph $H = (V, F)$ is called an *oriented* graph if its adjacency relation is antisymmetric, i.e., if

$$F \cap F^{-1} = \emptyset.$$

If, in addition, $F + F^{-1} = E$, then H (or F) is called an *orientation* of G (or E). The four nonisomorphic orientations of the pentagon are given in Figure 1.2.

Let $G = (V, E)$ be an undirected graph. We define the *complement* of G to be the graph $\bar{G} = (V, \bar{E})$, where

$$\bar{E} = \{(x, y) \in V \times V \mid x \neq y \text{ and } (x, y) \notin E\}.$$

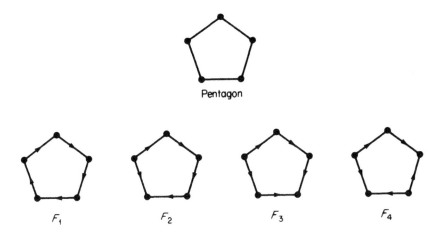

Figure 1.2.   The four nonisomorphic orientations of the pentagon.

Intuitively, the edges of G become the nonedges of $\bar{G}$ and vice versa. A graph is *complete* if every pair of distinct vertices is adjacent. Thus, the complement $\bar{G} = (V, \bar{E})$ of G could equivalently be defined as that set $\bar{E}$ satisfying $E \cap \bar{E} = \varnothing$ and $E + \bar{E}$ complete. The complete graph on n vertices is usually denoted by K_n (see Figure 1.3).

A *(partial) subgraph* of a graph $G = (V, E)$ is defined to be any graph $H = (V', E')$ satisfying $V' \subseteq V$ and $E' \subseteq E$. Two types of subgraphs are of particular importance, namely, the subgraph spanned by a given subset of edges and the subgraph induced by a given subset of vertices. They will now be described.

A subset $S \subseteq E$ of the edges *spans* the subgraph $H = (V_S, S)$, where $V_S = \{v \in V \mid v$ is an endpoint of some edge of $S\}$. We call H the (partial) subgraph spanned by S.

Figure 1.3.   Some complete graphs.

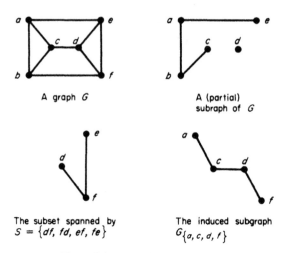

Figure 1.4. Examples of subgraphs.

Given a subset $A \subseteq V$ of the vertices, we define the *subgraph induced* by A to be $G_A = (A, E_A)$, where

$$E_A = \{xy \in E \,|\, x \in A \text{ and } y \in A\}.$$

For $v \in A$ we denote $\mathrm{Adj}_A(v) = \mathrm{Adj}(v) \cap A$. Obviously not every subgraph of G is an induced subgraph of G (Figure 1.4).

Let $G = (V, E)$ be an undirected graph. Consider the following definitions.

Clique: A subset $A \subseteq V$ of r vertices is an *r-clique* if it induces a complete subgraph, i.e., if $G_A \cong K_r$. A single vertex is a 1-clique. A clique A is *maximal* if there is no clique of G which properly contains A as a subset. A clique is *maximum* if there is no clique of G of larger cardinality. Some authors use the term *complete set* to indicate a clique.

$\omega(G)$ is the number of vertices in a maximum clique of G; it is called the *clique number* of G.

A *clique cover* of size k is a partition of the vertices $V = A_1 + A_2 + \cdots + A_k$ such that each A_i is a clique.

$k(G)$ is the size of a smallest possible clique cover of G; it is called the *clique cover number* of G.

A *stable set* is a subset X of vertices no two of which are adjacent. Some authors use the term *independent set* to indicate a stable set.

$\alpha(G)$ is the number of vertices in a stable set of maximum cardinality; it is called the *stability number* of G.

A *proper c-coloring* is a partition of the vertices $V = X_1 + X_2 + \cdots + X_c$ such that each X_i is a stable set. In such a case, the members of X_i are "painted" with the color i and adjacent vertices will receive different colors. We say that G is c-colorable. It is common to omit the word proper; a *coloring* will always be assumed to be a proper coloring.

$\chi(G)$ is the smallest possible c for which there exists a proper c-coloring of G; it is called the *chromatic number* of G.

It is easy to see that

$$\omega(G) \leq \chi(G) \quad \text{and} \quad \alpha(G) \leq k(G),$$

since every vertex of a maximum clique (maximum stable set) must be contained in a different partition segment in any minimum proper coloring (minimum clique cover). There is an obvious duality to these notions, namely,

$$\omega(G) = \alpha(\bar{G}) \quad \text{and} \quad \chi(G) = k(\bar{G}).$$

Let $G = (V, E)$ be an arbitrary graph. The *out-degree* of a vertex x, denoted by $d^+(x)$, is defined as $d^+(x) = |\text{Adj}(x)|$. The *in-degree* $d^-(x)$ of x is defined similarly:

$$d^-(x) = |\{y \in V \mid x \in \text{Adj}(y)\}|.$$

Although in general $d^+(x)$ and $d^-(x)$ will not be equal, we do have

$$\sum_{x \in V} d^+(x) = \sum_{x \in V} d^-(x) = |E|,$$

each ordered pair in E contributing 1 to both summands. A vertex whose out-degree (in-degree) equals zero is called a *sink* (*source*). If both $d^+(x) = 0$ and $d^-(x) = 0$, then x is an *isolated vertex*.

When G is an undirected graph the situation is somewhat special. In such a case $d^+(x) = d^-(x)$ for each $x \in V$, and we call this number simply the *degree* of x, denoted $d(x)$. That is, the degree of x in an undirected graph is the size of its adjacency set. Finally, defining $\|E\| = \frac{1}{2}|E|$ we obtain the familiar formula

$$\frac{1}{2} \sum_{x \in V} d(x) = \|E\|.$$

Let $G = (V, E)$ be an arbitrary graph. We present some fairly standard definitions.

Chain: A sequence of vertices $[v_0, v_1, v_2, \ldots, v_l]$ is a *chain of length l* in G if $v_{i-1}v_i \in E$ or $v_i v_{i-1} \in E$ for $i = 1, 2, \ldots, l$.

Path: A sequence of vertices $[v_0, v_1, v_2, \ldots, v_l]$ is a *path from v_0 to v_l of length l* in G provided that $v_{i-1}v_i \in E$ for $i = 1, 2, \ldots, l$.

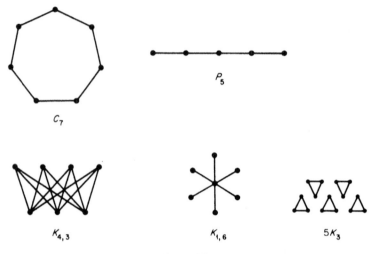

C_7

P_5

$K_{4,3}$

$K_{1,6}$

$5K_3$

Figure 1.5.

A path or chain in G is called *simple* if no vertex occurs more than once. It is called *trivial* if $l = 0$.

Connected graph: A graph G is *connected* if between any two vertices there exists a chain in G joining them.

Strongly connected graph: A graph G is *strongly connected* if for any two vertices x and y there exists a path in G from x to y.

Remark.   The notions of chain and path coincide when G is an undirected graph.

Cycle: A sequence of vertices $[v_0, v_1, v_2, \ldots, v_l, v_0]$ is called a *cycle* of length $l + 1$ (or closed path) if $v_{i-1}v_i \in E$ for $i = 1, 2, \ldots, l$ and $v_l v_0 \in E$.

Simple cycle: A cycle $[v_0, v_1, v_2, \ldots, v_l, v_0]$ is a *simple cycle* if $v_i \neq v_j$ for $i \neq j$.

Chordless cycle: A simple cycle $[v_0, v_1, v_2, \ldots, v_l, v_0]$ is *chordless* if $v_i v_j \notin E$ for i and j differing by more than 1 mod $l + 1$.

Bipartite graph: An undirected graph $G = (V, E)$ is *bipartite* if its vertices can be partitioned into two disjoint stable sets $V = S_1 + S_2$, i.e., every edge has one endpoint in S_1 and the other in S_2. Equivalently, G is bipartite if and only if it is 2-colorable. It is customary to use the notation $G = (S_1, S_2, E)$, which emphasizes the partition. Vertices $x \in S_i$ and $y \in S_j$ are of the *same parity* if $i = j$ and are of *opposite parity* if $i \neq j$.

Complete bipartite graph: A bipartite graph $G = (S_1, S_2, E)$ is *complete* if for every $x \in S_1$ and $y \in S_2$ we have $xy \in E$, i.e., every possible edge that could exist does exist.

Throughout the text certain graphs will occur many times. We give names to some of them (see Figure 1.5).

K_n: the *complete graph* on n vertices or *n-clique*.
C_n: the *chordless cycle* on n vertices or *n-cycle*.
P_n: the *chordless path graph* on n vertices or *n-path*.
$K_{m,n}$: the *complete bipartite graph* on $m + n$ vertices partitioned into an *m-stable set* and an *n-stable set*.
$K_{1,n}$: the *star graph* on $n + 1$ vertices.
mK_n: m disjoint copies of K_n.

There is obviously some overlap with these names. For example, $K_3 = C_3$ is called a *triangle*. Notice also that $\bar{C}_4 = 2K_2$ and $K_{n,n} = \overline{2K_n}$.

2. Intersection Graphs

Let $\mathscr{F}$ be a family of nonempty sets. The *intersection graph* of $\mathscr{F}$ is obtained by representing each set in $\mathscr{F}$ by a vertex and connecting two vertices by an edge if and only if their corresponding sets intersect. When $\mathscr{F}$ is allowed to be an arbitrary family of sets, the class of graphs obtained as intersection graphs is simply all undirected graphs (Marczewski [1945]). The problem of characterizing the intersection graphs of families of sets having some specific topological or other pattern is often very interesting and frequently has applications to the real world.

The intersection graph of a family of intervals on a linearly ordered set (like the real line) is called an *interval graph*. If these intervals are required to have unit length, then we have a *unit interval graph*; a *proper interval graph* is constructed from a family of intervals on a line such that no interval properly contains another. Roberts [1969a] showed that the classes of unit interval graphs and proper interval graphs coincide. Interval graphs are discussed in Section 1.3 and in Chapter 8.

Consider the following relaxation of the notion of intervals on a line. If we join the two ends of our line, thus forming a circle, the intervals will become arcs on the circle. Allowing arcs to slip over and include the point of connection, we obtain a class of intersection graphs called the *circular-arc graphs*, which properly contains the interval graphs. Circular-arc graphs have been extensively studied by A. C. Tucker and others. We will survey these results in Section 8.6. There are a number of interesting applications of circular-arc graphs, including computer storage allocation and the phasing of traffic lights. Let us look at an example of the latter application.

Example.   The traffic flow at the corner of Holly, Vood, and Wine is pictured in Figure 1.6. Certain lanes are compatible with one another, such as c and j, or d and k, while others are incompatible, such as b and f. In order to avoid collisions, we wish to install a traffic light system to control the flow of vehicles. Each lane will be assigned an arc on a circle representing the time interval during which it has a green light. Incompatible lanes must be assigned disjoint arcs. The circle may be regarded as a clock representing an entire cycle which will be continually repeated. An arc assignment for our example is given in Figure 1.7. In general, if G is the intersection graph of the arcs of such an assignment (see Figure 1.8), and if H is the compatibility relation defined on the pairs of lanes, then clearly G is a (partial) subgraph of H. In our example, the compatible pairs (d, k), (h, j), and (i, j) are in H but are not in G. Additional aspects of this problem, such as how to choose an arc assignment which minimizes waiting time, can also be incorporated into the model. The reader is referred to Stoffers [1968] and Roberts [1976, pp. 129–134; 1978, Section 3.6] for more details.

A *proper circular-arc* graph is the intersection graph of a family of arcs none of which properly contains another. It can be shown (Theorem 8.18)

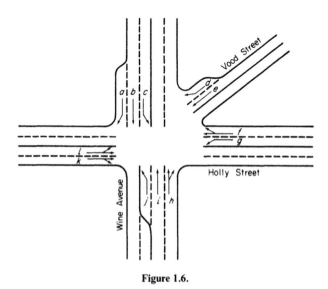

Figure 1.6.

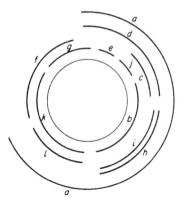

Figure 1.7. The clock cycle.

that every proper circular-arc graph has a representation as intersecting arcs of a circle in which not only is no arc properly contained in another but also no pair of arcs together cover the entire circle.

In a different generalization of interval graphs, Renz [1970] characterized the intersection graphs of paths in a tree, and Gavril [1978] gives a recognition algorithm for them. Walter [1972], Buneman [1974], and Gavril [1974] carried this idea further and showed that the intersection graphs of subtrees of a tree are exactly the triangulated graphs of Chapter 4. All of this is summarized in Figure 1.9.

A permutation diagram consists of n points on each of two parallel lines and n straight line segments matching the points. The intersection graph of the line segments is called a *permutation graph*. These graphs will be discussed

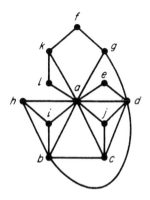

Figure 1.8. G, the circular-arc graph.

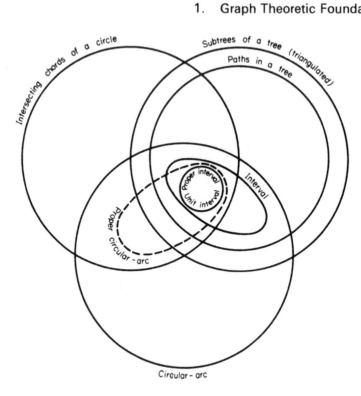

Figure 1.9.

in Chapter 7. If the 2*n* points are located randomly around a circle, then the matching segments will be *chords of the circle* and the resulting class of intersection graphs, studied in Chapter 11, properly contains the permutation graphs. A simple argument shows that *every proper circular-arc graph is also the graph of intersecting chords of a circle*: We may assume that no pair of arcs together covers the entire circle (Theorem 8.18). For each arc on the circle, draw the chord connecting its two endpoints. Clearly, two arcs overlap if and only if their corresponding chords intersect.

There are many other interesting classes of intersection graphs. We have introduced you to only some of them, specifically those classes which will be developed further in the text. To the reader who wishes to investigate other intersection graphs we offer the following references:

Cubes and boxes in *n*-space: Danzer and Grunbaum [1967],
                                 Roberts [1969b].
Convex sets in *n*-space: Wegner [1967],
                                 Ogden and Roberts [1970].

3. Interval Graphs—A Sneak Preview of the Notions Coming Up

Our intention in this section is to arouse the reader's curiosity by presenting some basic ideas that will be pursued in greater detail in later chapters. We also hope to imbue the reader with a sense of how the subject matter is relevant to applied mathematics and computer science.

An undirected graph G is called an *interval graph* if its vertices can be put into one-to-one correspondence with a set of intervals $\mathscr{I}$ of a linearly ordered set (like the real line) such that two vertices are connected by an edge of G if and only if their corresponding intervals have nonempty intersection. We call $\mathscr{I}$ an *interval representation* for G. (It is unimportant whether we use open intervals or closed intervals; the resulting class of graphs will be the same.) An interval representation of the windmill graph is given in Figure 1.10.

Let us discuss one application of interval graphs. Many other such applications will be presented in Section 8.4.

Application to Scheduling

Consider a collection $C = \{c_i\}$ of courses being offered by a major university. Let T_i be the time interval during which course c_i is to take place. We would like to assign courses to classrooms so that no two courses meet in the same room at the same time.

This problem can be solved by properly coloring the vertices of the graph $G = (C, E)$ where

$$c_i c_j \in E \Leftrightarrow T_i \cap T_j \neq \varnothing.$$

Each color corresponds to a different classroom. The graph G is obviously an interval graph, since it is represented by time intervals.

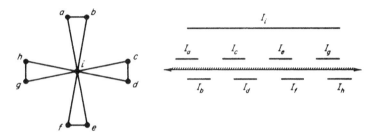

Figure 1.10.   An interval graph—the windmill graph (at left)—and an interval representation for it.

This example is especially interesting because efficient, linear-time algorithms are known for coloring interval graphs with a minimum number of colors. (The minimum coloring problem is NP-complete for general graphs, Section 2.1.) We will discuss these algorithms in subsequent chapters.

Remark.   The determination of whether a given graph is an interval graph can also be carried out in linear time (Section 8.3).

We have chosen interval graphs as an introduction to our studies because they satisfy so many interesting properties. The first fact that we notice is that being an interval graph is a *hereditary property*.

Proposition 1.1.   An induced subgraph of an interval graph is an interval graph.

Proof.   If $\{I_v\}_{v \in V}$ is an interval representation for a graph $G = (V, E)$, then $\{I_v\}_{v \in X}$ is an interval representation for the induced subgraph $G_X = (X, E_X)$.   ∎

Hereditary properties abound in graph theory. Some of our favorites include planarity, bipartiteness, and any "forbidden subgraph" characterization. The next property of interval graphs is also a hereditary property.

Triangulated graph property.   Every simple cycle of length strictly greater than 3 possesses a chord.

Graphs which satisfy this property are called *triangulated graphs*. The graph in Figure 1.10 is triangulated, but the house graph in Figure 1.11 is not triangulated because it contains a chordless 4-cycle.

Proposition 1.2 (Hajös [1958]).   An interval graph satisfies the triangulated graph property.

Figure 1.11.   A graph which is not triangulated: The house graph.

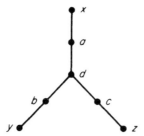

Figure 1.12.   A triangulated graph which is not an interval graph.

Proof. Suppose the interval graph G contains a chordless cycle $[v_0, v_1, v_2, \ldots, v_{l-1}, v_0]$ with $l > 3$. Let I_k denote the interval corresponding to v_k. For $i = 1, 2, \ldots, l - 1$, choose a point $p_i \in I_{i-1} \cap I_i$. Since I_{i-1} and I_{i+1} do not overlap, the p_i constitute a strictly increasing or strictly decreasing sequence. Therefore, it is impossible for I_0 and I_{l-1} to intersect, contradicting the criterion that $v_0 v_{l-1}$ is an edge of G.   ∎

Not every triangulated graph is an interval graph. Consider the tree T given in Figure 1.12, which certainly has no chordless cycles. The intervals $I_a, I_b,$ and I_c of a representation for T would have to be disjoint, and I_d would properly include the middle interval, say I_b. Where, then, could we put I_y so that it intersects I_b but not I_d? Clearly we would be stuck. So there must be more to the story of interval graphs than we have told so far.

Transitive orientation property.   Each edge can be assigned a one-way direction in such a way that the resulting oriented graph (V, F) satisfies the following condition:

$$ab \in F \text{ and } bc \in F \quad \text{imply} \quad ac \in F \qquad (\forall a, b, c \in V). \qquad (1)$$

An undirected graph which is transitively orientable is sometimes called a *comparability graph*. Figure 1.13 shows a transitive orientation of the A graph and of the suspension bridge graph. The odd length chordless cycles $C_5, C_7, C_9, \ldots$ and the bull's head graph (see Figure 1.14) cannot be transitively oriented.

Proposition 1.3 (Ghouila-Houri [1962]).   The complement of an interval graph satisfies the transitive orientation property.

Proof. Let $\{I_v\}_{v \in V}$ be an interval representation for $G = (V, E)$. Define an orientation F of the complement $\bar{G} = (V, \bar{E})$ as follows:

$$xy \in F \Leftrightarrow I_x < I_y \qquad (\forall xy \in \bar{E}).$$

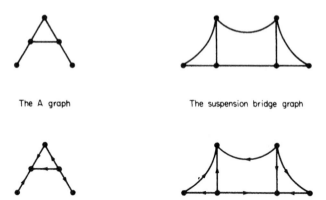

The A graph   The suspension bridge graph

Figure 1.13.   Transitive orientations of two comparability graphs.

Here $I_x < I_y$ means that the interval I_x lies entirely to the left of the interval I_y. (Remember, they are disjoint.) Clearly (1) is satisfied, since $I_x < I_y < I_z$ implies $I_x < I_z$. Thus, F is a transitive orientation of $\bar{G}$.   ∎

As in the case of triangulated graphs, there are graphs whose complements are comparability graphs but which fail to be interval graphs. So it seems that Propositions 1.2 and 1.3 simply provide necessary (but not sufficient) conditions for interval graphs. Rather than wait any longer, we state an important result that says, if we put these two properties together, we get (drum roll, please) exactly all interval graphs.

Theorem 1.4 (Gilmore and Hoffman [1964]).   An undirected graph G is an interval graph if and only if G is a triangulated graph and its complement $\bar{G}$ is a comparability graph.

The proof of sufficiency is postponed until Chapter 8, primarily because this is a "getting acquainted with" section.

Looking back, each of the graphs in Figures 1.10, 1.11, and 1.13 can be properly colored using three colors and each contains a triangle. Therefore,

C_5 The bull's head graph

Figure 1.14.   Two graphs which are not transitively orientable. Why?

for these graphs, their chromatic number equals their clique number. This is not an accident. In Chapters 4 and 5 we will show that any triangulated graph and any comparability graph also satisfies the following properties.

χ-Perfect property.   For each induced subgraph G_A of G,

$$\chi(G_A) = \omega(G_A).$$

The chordless cycles C_5, C_7, C_9 are not χ-perfect. A dual notion of χ-perfection is the following:

α-Perfect property.   For each induced subgraph G_A of G,

$$\alpha(G_A) = k(G_A).$$

A very important theorem in Chapter 3 states that a graph is χ-perfect if and only if it is α-perfect. This equivalence was originally conjectured by Claude Berge, and it was proved some ten years later by László Lovász.

4. Summary

   The reader has been introduced to the graph theoretic foundations needed for the remainder of the book. In addition, he has had a taste of some of the particular notions that we intend to investigate further. Returning to the table of contents at this point, he will recognize many of the topics listed. The chapter dependencies are given in Figure 1.15.

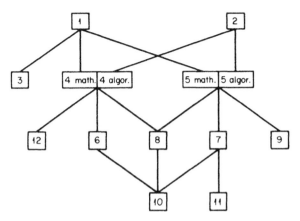

Figure 1.15.   The chapter dependencies. The reader may wish to read Chapters 1 and 2 quickly and refer back to them as needed.

In the next chapter we will present the foundations of algorithmic design and analysis. As was the case in this chapter, many examples will be given which will introduce the reader to the ideas and techniques that he will find helpful in subsequent chapters.

EXERCISES

1. Show that the graphs in Figures 1.16 and 1.17 are both intersection graphs of a family of chords of a circle but that neither is a circular-arc graph.

Figure 1.16.   Figure 1.17.

2. Can you find graphs for each zone of the Venn diagram in Figure 1.9?
3. Let $\mathscr{F}$ be a family of intervals on a line such that no interval contains another. Show that none of the left endpoints coincide. Give a procedure which constructs a family $\mathscr{F}'$ of unit intervals such that the intersection graphs of $\mathscr{F}$ and $\mathscr{F}'$ are isomorphic.
4. Let $G = (V, E)$ be any undirected graph. Show that there is a family $\mathscr{F}$ of subsets of V such that G is the intersection graph of $\mathscr{F}$.
5. Let G be the intersection graph of a family of paths in a tree and let v be a vertex of G. Show that the induced subgraph $G_{\{v\} + \mathrm{Adj}(v)}$ is an interval graph.
6. Prove directly (using only the definition) that the graph in Figure 1.17 does not have an interval representation and is therefore not an interval graph.
7. Give an interval representation for the graph in Figure 1.18. Show that it is not a comparability graph. Why is this not in conflict with the Gilmore–Hoffman theorem?

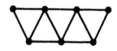

Figure 1.18.

8. Give a graph theoretic solution to the following problem: A group of calculus teaching assistants each gives two office hours weekly which are chosen in advance. Because of budgetary reasons, the TAs must share offices. Since each office has only one blackboard, how can office space be assigned so that at any particular time no more than one TA is meeting with students?

9. Give an example to show that the graph you obtain in Exercise 8 is not necessarily an interval graph. How could we alter the problem so that we would obtain only interval graphs?

10. Is the bull's head graph (Figure 1.14) an interval graph? Is the complement of the suspension bridge graph (Figure 1.13) an interval graph? What is a good name for this last graph?

11. An undirected graph is *self-complementary* if it is isomorphic to its complement. Show that there are exactly two self-complementary graphs having five vertices. How many are there for four vertices? Six vertices?

12. Let $G = (V, E)$ be an undirected graph. A subset $A \subseteq V$ is called an *edge cover* of G if for every edge $xy \in E$, either $x \in A$ or $y \in A$ or both. Prove that A is a minimum edge cover if and only if $V - A$ is a maximum stable set.

13. Let $\mathscr{F} = \{S_x\}_{x \in V}$ be a family of subsets of a set. Two members S_x and S_y of $\mathscr{F}$ *overlap*, denoted $S_x \between S_y$, if $S_x \cap S_y \neq \varnothing$, $S_x \nsubseteq S_y$, and $S_y \nsubseteq S_x$. The *overlap graph* of $\mathscr{F}$ is the undirected graph $G = (V, E)$ where

$$xy \in E \quad \text{if and only if} \quad S_x \between S_y \qquad (x, y \in V).$$

(i) Show that if x and y are in separate connected components G_A and G_B of G, then

$$S_x \subseteq S_y \Rightarrow S_a \subseteq S_y \qquad (a \in A).$$

(ii) Let $\mathscr{G}$ be the collection of (maximal) connected components of G. Show that the relation $<$, defined for all $G_A, G_B \in \mathscr{G}$ as

$$G_A < G_B \Leftrightarrow \exists x \in A, y \in B \text{ such that } S_X \subseteq S_Y,$$

is a strict partial order of $\mathscr{G}$.

14. A family $\mathscr{F}$ of distinct nonempty subsets of a set S is a *representation* of a graph G if the intersection graph of $\mathscr{F}$ is isomorphic to G. A representation is *minimum* if the set S is of smallest possible cardinality over all representations of G. A graph G is *uniquely intersectable* if for all minimum representations $\mathscr{F}_1$ and $\mathscr{F}_2$ of G, $\mathscr{F}_1$ and $\mathscr{F}_2$ are isomorphic.

(i) Prove that every triangle-free graph is uniquely intersectable.

A *star n-gon* is constructed from the cycle C_n by adjoining new vertices to the endpoints of each edge. Figure 1.19 illustrates a star 7-gon.

(ii) Verify that the family $\mathscr{F} = \{S_0, S_1, \ldots, S_{n-1}, D_0, D_1, \ldots, D_{n-1}\}$ is a minimum representation of the star *n*-gon, where $S_i = \{i\}$ and $D_i = \{i, i + 1 \bmod n\}$.

(iii) Prove that every star *n*-gon is uniquely intersectable (Alter and Wang [1977]).

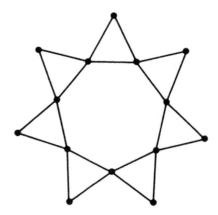

Figure 1.19.   A star 7-gon.

15.   *The Berge mystery story.* Six professors had been to the library on the day that the rare tractate was stolen. Each had entered once, stayed for some time, and then left. If two were in the library at the same time, then at least one of them saw the other. Detectives questioned the professors and gathered the following testimony: Abe said that he saw Burt and Eddie in the library; Burt said that he saw Abe and Ida; Charlotte claimed to see Desmond and Ida; Desmond said that he saw Abe and Ida; Eddie testified to seeing Burt and Charlotte; Ida said that she saw Charlotte and Eddie.
   One of the professors lied!! Who was it?

Research Problem.   Characterize uniquely intersectable graphs and/or give a recognition algorithm.

Bibliography

Alter, R., and Wang, C. C.
   [1977]   Uniquely intersectable graphs, *Discrete Math.* **18**, 217–226.
Berge, Claude
   [1973]   "Graphs and Hypergraphs," Chapter 16, North-Holland. Amsterdam, 1973. MR50 #9640.
   [1975]   Perfect graphs, *in* "Studies in Graph Theory," Part I (D. R. Fulkerson, ed.), pp. 1–22. M.A.A. Studies in Mathematics Vol. 11, Math. Assoc. Amer., Washington, D.C. MR53 #10585.
Buneman, Peter
   [1974]   A characterization of rigid circuit graphs, *Discrete Math.* **9**, 205–212. MR50 # 9686.
Danzer, L., and Grunbaum, B.
   [1967]   Intersection properties of boxes in $\mathbb{R}^d$, unpublished mimeograph, Univ. of Washington, Seattle.

Erdös, P., Goodman, A., and Pósa, L.,

[1966] The representation of a graph by set intersections, *Canad. J. Math.* **18**, 106–112. MR32 #4034.

Gavril, Fanica

[1974] The intersection graphs of subtrees in trees are exactly the chordal graphs, *J. Combinatorial Theory B* **16**, 47–56. MR48 #10868.

[1978] A recognition algorithm for the intersection graphs of paths in trees, *Discrete Math.* **23**, 211–227.

Ghouilà-Houri, Alain

[1962] Caractérisation des graphes non orientés dont on peut orienter les arrêtes de manière à obtenir le graphe d'une relation d'ordre, *C. R. Acad. Sci. Paris* **254**, 1370–1371. MR30 #2495.

Gilmore, Paul C., and Hoffman, Alan J.

[1964] A characterization of comparability graphs and of interval graphs, *Canad. J. Math.* **16**, 539–548. MR31 #87.

Hajös, G.

[1957] Über eine Art von Graphen, *Intern. Math. Nachr.* **11**, Problem 65. First posed the problem of characterizing interval graphs.

Marczewski, E.

[1945] Sur deux propriétés des classes d'ensembles, *Fund. Math.* **33**, 303–307.

Ogden, W. F., and Roberts, Fred S.

[1970] Intersection graphs of families of convex sets with distinguished points, *in* "Combinatorial Structures and Their Applications" (R. Guy, H. Hanani, N. Sauer, and J. Schönheim, eds.), pp. 311–313. Gordon and Breach, New York.

Renz, P. L.

[1970] Intersection representations of graphs by arcs, *Pacific J. Math.* **34**, 501–510. MR42 #5839.

Roberts, Fred S.

[1969a] Indifference graphs, *in* "Proof Techniques in Graph Theory" (F. Harary, ed.), pp. 139–146. Academic Press, New York. MR40 #5488.

[1969b] On the boxity and cubicity of a graph, *in* "Recent Progress in Combinatorics" (W. T. Tutte, ed.), pp. 301–310. Academic Press, New York.

[1976] "Discrete Mathematical Models, with Applications to Social, Biological, and Environmental Problems." Prentice-Hall, Englewood Cliffs, New Jersey.

[1978] Graph Theory and its Applications to Problems of Society, NFS-CBMS Monograph No. 29. SIAM, Philadelphia, Pennsylvania.

Stoffers, K. E.

[1968] Scheduling of traffic lights—a new approach, *Transportation Res.* **2**, 199–234.

Walter, J. R.

[1972] Representations of rigid cycle graphs, Ph.D. thesis, Wayne State Univ.

Wang, D. L.

[1976] A note on uniquely intersectable graphs, *Studies in Appl. Math.* **55**, 361–363.

Wegner, G.

[1967] Eigenschaften der Nervan Homologische-einfacher Familien in $\mathbb{R}^n$, Ph.D. thesis, Göttingen.

CHAPTER 2

The Design of
Efficient Algorithms

1. The Complexity of Computer Algorithms

With the advent of the high-speed electronic computer, new branches of applied mathematics have sprouted forth. One area that has enjoyed a most rapid growth in the past decade is the complexity analysis of computer algorithms. At one level, we may wish to compare the relative efficiencies of procedures which solve the same problem. At a second level, we can ask whether one problem is intrinsically harder to solve than another problem. It may even turn out that a task is too hard for a computer to solve within a reasonable amount of time. Measuring the costs to be incurred by implementing various algorithms is a vital necessity in computer science, but it can be a formidable challenge.

Let us reflect for a moment on the differences between *computability* and *computational complexity*. These two topics, along with formal languages, become the pillars of the theory of computation. Computability addresses itself mostly to questions of existence: Is there an algorithm which solves problem Π? An early surprise for many math and computer science students is that one can prove mathematically that computers cannot do everything. A standard example is the unsolvability of the halting problem. Loosely stated, this says that *it is impossible for a professor to write a computer program which will accept as data any student's programming assignment and will return either the answer "yes, this student's program will halt within finite time" or "no, this student's program (has an infinite loop and) will run forever."* Proving that a problem *is* computable usually, but not always, consists of demonstrating an actual algorithm which will terminate with a correct answer for every input. The amount of resources (time and space) used in the calculation, although finite, is unlimited. Thus, computability gives us an understanding of the capabilities and limitations of the machines that mankind can build, but without regard to resource restrictions.

In contrast to this, computational complexity deals precisely with the quantitative aspects of problem solving. It addresses the issue of what can be computed within a *practical* or *reasonable* amount of time and space by measuring the resource requirements exactly or by obtaining upper and lower bounds. Complexity is actually determined on three levels: the problem, the algorithm, and the implementation. Naturally, we want the best algorithm which solves our problem, and we want to choose the best implementation of that algorithm.

A *problem* consists of a question to be answered, a requirement to be fulfilled, or a best possible situation or structure to be found, called a *solution*, usually in response to several input *parameters* or *variables*, which are described but whose values are left unspecified. A *decision problem* is one which requires a simple "yes" or "no" answer. An *instance* of a problem Π is a specification of particular values for its parameters. An *algorithm* for Π is a step-by-step procedure which when applied to any instance of Π produces a solution.

Usually we can rewrite an optimization problem as a decision problem which at first seems to be much easier to solve than the original but turns out to be just about as hard. Consider the following two versions of the graph coloring problem.

GRAPH COLORING (optimization version)
Instance: An undirected graph G.
Question: What is the smallest number of colors needed for a proper coloring of G?

GRAPH COLORING (decision version)
Instance: An undirected graph G and an integer $k > 0$.
Question: Does there exist a proper k coloring of G?

The optimization version can be solved by applying an algorithm for the decision version n times for an n-vertex graph. If the n decision problems are solved sequentially, then the time needed to solve the optimization version is larger than that for the decision version by at most a factor of n. However, if they can be solved simultaneously (in parallel), then the time needed for both versions is essentially the same.

It is customary to express complexity as a function of the size of the input. We say that an algorithm $\mathscr{A}$ for Π runs in time $O(f(m))$ if for some constant

$c > 0$ there exists an implementation of $\mathscr{A}$ which terminates after at most $cf(m)$ (computational) steps for all instances of size m. The complexity of an algorithm $\mathscr{A}$ is the smallest function f such that $\mathscr{A}$ runs in $O(f(m))$. The complexity of a problem Π is the smallest f for which there exists an $O(f(m))$-time algorithm $\mathscr{A}$ for Π, i.e., the minimum complexity over all *possible* algorithms solving Π. Thus, demonstrating and analyzing the complexity of a particular algorithm for Π provides us with an *upper bound* on the complexity of Π.*

By presenting faster and more efficient algorithms and implementations of algorithms, successive researchers have improved the complexity upper bounds (i.e., lowered them) for many problems in recent years. Consider the example of testing a graph for planarity. A graph is *planar* if it can be drawn on the plane (or on the surface of a sphere) such that no two edges cross one another. Kuratowski's [1930] characterization of planar graphs in terms of forbidden configurations provides an obvious exponential-time planarity algorithm, namely, verify that no subset of vertices induces a subgraph homeomorphic to K_5 or $\overline{2K_3}$. Auslander and Parter [1961] gave a planar embedding procedure, which Goldstein [1963] was able to formulate in such a way that halting was guaranteed. Shirey [1969] implemented this algorithm to run in $O(n^3)$ time for an n-vertex graph. In the meantime, Lempel, Even, and Cederbaum [1967] gave a different planarity algorithm, and, although they did not specify a time bound, an easy $O(n^2)$ implementation exists. Hopcroft and Tarjan [1972, 1974] then improved the Auslander–Parter method first to $O(n \log n)$ and finally to $O(n)$, which is the best possible. Booth and Leuker showed that the Lempel–Even–Cederbaum method could also be implemented to run in $O(n)$ time. Table 2.1 shows the stages of improvement for the planarity problem and for the maximum-network-flow problem. Tarjan [1978] summarizes the progress on a number of other problems.

Determining the complexity of a problem Π requires a two-sided attack:

(1)   *The upper bound*—the minimum complexity over all *known* algorithms solving Π.

(2)   *The lower bound*—the largest function f for which it has been proved (mathematically) that all *possible* algorithms solving Π are required to have complexity at least as high as f.

Our ultimate goal is to make these bounds coincide. A gap between (1) and (2) tells us how much more research is needed to achieve this goal. For many

---

* We have just described the worst-case complexity analysis. One may also formulate complexity according to the average case. A good discussion of the pros and cons of average-case analysis can be found in Weide [1977, Section 4].

Table 2.1

Progress on the complexity of two combinatorial problems

Planarity: A graph with n vertices		Maximum network flow: A network with n vertices and e edges	
exp	Kuratowski [1930]	Nonterminating under certain conditions	Ford and Fulkerson [1962]
↓		↓	
		ne^2	Edmonds and Karp [1972][a]
n^3	Auslander and Parter [1961] Goldstein [1963] Shirey [1969]	↓	
		n^2e	Dinic [1970][a]
↓		↓	
n^2	Lempel, Even, and Cederbaum [1967]	n^3	Karzanov [1974]
↓		↓	
$n \log n$	Hopcroft and Tarjan [1972]	$n^2e^{1/2}$	Cherkasky [1977]
↓		↓	
n	Hopcroft and Tarjan [1974] Booth and Leuker [1976]	$n^{5/3}e^{2/3}$ ↓ ?	$ne \log^2 n$ Galil [1978] Galil and Naamad [1979]

[a] Done independently.

problems this gap is stubbornly large. An example of this is the problem of matrix multiplication.

In Strassen [1969] an algorithm is presented for multiplying a pair of 2×2 matrices using only seven scalar multiplications. It is now known that seven multiplications is the best possible. For arbitrary $n \times n$ matrices Strassen's algorithm may be applied recursively (by first embedding the matrices into the next larger power of 2 in size) to obtain a general algorithm whose complexity is $O(n^{\log_2 7}) \approx O(n^{2.81})$. Until recently, $O(n^{2.81})$ was the best result known. The best algorithm known for the case of 3×3 matrices is given by Laderman [1976]; it uses 23 scalar multiplications. By appropriately composing these two methods with themselves or each other, we can obtain the best algorithms known for $n = 4, 6, 7, 8, 9$, and many other values. For

$n = 5$, G. A. Schachtel has an algorithm using 103 multiplications, an improvement of one given by O. Sýkora, which used 105. Asymptotically, however, in order to improve Strassen's general bound, one would need an algorithm for $n = 3$ using 21 or fewer multiplications (since $\log_3 21 < \log_2 7 < \log_3 22$) or an algorithm for $n = 5$ using 91 or fewer multiplications, etc. Amazingly, Pan [1978, 1979a] has discovered a collection of algorithms which do improve upon Strassen's bound. The best of these is an algorithm for $n = 48$ which uses 47,216 multiplications. Since $\log_{48} 47{,}216 \approx 2.78$, Pan's algorithm has a complexity of $O(n^{2.78})$. In very recent work Pan [1979b] has reduced the complexity down to $O(n^{2.6054})$ for very large n. This is currently the upper bound for the matrix multiplication problem. On the other hand, the tightest lower bound known to date for this problem is only $O(n^2)$ [see Aho, Hopcroft, and Ullman, 1974, p. 438].

The biggest open question involving the gap between upper and lower complexity bounds involves the so called NP-complete problems (discussed below). For each of the problems in this class only exponential-time algorithms are known, yet the best lower bounds proven so far are polynomial functions. Furthermore, if a polynomial-time algorithm exists for one of them, then such an algorithm exists for all of them. Included among the NP-complete problems on graphs are finding a Hamiltonian circuit, a minimum coloring, or a maximum clique. Appendix A contains a small collection of NP-complete problems which will suffice for the purposes of this book. For a more comprehensive list, the reader is referred to Garey and Johnson [1978]. Let us discuss the basics of this theory.

The *state* of an algorithm consists of the current values of all variables and the location of the current instruction to be executed. A *deterministic algorithm* is one for which each state upon execution of the instruction uniquely determines at most one next state. Virtually all computers, as we know them, run deterministically. A problem Π is in the class P if there exists a deterministic polynomial-time algorithm which solves Π.

A *nondeterministic algorithm* is one for which a state may determine many next states and which follows up on each of the next states *simultaneously*. We may regard a nondeterministic algorithm as having the capability of branching off into many copies of itself, one for each next state. Thus, while a deterministic algorithm must explore a set of alternatives one at a time, a nondeterministic algorithm examines all alternatives at the same time.

Following Reingold, Nievergelt, and Deo [1977], three special instructions are used in writing nondeterministic algorithms for decision problems:

$x \leftarrow$ **choice**(S) creates $|S|$ copies of the algorithm, and assigns every member of the set S to the variable x in one of the copies.

failure causes that copy of the algorithm to stop execution.

success causes all copies of the algorithm to stop execution and indicates a "yes" answer to that instance of the problem.

A nondeterministic polynomial-time algorithm for the decision version of the CLIQUE problem is the following: Let $G = (V, E)$ be an undirected graph and let $k \geq 0$.

```
procedure CLIQUE(G, k):
begin
1.   A ← ∅;
2.   for all v ∈ V do A ← choice({A + {v}, A});;
3.   if |A| < k then failure;
4.   for all v, w ∈ A, v ≠ w do
5.       if vw ∉ E then failure;;
6.   success;
end
```

The loop in line 2 nondeterministically selects a subset of vertices $A \subseteq V$; lines 4–6 decide if A is a complete set. If **success** is reached in one of the copies, then the final value of A in that copy is a clique of size at least k. Using the above procedure we obtain a nondeterministic polynomial-time algorithm for the optimization version of the CLIQUE problem as follows: Let G be an undirected graph with n vertices.

```
procedure MAXCLIQUE(G):
begin
   for k ← n to 1 step −1 do
       if CLIQUE(G, k) then return k;
end
```

A problem Π is in the class NP if there exists a nondeterministic polynomial-time algorithm which solves Π. We have just demonstrated that CLIQUE $\in$ NP by presenting an appropriate algorithm. Clearly, P $\subseteq$ NP. An important open question in the theory of computation is whether the containment of P in NP is proper; i.e., is P $\neq$ NP?

One problem Π_1 is *polynomially transformable* to another problem Π_2, denoted $\Pi_1 \preccurlyeq \Pi_2$, if there exists a function f mapping the instances of Π_1 into the instances of Π_2 such that

(i) f is computable deterministically in polynomial time, and
(ii) a solution to the instance $f(I)$ of Π_2 gives a solution to the instance I of Π_1, for all I.

Intuitively this means that Π_1 is no harder to solve than Π_2 up to an added polynomial term, for we could solve Π_1 by combining the transformation f with the best algorithm for solving Π_2. Thus, if $\Pi_1 \preccurlyeq \Pi_2$, then

$$\text{COMPLEXITY}(\Pi_1) \leq \text{COMPLEXITY}(\Pi_2) + \text{POLYNOMIAL.}$$

If Π_2 has a deterministic polynomial-time algorithm, then so does Π_1; if every deterministic algorithm solving Π_1 requires at least an exponential amount of time, then the same is true of Π_2.

A problem Π is *NP-hard* if any one of the following equivalent conditions holds:

(H_1) $\Pi' \leqslant \Pi$ for all $\Pi' \in NP$;

(H_2) $\Pi \in P \Rightarrow P = NP$;

(H_3) the existence of a deterministic polynomial-time algorithm for Π would imply the existence of a polynomial-time algorithm for every problem in NP.

A problem Π is *NP-complete* if it is both a member of NP and is NP-hard (see Figure 2.1). The NP-complete problems are the most difficult of those in the "zone of uncertainty."

The topic of NP-completeness was initiated by Cook [1971]. Emphasizing the significance of polynomial-time reducibility, he focused attention on NP decision problems. He proved that the SATISFIABILITY problem of

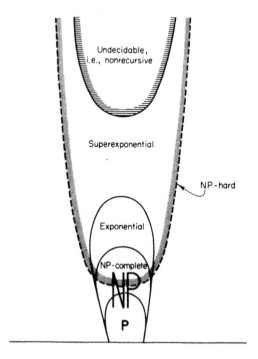

Figure 2.1. The hierarchy of complexities. The big open question is whether or not the "zone of uncertainty," NP-P, is empty.

mathematical logic is NP-complete (Cook's theorem), and he suggested other problems which might be NP-complete. Karp [1972] presented a large collection of NP-complete problems (about two dozen) arising from combinatorics, logic, set theory, and other areas of discrete mathematics. In the next few years, hundreds of problems were shown to be NP-complete. The standard technique employed with NP-completeness is as follows: First, by Cook's theorem, place SATISFIABILITY in the bag of NP-complete problems. Next, repeat the following sequence of instructions a few hundred times:

Find a candidate Π which might be NP-complete. Select an appropriate Π' from the bag of NP-complete problems. Show that $\Pi \in$ NP and $\Pi' \leqslant \Pi$. Add Π to the bag.

An amount of cleverness is needed in selecting Π' and finding a transformation from Π' to Π. By way of illustration we will demonstrate such a reduction in the proof of the next theorem. For a more complete treatment of Cook's theorem and the reductions following from it, see (in increasing level of scope) Reingold, Nievergelt, and Deo [1977], Aho, Hopcroft, and Ullman [1974], and Garey and Johnson [1978].

To illustrate the technique of reduction, we present the following result.

Theorem 2.1 (Poljak [1974]). (i) STABLE SET $\leqslant$ STABLE SET ON TRIANGLE-FREE GRAPHS;
   (ii)   STABLE SET $\leqslant$ GRAPH COLORING.

Proof. (i)   Let G be an undirected graph with n vertices and e edges. The idea of our proof will be to construct from G a certain triangle-free graph H with the property that knowing $\alpha(H)$ will immediately give us $\alpha(G)$. Subdivide each edge of G into a path of length 3; call the resulting graph H. Clearly, H is a triangle-free graph with $n + 2e$ vertices and $3e$ edges. Also, H can be constructed from G in $O(n + e)$ steps. Finally, since $\alpha(H) = \alpha(G) + e$, a deterministic polynomial time algorithm which solves for $\alpha(H)$ yields a solution to $\alpha(G)$.

   (ii)   Let G be an undirected graph and construct H as in part (i). Next we construct H' from H as follows. The vertices of H' correspond to the edges of H, and we connect two vertices of H' if their corresponding edges in H do not share a common vertex. This construction can be easily carried out in $O(e^2)$ steps. Since H is triangle-free, $\chi(H') = (2e + n) - \alpha(H) = e + n - \alpha(G)$. Thus, $\alpha(G)$ can be determined from $\chi(H')$. ∎

Since it is well known that STABLE SET is NP-complete, we obtain the following lesser known result.

Corollary 2.2. STABLE SET ON TRIANGLE-FREE GRAPHS is NP-complete.

A graph theoretic or other type of problem Π which is normally hard to solve in the general case may have an efficient solution if the input domain is suitably restricted. The HAMILTONIAN CIRCUIT problem, for example, is trivial if the only graphs considered are trees. However, we have seen that restricting the STABLE SET problem to triangle-free graphs is not sufficient to allow fast calculation (until someone proves that P = NP). Research has found interesting families $\mathscr{F}$ of graphs for which certain hard problems Π when restricted to $\mathscr{F}$ are nontrivial and tractable (i.e., in P). In this book we will consider this situation for various families of perfect graphs and some not so perfect graphs. A more perplexing topic currently under investigation by many complexity theorists is that of finding and understanding the cause of the *boundary* between the tractability and intractability of various problems Π.

One final note: Our definition of complexity suppressed one fundamental point. An implementation of an algorithm is always taken relative to some specified type of machine. As an underlying assumption throughout this book we will take the random access machine (RAM), introduced by Cook and Reckhow [1973], as our model of computation. The RAM is an abstraction of a general-purpose digital computer in which each storage cell has a unique address, allowing it to perform in one computational step an access to any cell, an arithmetic or Boolean operation, or a comparison. A computation is performed sequentially by a RAM, one step at a time. The theory of NP-complete problems is usually formulated using the Turing machine model rather than the RAM. This presents no difficulty, however, since any RAM can be simulated on a deterministic Turing machine with only a polynomial increase in running time.

Summary

Besides providing a basis for comparing algorithms which solve the same problem, algorithmic analysis has other practical uses. Most importantly, it affords us the opportunity to know *in advance of the computation* an estimate or a bound on the storage and run time requirements. Such advance knowledge would be essential when designing a computer system for a manned spacecraft in which the ability to calculate trajectories and fire the guidance rockets appropriately within tight constraints had better be guaranteed. Even in less urgent situations, having advanced estimates allows a programmer to set job card limits to abort those runs which exceed the expected bounds and hence probably contain errors, and to avoid aborting correct programs. Also such estimates are needed by the person who must decide whether or not it is worthwhile spending the necessary funds on computer time to carry out a certain (very large) computation.

2. Data Structures

As the name suggests, data structures provide a systematic framework in which the variables being processed (both input and internal) can be organized. Data structures are really mathematical objects, but we will usually refer to their computer implementations by the same names. The most familiar data structure is the *array*, which is used in conjunction with subscripted variables. A *0-dimensional array* is a single variable or storage location. A *d-dimensional array* can be defined recursively as a finite sequence of $(d - 1)$-dimensional arrays all of the same size. A *vector* is usually stored as a 1-dimensional array and a *matrix* as a 2-dimensional array. It is generally accepted that the entries of an array must be homogeneous (i.e., all of the same type and all requiring the same amount of space).

The main feature of an array is its indexing capability. The subscripts should uniquely determine the location of each data item. The entries of an array are stored consecutively, and an addressing scheme using *multipliers* allows access to any entry in a constant amount of time, independent of the size of the array, on a random access machine. Thus, a query of the form "Is $A_{5, 12} > 0$?" can be executed in essentially one step.

For those unfamiliar with the use of multipliers, the technique will be illustrated for an $m_1 \times m_2$ matrix A. Let us assume that the entries of A are stored sequentially in locations of size s in the order $A_{1, 1}, A_{1, 2}, \ldots, A_{1, m_2}, A_{2, 1}, A_{2, 2}, \ldots, A_{2, m_2}, \ldots, A_{m_1, 1}, A_{m_1, 2}, \ldots, A_{m_1 m_2}$ (row-major ordering). Then the space used by each row of A equals $m_2 s$. Now $A_{i, j}$ could be accessed by starting at $A_{1, 1}$, jumping down $i - 1$ rows, and then moving over $j - 1$ columns. Thus, if $B = \text{ADDRESS}(A_{1, 1})$, then we have the formula

$$\text{ADDRESS}(A_{i, j}) = B + (i - 1)m_2 s + (j - 1)s.$$

An analogous formula can be obtained for column-major ordering. This idea easily extends to *d*-dimensional arrays (Exercise 14).

A *list* is a data structure which consists of homogeneous *records* which are linked together in a linear fashion. Each record will contain one field or more of data and one field or more of pointers. Figure 2.2 shows two singly linked lists; each record has a single forward pointer. Unlike an array, in which the

COURANT

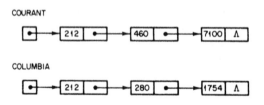

COLUMBIA

Figure 2.2. Two singly linked lists.

data is stored sequentially in memory, the records of a list can be scattered throughout memory. The pointers maintain law and order. This allows the flexibility of changing the size of the data structure, inserting and deleting items, by simply changing the values of a few pointers rather than shifting large blocks of data. An implementation of our examples is given in Figure 2.3. It uses two arrays and two single variables. The Λ is a special symbol indicating undefined. The list COURANT can be printed out by the following program:

```
begin
    P ← COURANT;
    while P ≠ Λ
        print DATA(P);
        P ← POINTER(P);;
end
```

This is an example of *scanning* a list. Scanning takes time proportional to the length of the list.

Two special types of lists should be mentioned here because of their usefulness in computer science. A *stack* is a list in which we are only permitted to insert and delete elements at one end, called the *top* of the stack. A *queue* is a list in which we are only permitted to insert at one end, called the *tail* of the queue, and delete from the other end, called the *head* of the queue.

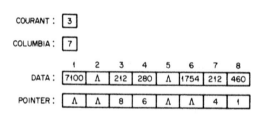

Figure 2.3. An implementation of the lists COURANT and COLUMBIA using arrays. (In what year was Columbia founded?)

The Adjacency Matrix of a Graph

Let $G = (V, E)$ be a graph whose vertices have been (arbitrarily) ordered $v_1, v_2, \ldots, v_n$. The adjacency matrix $\mathbf{M} = (m_{i,j})$ of G is an $n \times n$ matrix with entries

$$m_{i,j} = \begin{cases} 0 & \text{if } v_i v_j \notin E, \\ 1 & \text{if } v_i v_j \in E \end{cases}$$

(see Figure 2.4b). By definition, the main diagonal of $\mathbf{M}$ is all zeros, and M is symmetric about the main diagonal if and only if G is an undirected graph. If $\mathbf{M}$ is stored in a computer as a 2-dimensional array, then only one step (more precisely $O(1)$ time) is required for the statements "Is $v_i v_j \in E$?" or "Erase the edge $v_i v_j$." An instruction such as "mark each vertex which is adjacent to v_j" requires scanning the entire column j and hence takes n steps. Similarly, "mark each edge" takes n^2 steps. The space requirement for the array representation is $O(n^2)$. A graph whose edges are weighted can be represented in the same fashion. In this more general case $m_{i,j}$ will equal the weight of $v_i v_j$; a nonedge will have weight either zero or infinity depending upon the application.

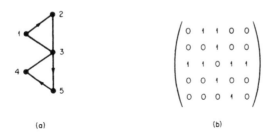

(a)   (b)

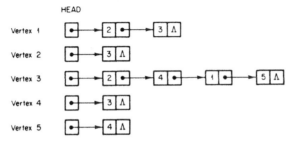

(c)

Figure 2.4. (a) The graph G. (b) The adjacency matrix of G. (c) The adjacency lists of G.

Some of the performance figures above can be improved upon when the density of **M** is low. We use the term *sparse* to indicate that $|E| \ll n^2$, i.e., the number of edges is *much* less than n^2. One of the most talked about classes of sparse graphs are the planar graphs for which Euler proved that $\|E\| \leq 3n - 6$.

The Adjacency Lists of a Graph

For each vertex v_i of G we create a list $\text{Adj}(v_i)$ containing those vertices adjacent to v_i. The adjacency lists are not necessarily sorted although one might wish them to be (see Figure 2.4c). The space requirement for the adjacency list representation of a graph with n vertices and e edges is

$$O\left(\sum_i [1 + d_i]\right) = O(n + e),$$

where d_i denotes the degree of v_i (see Figure 2.5). Thus, from *storage* considerations, it is usually more advantageous to use adjacency lists than the

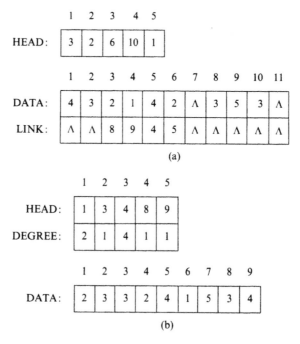

Figure 2.5.   Two implementations of Figure 2.4c. (a) An implementation of the adjacency sets of G as linked lists. (b) An implementation of the adjacency sets of G in sequential storage.

Table 2.2

Some typical graph operations and their complexity with respect to three data structures[a]

	Adjacency matrix stored as an array	Adjacency sets stored as lists	Adjacency sets stored sequentially
Is v_iv_j an edge?	$O(1)$	$O(d_i)^*$	$O(d_i)^*$
Mark each vertex which is adjacent to v_i	$O(n)$	$O(d_i)$	$O(d_i)$
Mark each edge	$O(n^2)$	$O(e)$	$O(e)$
Add an edge v_iv_j	$O(1)$	$O(1)^{**}$	$O(e)$
Erase an edge v_iv_j	$O(1)$	$O(d_i)^*$	$O(e)$

[a] If the adjacency sets are sorted, then the starred entries can be reduced to $O(\log d_i)$ using a binary search, but the double starred entry will increase to $O(d_i)$.

adjacency matrix to store a sparse graph. Often, it is also advantageous from *time* considerations to store a sparse graph using adjacency lists. For example, the instruction "mark each vertex which is adjacent to v_j" requires scanning the list Adj(v_j) and hence takes d_j steps. Similarly, "mark each edge" takes $O(e)$ steps using adjacency lists, a substantial saving over the adjacency matrix for a sparse graph. However, erasing an edge is more complex with lists than with the matrix (see Table 2.2). Thus *there is no representation of a graph that is best for all operations and processes.* Since the selection of a particular data structure can noticeably affect the speed and efficiency of an algorithm, decisions about the representation must incorporate a knowledge of the algorithms to be applied. Conversely, the choice of an algorithm may depend on how the data is initially given. For example, an algorithm to set up the adjacency lists of a sparse graph will take longer if we are initially given its adjacency matrix as an $n \times n$ array rather than as a collection of ordered pairs representing the edges.

A graph problem is said to be *linear in the size of the graph*, or simply *linear*, if it has an algorithm which can be implemented to run in $O(n + e)$ steps on a graph with n vertices and e edges. This is usually the best that one could expect for a graph problem. By a careful choice of algorithm and data structure a number of simple problems can be solved in linear time; these include testing for connectivity (Section 2.3), biconnectivity (Exercise 5), and planarity (Table 2.1). We will illustrate this on the problem of converting the adjacency lists of a graph into *sorted adjacency lists.*

It is by now a well-known fact that any algorithm which correctly sorts a set of k numbers using comparisons will require at least $k \log k$ comparisons both in the worst case and in the average case.* Furthermore, many

* All logarithms will be base 2.

$O(k \log k)$-time algorithms for sorting by comparisons are available: HEAP-SORT, BINARY INSERTION, MERGESORT, etc. This might suggest that sorting the adjacency list of v_i requires $d_i \log d_i$ steps, so that sorting *all* the adjacency lists would take $\sum_{i=1}^{n} d_i \log d_i$ steps, which is superlinear, i.e., greater than $O(n + e)$. As an alternative to comparison sorting, $Adj(v_i)$ could be put into order using a radix or bucket sort. This method takes $O(n + d_i)$ moves and is executed as follows:

1. Initialize bit vector: $\langle b_1, b_2, \ldots, b_n \rangle \leftarrow \langle 0, 0, \ldots, 0 \rangle$
2. Scan $Adj(v_i)$ assigning: $b_j \leftarrow 1$ for each $v_j \in Adj(v_i)$
3. Set $SortedAdj(v_i) \leftarrow \emptyset$
4. Scan bit vector: **for** $j \leftarrow 1$ **to** n **do**
   if $b_j = 1$ **then** CONCATENATE v_j to $SortedAdj(v_i)$

If this were done for all adjacency sets, it would require $O(n^2)$ steps which is superlinear for sparse graphs. Happily, there is yet another method for ordering the adjacency lists, which turns out to be linear. It is conceptually very simple and differs from the above in that the $SortedAdj(v_i)$ are not created separately, but rather, simultaneously.

Algorithm 2.1. Sorting the adjacency lists of a graph.

Input: The unsorted adjacency lists of a graph $G = (V, E)$ whose vertices are numbered $v_1, v_2, \ldots, v_n$.
Output: The sorted adjacency lists of the reversal $G^{-1} = (V, E^{-1})$. (If G is undirected, then $G = G^{-1}$; otherwise run the algorithm a second time on G^{-1}.)
Method: The algorithm is as follows:

```
    begin
1.      for i ← 1 to n do SortedAdj(vᵢ) ← empty list;
2.      for i ← 1 to n do
3.          for each vⱼ ∈ Adj(vᵢ) do
4.              CONCATENATE vᵢ to SortedAdj(vⱼ);
    end
```

Theorem 2.3. Algorithm 2.1 runs in $O(n + e)$ time.

Proof. Line 1 is a loop which takes $O(n)$ steps. Concatenation is independent of the length of a list provided that a pointing variable is used to remember the address of the end of the list. Thus line 4 takes $O(1)$ steps, and the loop 3–4 takes $O(d_i)$ steps. Therefore, the nested loops 2–4 require a total of $\sum_{i=1}^{n} O(d_i) = O(e)$ steps, which proves the theorem. ∎

The usual implementation of adjacency sets as linked lists is illustrated in Figure 2.5a. There is an alternate way of storing the adjacency sets when no inserting or deleting is anticipated. Under these circumstances sequential storage can be used to eliminate the links that were present in the list representation and thus save space. In both implementations HEAD(i) points to the first member of Adj(v_i), but Adj(v_i) is now stored in *consecutive locations* DATA(HEAD(i)), ..., DATA(HEAD(i) + DEGREE(i) − 1) (see Figure 2.5b) and Exercise 9).

For further reading on data structures and their uses see Knuth [1969], Aho, Hopcroft, and Ullman [1974], Horowitz and Sahni [1976], Lewis and Smith [1976], Wirth [1976], Goodman and Hedetniemi [1977], Reingold, Nievergelt, and Deo [1977], and Gotleib and Gotleib [1978].

3.   How to Explore a Graph

In designing algorithms we frequently require a mechanism for exploring the vertices and edges of a graph. Having the adjacency sets at hand allows us to repeatedly pass from a vertex to one of its neighbors and thus "walk" through the graph. Typically, in the midst of such a searching algorithm, some of the vertices will have been visited, the remainder not yet visited. A decision will have to be made as to which vertex x is being visited next. Since, in general, there will be many eligible candidates for x, we may want to establish some sort of priority among them.

Two criteria of priority which prove to be especially useful in exploring a graph are discussed in this section. They are depth-first search (DFS) and breadth-first search (BFS). In both methods each edge is traversed exactly once in the forward and reverse directions and each vertex is visited. By examining a graph in such a structured way, some algorithms become easier to understand and faster to execute. The choice of which method to use will often affect the efficiency of the algorithm. Thus, simply selecting a clever data structure is not sufficient to insure a good implementation. A carefully chosen search technique is also needed.

Depth-First Search

In DFS we select and visit a vertex a, then visit a vertex b adjacent to a, continuing with a vertex c adjacent to b (but different from a), followed by an "unvisited" d adjacent to c, and so forth. As we go deeper and deeper into the graph, we will eventually visit a vertex y with no unvisited neighbors; when this happens, we return to the vertex x immediately preceeding y in the search and revisit x. Note that if G is a connected undirected graph, then

procedure DFSEARCH(v):
begin
1.   mark v "visited"; $i \leftarrow i + 1$; DFSNUMBER(v) $\leftarrow i$;
2.   **for** each $w \in$ Adj(v) **do**
3.   **if** w is marked "unvisited" **then**
         begin
4.   add the edge vw to T; FATHER(w) $\leftarrow v$;
5.   DFSEARCH(w);
         end
end

Figure 2.6. Depth-first search.

each vertex will be visited and every edge will be explored once in both directions. If G is not connected, then such a search is carried out for each connected component of G.

A depth-first search of an undirected graph $G = (V, E)$ partitions the edge set into two classes T and B where T comprises a *spanning forest* of G with one spanning tree for each component of G. The edge xy is placed into T if vertex y was visited for the first time immediately following a visit to x. In this case x is called the *father* of y and y is the *son* of x. The origin of this male-dominated nomenclature appears to be biblical. The edges in T are called *tree edges*. The remaining edges, called *back edges*, are placed into B; they are also called *fronds* by an(n) arborist graph theorist. If G is connected then (V, T) is called a *depth-first spanning tree*. We consider each tree of the depth-first spanning forest to be *rooted* at the vertex at which the DFS of that tree was begun.

An algorithm for depth-first search is given below.

Algorithm 2.2. Depth-first search of a graph.

Input: An undirected graph $G = (V, E)$ represented by adjacency sets Adj(v), for $v \in V$.
Output: A partition of E into a set T of tree edges and a set B of back edges.
Method: All vertices are initially marked "unvisited." The procedure DFSEARCH in Figure 2.6 is used recursively. All edges in E not placed into T are assumed to be in B. In addition, the vertices are numbered from 1 to n according to the order in which they are first visited; DFSNUMBER(v) denotes this number for a vertex v. The algorithm is as follows:

begin
6.   initialize: $T \leftarrow \emptyset$; $i \leftarrow 0$;
7.   **for** all $v \in V$ **do** mark v "unvisited";
8.   **while** there exists $v \in V$ marked "unvisited" **do**
9.   DFSEARCH(v);
end

In general a graph may have many depth-first spanning forests. Indeed there is quite a bit of freedom in choosing the vertices in lines 2 and 8. Nonetheless, a depth-first spanning forest T has some important and useful properties, which we now state.

(D1) If v is a proper ancestor of w in T, then DFSNUMBER(v) < DFSNUMBER(w).

(D2) For every edge of G, whether tree or back edge, one of its endpoints is an ancestor of the other endpoint, that is, there are no "cross edges."

We leave the proof of properties (D1) and (D2) as an exercise.

DFSEARCH(v) is an example of a *recursive procedure*, that is, it calls itself. Such a procedure is implemented using a stack. When a call to itself is made, the current values of all variables local to the procedure and the line of the procedure which made the call are stored at the top of the stack. In this way when control is returned the computation can continue where it had left off. Some computer languages, like ALGOL, PL/I, PASCAL, and SETL, allow recursive subroutines and set up the stack automatically for you. Other languages, like FORTRAN, COBOL, or BASIC, do not have this feature, so that the programmer must set up his own stack to simulate the recursion.

Breadth-First Search

In BFS we select a vertex and put it on an initially empty queue of vertices *to be visited*. We repeatedly remove the vertex x at the head of the queue and then place onto the queue all vertices adjacent to x which have never been enqueued. As in the case of depth-first search, BFS is carried out once for each connected component of the graph. However, in BFS each vertex is visited only once (and is thus exhausted, having produced all its offspring in one visit).

A breadth-first search of an undirected graph $G = (V, E)$ also partitions the edge set into two classes: the tree edges in T and the back edges in B. Here an edge xy is placed into T if vertex y is enqueued during the visit to x. The (partial) subgraph (V, T) is called a *breadth-first spanning forest*.

An algorithm for breadth-first search is given below.

Algorithm 2.3. Breadth-first search of a graph.

Input: An undirected graph $G = (V, E)$ represented by adjacency sets Adj(v), for $v \in V$.

Output: A partition of E into a set T of tree edges and a set B of back edges.

Method: All vertices are initially marked "never enqueued." The procedure BFSEARCH in Figure 2.7 is used to visit a vertex. All edges in E not placed

procedure BFSEARCH(x):
begin
1. $i \leftarrow i + 1$; BFSNUMBER(x) $\leftarrow i$;
2.   **for** each $y \in \text{Adj}(x)$ **do**
3.   **if** y is marked "never enqueued" **then**
          begin
4.   add the edge xy to T; FATHER(y) $\leftarrow x$;
5.   add y to Q; mark y "enqueued";
          end
end

Figure 2.7.   Breadth-first search.

into T are assumed to be in B. An array BFSNUMBER records the order in which the vertices are enqueued and visited. The algorithm is as follows:

begin
6.   initialize: $T \leftarrow \varnothing$; $Q \leftarrow$ empty queue; $i \leftarrow 0$;
7.   **for** all $v \in V$ **do** mark v "never enqueued";
8.   **while** Q is empty and there exists $v \in V$ marked "never enqueued"
9.   add v to Q; mark v "enqueued";
10.   **while** Q is nonempty
11. $x \leftarrow$ head of Q; $Q \leftarrow Q - x$;
12.   BFSEARCH(x);
        end
      end
end

Let T be a breadth-first spanning forest of an undirected graph $G = (V, E)$. As was the case for DFS, a graph may have many breadth-first spanning forests. The *level* (in T) of a vertex v is defined inductively:

$$\text{LEVEL}(v) = \begin{cases} 0, & \text{if } v \text{ is a root of a tree in } T \\ 1 + \text{LEVEL(FATHER}(v)), & \text{otherwise.} \end{cases}$$

A breadth-first spanning forest T satisfies the following properties.

(B1)   If v is a proper ancestor of w in T, then BFSNUMBER(v) < BFSNUMBER(w).

(B2)   Every edge of G, whether tree or back edge, connects two vertices whose level in T differs by at most 1.

(B3)   If v is a vertex in the connected component of G whose root in T is r, then the level of v equals the length of the shortest path from r to v in G.

In Section 4.3 we will discuss a variant of the process described here, called lexicographic breadth-first search, in which the vertices of a given level are not searched in the same order as they are enqueued, but rather according to a priority which depends on their ancestors.

Implementation and Complexity

Let $G = (V, E)$ be a graph. Both Algorithms 2.2 and 2.3 can be implemented to run in time and space proportional to $|V| + |E|$. Such an implementation is said to be *linear* in the size of G. This is usually the best that one can expect from a graph algorithm, since it is reasonable to assume each vertex and each edge must be processed. Let us describe in detail a linear implementation of Algorithm 2.2 and leave Algorithm 2.3 as an exercise.

The adjacency sets of G can be stored either as singly linked lists or by using sequential allocation*; thus, the input can be entered in $O(|V| + |E|)$ time and space. A Boolean array VISITED of size $|V|$ can serve to mark each vertex v *unvisited* if VISITED$(v) = 0$ and *visited* if VISITED$(v) = 1$. Thus, line 7 of Algorithm 2.2 can be executed in $O(|V|)$ time, and the tests in lines 3 and 8 can be done in constant time. The set T can be a singly linked list, while FATHER and DFSNUMBER will be arrays of size $|V|$. Hence statements 1, 4–6, and 9 can each be done in constant time. Now comes the crucial part of the complexity analysis. Statement 8 requires a pointing variable which will scan, or run through, all the vertices exactly once. That is, when this pointer finds an unvisited vertex, the pointer's value will be saved, so that the next time statement 8 is required the search for an unvisited vertex can resume at the spot where it had last left off (rather than starting at the beginning of V each time). Therefore, the total number of operations summed over all executions of statement 8 is proportional to $|V|$. Exactly the same technique is used in statement 2 to scan Adj(v) which, together with our previous comments, implies that the entire procedure DFSEARCH(v) takes $O(|Adj(v)|)$ time. Finally, the procedure is called once for each vertex, so the total time (and space) complexity of our implementation is

$$O(|V|) + \sum_{v \in V} O(|Adj(v)|),$$

which equals $O(|V| + |E|)$.

As we mentioned in the opening paragraphs of this section, one search technique may be preferable over another, that is, it may give us a more efficient implementation. We list some instances of problems for which DFS and BFS are most effective, respectively.

DFS—planarity testing; certain connectivity related problems (biconnectivity, triconnectivity); topological sorting; testing for cycles in an oriented graph.

---

* If inserting or deleting of edges were required in the algorithm, then sequential allocation would not be advisable.

BFS—shortest-path problems; testing for chordless cycles (Sections 4.3 and 4.4); network flow problems.

In the next section we will discuss one of these problems—topological sorting. Upon completing that section the reader will have been exposed to all the algorithmic tools needed for the remainder of the book. For additional reading in this area see Aho, Hopcroft, and Ullman [1974], Goodman and Hedetniemi [1977], and Reingold, Nievergelt, and Deo [1977].

4. Transitive Tournaments and Topological Sorting

Let F be an orientation of the complete graph K_n on n vertices. Each edge xy of F may be regarded as the outcome of a contest between the vertices x and y, where x was the loser and y the winner. We call F a *transitive tournament* if, for all triples of vertices,

$$xy \in F \quad \text{and} \quad yz \in F \quad \text{implies} \quad xz \in F. \tag{1}$$

Condition (1) simply says that F has no 3-cycles. A stronger statement can be made.

Theorem 2.4. Let F be an orientation of the complete graph K_n. The following statements are equivalent.

(i) F is a transitive tournament.
(ii) F is acyclic.

Moreover, the vertices can be linearly ordered $[v_1, v_2, \ldots, v_n]$ such that
(iii) v_i has in-degree $i - 1$ in F, for all i, and
(iv) $v_i v_j \in F$ if and only if $i < j$.

This linear ordering of the vertices is unique. Figure 2.8 shows a transitive tournament and the linear ordering of its vertices.

Proof. (i) $\Rightarrow$ (ii) Since F is transitive, it has no 3-cycle. Suppose F has an l-cycle ($l > 3$) where l is smallest possible. But this l-cycle has a chord which shortcuts it, producing a cycle of shorter length and thus contradicting the minimality of l. Hence, F is acyclic.

(ii) $\Rightarrow$ (iii) If F is acyclic, then it has a sink (a vertex of out-degree zero). Call the sink v_n. Clearly v_n has in-degree $n - 1$. Deleting v_n from the graph, we obtain a smaller acyclic oriented graph, and the conclusion follows by induction.

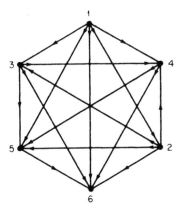

Figure 2.8.   A transitive tournament.

(iii) ⇒ (iv)   By induction.
(iv) ⇒ (i)   Obvious.

This theorem provides us with a linear time algorithm for recognizing transitive tournaments. First, calculate the in-degree of each vertex; then, using a Boolean vector, verify that there are no duplicates among the in-degrees. The technique of recognizing a class of graphs *solely on the basis of the degrees of their vertices* will be seen again when we study threshold graphs (Chapter 10) and split graphs (Chapter 6).

A slightly more general problem than recognizing transitive tournaments is that of topologically sorting an arbitrary acyclic oriented graph $G = (V, F)$. What we seek is a linear ordering of the vertices $[v_1, v_2, \ldots, v_n]$ which is consistent with the edges of G; that is,

$$v_i v_j \in F \Rightarrow i < j \qquad \text{(for all } i, j\text{).} \tag{2}$$

An ordering which satisfies (2) is called a *topological sorting* of G. If G had a cycle, then a topological sorting would clearly be impossible. Why? But, if G is acyclic, then it is always possible. One method for finding an ordering satisfying (2) is the following:

for $j \leftarrow |V|$ **to** 1 **step** -1
  Locate a sink v of the remaining graph and call it v_j;
  Delete v and all edges incident on v from the graph;   (3)
next j;

The correctness of this method is left as an exercise. In practice, we can implement (3) without actually deleting anything from our data structures.

Rather, we employ a depth-first search and some clever labeling. The algorithm is presented below.

Algorithm 2.4.　Topological sorting.

Input: An acyclic oriented graph $G = (V, F)$ stored as adjacency lists.
Output: A DFS numbering of the vertices called DFSNUMBER and a topological sort numbering of the vertices called TSNUMBER. The algorithm also tests to make sure that G is acyclic.
Method: To find the jth vertex of the desired ordering, the depth-first search procedure TOPSORT in Figure 2.9 locates a vertex v all of whose successors in G have already been searched and numbered and are therefore considered as having been deleted. This vertex v is then numbered. The entire algorithm is as follows:

```
begin
   for each x ∈ V do
      DFSNUMBER(x) ← 0;
      TSNUMBER(x) ← 0;;
   j ← |V|;
   i ← 0;
   for each x ∈ V do
      if DFSNUMBER(x) = 0 then
         TOPSORT(x);
end
```

Algorithm 2.4 is illustrated in an example in Appendix C.

```
procedure TOPSORT(v):
   i ← i + 1;
   DFSNUMBER(v) ← i;
   for all w ∈ Adj(v) do
      begin
         if DFSNUMBER(w) = 0 then
            TOPSORT(w);
         else if TSNUMBER(w) = 0 then
            "G is not acyclic";
      end
   comment: We now label v with a value smaller than
      the value assigned to any descendant.
   TSNUMBER(v) ← j;
   j ← j - 1;
   return
```

Figure 2.9.

EXERCISES

1. (a) Show that a spanning tree of the complete graph K_4 is either a depth-first spanning tree or a breadth-first spanning tree.
(b) Find a spanning tree of the complete graph K_5 which is neither a depth-first nor a breadth-first spanning tree.
2. Modify the DFS and BFS Algorithms 2.2 and 2.3 to count the number of connected components of an undirected graph G.
3. Prove properties (D1) and (D2) for any depth-first search spanning forest T.
4. A vertex x is an *articulation vertex* of G if deleting x and all edges incident on it increases the number of connected components. Let G be a connected undirected graph, and let T be a DFS spanning tree of G. Prove that a vertex x is an articulation vertex of G if and only if one of the following holds:
   (i) x is the root of T, and x has more than one son;
   (ii) x is not the root of T, and for some son s of x there is no back edge between any descendent of s (including s itself) and a proper ancestor of x.

   Remark. A connected undirected graph G is *biconnected* (there are two vertex-disjoint paths between every pair of vertices) if and only if G has no articulation vertex.

5 (Biconnectivity). Let T be a DFS spanning tree of an undirected graph G. Assume that the vertices are numbered consecutively as they are first visited during depth-first search, and let $\pi(v)$ denote this number. For each vertex x, define

$$\mathrm{LOW}(x) = \mathrm{MIN}\{\pi(x), \pi(w)\},$$

where w runs over all proper ancestors of x accessible from a son of x by going down some tree edges and then up *one* back edge.
   (a)   Write a depth-first search algorithm which assigns the values $\pi(x)$ and calculates the values $\mathrm{LOW}(x)$ for all vertices x.
   (b)   Prove that your algorithm can run in $O(|V| + |E|)$ time for an arbitrary graph $G = (V, E)$.
   (c)   Show how your algorithm can detect articulation vertices using the function LOW.
6. Describe an efficient implementation of Algorithm 2.3 and prove that it is linear in the size of the graph.
7. Let S and T be subsets of the integers $1, 2, 3, \ldots, n$, and let A be a one-dimensional array of size n whose values have been initialized $A(1) = A(2) = \cdots = A(n) = 0$. Write subroutines which calculate $S \cup T$ and $S \cap T$ in time proportional to $|S| + |T|$. Assume that S and T are stored as (unordered) singly linked lists. (The answer appears in Appendix B.)

8.   Let $H = (V, F)$ be an acyclic oriented graph. A height function h is defined on the vertices inductively:

$$h(v) = \begin{cases} 0 & \text{if } v \text{ is a sink,} \\ 1 + \max\{h(w)\,|\,w \in \text{Adj}(v)\} & \text{otherwise.} \end{cases}$$

Write a DFS algorithm which assigns a height function h to the vertices. Prove that your algorithm can be implemented to run in $O(|V| + |F|)$ time.

9.   Let $V = \{1, 2, \ldots, n\}$ and let E be a collection of m ordered pairs representing the edges of a graph $G = (V, E)$. Write a FORTRAN program which allocates *sequential space* in an array A of size m to store the adjacency sets of G, where $\text{Adj}(1)$ is followed by $\text{Adj}(2)$, etc. Let b_i denote the location in A of the beginning of $\text{Adj}(i)$, and let d_i denote the out-degree of vertex i (i.e., the number of ordered pairs in which it is the first coordinate). You are permitted exactly two scans of E, one to calculate the out-degrees of the vertices and one to fill the array. For example, if

$$E = \{(4, 5), (1, 4), (6, 7), (3, 2), (4, 1), (5, 4), (8, 2),$$
$$(7, 6), (2, 3), (2, 8), (9, 4), (1, 6), (4, 9), (6, 1),$$
$$(5, 7), (4, 6), (4, 7), (7, 5), (6, 4), (7, 4)\},$$

then the array A should look as indicated in Figure 2.10. Note that

$$b_i = 1 + \sum_{j < i} d_j.$$

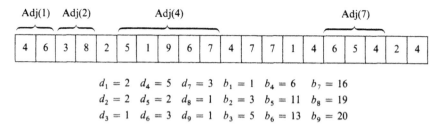

Adj(1)   Adj(2)   Adj(4)   Adj(7)

| 4 | 6 | 3 | 8 | 2 | 5 | 1 | 9 | 6 | 7 | 4 | 7 | 7 | 1 | 4 | 6 | 5 | 4 | 2 | 4 |

$d_1 = 2$	$d_4 = 5$	$d_7 = 3$	$b_1 = 1$	$b_4 = 6$	$b_7 = 16$
$d_2 = 2$	$d_5 = 2$	$d_8 = 1$	$b_2 = 3$	$b_5 = 11$	$b_8 = 19$
$d_3 = 1$	$d_6 = 3$	$d_9 = 1$	$b_3 = 5$	$b_6 = 13$	$b_9 = 20$

Figure 2.10.

10.   Using the data structure from Exercise 9 implement the algorithm from Exercise 8 and test it on some sample graphs.

11.   Using the data structure from Exercise 9 implement the algorithm from Exercise 5 to test some sample undirected graphs for biconnectivity.

12. Let $U(n)$ be the set of all upper triangular, $(0, 1)$-valued $n \times n$ matrices. That is, an $n \times n$ matrix $\mathbf{M} = [m_{ij}]$ is in $U(n)$ if

$$m_{ij} = \begin{cases} 0 & i > j, \\ 1 & i = j, \\ 0 \text{ or } 1 & i < j. \end{cases}$$

Show that $U(n)$ forms a group under matrix multiplication over the two element field GF(2) and that the identity matrix I is the identity element of this group. (In GF(2): $0 + 0 = 1 + 1 = 0$, $0 + 1 = 1 + 0 = 1$, $0 \cdot 0 = 0 \cdot 1 = 1 \cdot 0 = 0$, and $1 \cdot 1 = 1$.)

13. Let $G = (V, F)$ be an acyclic, topologically sorted, oriented graph; i.e., its vertices have been renamed such that $V = \{1, 2, \ldots, n\}$ and

$$ij \in F \quad \text{implies} \quad i < j \qquad (i, j \in V).$$

Clearly, this numbering implies that the adjacency matrix $\mathbf{M}(G)$ of G is upper triangular and is therefore in $U(n)$ (see Exercise 12). Consider the subset

$$U_G = \{\mathbf{M} \in U(n) | m_{ij} = 1, i < j \Rightarrow ij \in F\}$$

consisting of those matrices in $U(x)$ with nonzeros only in the positions determined by the nonzeros of $\mathbf{M}(G)$.

    (i)   Show that the elements of U_G can be ordered by set inclusion to form a complete, distributive lattice.

    (ii)   Show that U_G is a subgroup of $U(n)$ if and only if F is transitive (i.e., a strict partial order).

14. Let A be a d-dimensional array of size $m_1 \times m_2 \times \cdots \times m_d$. Discuss how A may be stored in consecutive storage locations of size s in a manner similar to row-major or column-major ordering. Give a formula for obtaining the address of $A_{i_1, i_2, \ldots, i_d}$.

15. In the proof of Theorem 2.1, show that the following claims are valid:

    (i)   The transformation $G \mapsto H$ is $O(n + e)$.

    (ii) $\alpha(H) = \alpha(G) + e$.

    (iii) The transformation $H \mapsto H'$ is $O(e^2)$. Can this be improved?

    (iv) $\chi(H') = e + n - \alpha(G)$.

16. Prove the following: If G has n vertices, then $\chi(G) \leq r$ if and only if $\alpha(G \times K_r) = r$, where $\times$ denotes the Cartesian product. (The Cartesian product of two graphs $G_1 = (V_1, E_1)$ and $G_2 = (V_2, E_2)$ is the graph $G = (V_1 \times V_2, E)$, where $E = \{((v_1, v_2), (v_1', v_2')) | \text{either } v_1 = v_1' \text{ and } (v_2, v_2') \in E_2 \text{ or } v_2 = v_2' \text{ and } (v_1, v_1') \in E_1\}$ (Chvátal [1973, p. 326]).

17. Using Exercise 16, show that GRAPH COLORING $\preccurlyeq$ STABLE SET.

18. Prove that assigning a minimum coloring to a bipartite graph has complexity which is linear in the size of the graph.

19. Prove that STABLE SET restricted to bipartite graphs has complexity which is polynomial in the size of the graph.

20. Prove that HAMILTONIAN CIRCUIT restricted to bipartite graphs is NP-complete.

21. If a positive integer m can be stored in $1 + [\log_2 m]$ space, show that the numbers $1, 2, 3, \ldots, n$ can be stored in a total of $O(n)$ space.

22. Algorithms $\mathcal{A}$ and $\mathcal{B}$ run in n^2 and 2^n steps, respectively, on an input of size n.

(i) If current computers can execute 10^9 steps/sec, what size input can be processed by each algorithm in one minute? In one hour? In one year?

(ii) Suppose that by the time this book reaches your university library the computer industry has a technological breakthrough, which increases the speed of execution by 100-fold. What will be the corresponding increased capability of algorithms $\mathcal{A}$ and $\mathcal{B}$?

Bibliography

Aho, A. V., Hopcroft, J. E., and Ullman, J. D.
[1974] "The Design and Analysis of Computer Algorithms." Addison-Wesley, Reading, Massachusetts.
Auslander, L., and Parter, S.
[1961] On embedding graphs in the sphere, *J. Math. Mech.* **10**, 517–523. MR25 #1548.
Booth, K. S., and Leuker, G. S.
[1976] Testing for the consecutive ones property, interval graphs, and graph planarity using PQ-tree algorithms, *J. Comput. System Sci.* **13**, 335–379. MR55 #6932.
Cherkasky, B. V.
[1977] Algorithm of construction of maximal flow in networks with complexity of $O(V^2 \sqrt{E})$ operations, (in Russian), *Math. Methods of Solution of Economical Problems* 7, 117–125.
Christofides, Nicos
[1975] "Graph Theory—An Algorithmic Approach." Academic Press, New York.
Chvátal, Václáv.
[1973] Edmonds polytopes and a hierarchy of combinatorial problems, *Discrete Math.* **4**, 305–337.
Cook, Stephen A.
[1971] The complexity of theorem-proving procedures, *Proc. 3rd Ann. ACM Symp. on Theory of Computing Machinery, New York*, pp. 151–158.
Cook, S. A., and Reckhow, R. A.
[1973] Time bounded random access machines, *J. Comput. System Sci.* **7**, 354–375.
Dinic, E. A.
[1970] Algorithm for solution of a problem of maximal flow in a network with proper estimation, *Soviet Math. Dokl.* **11**, 1277–1280. MR44 #5178.
Edmonds, Jack, and Karp, Richard M.
[1972] Theoretical improvements in algorithmic efficiency for network flow problems, *J. ACM* **19**, 248–264.

Ford, L. R., and Fulkerson, D. R.
  [1962] "Flows in Networks." Princeton Univ. Press, Princeton, New Jersey
Galil, Zvi
  [1978] A new algorithm for the maximum flow problem, *Proc. 19th IEEE Annu. Symp. on Foundations of Computer Science, Ann Arbor, Michigan, 16–18 October*, pp. 231–245.
Galil, Zvi, and Naamad, Amnon
  [1979] Network flow and generalized path compression, *Proc. 11th Annu. ACM Symp. on Theory of Computing*.
Garey, Michael R., and Johnson, David S.
  [1978] "Computers and Intractability: A Guide to the Theory of NP-completeness." Freeman, San Francisco, California.
Goldstein, A. J.
  [1963] An efficient and constructive algorithm for testing whether a graph can be embedded in the plane, *Graph and Combinatorics Conf.*, Office of Naval Research Logistics Proj., Dept. of Math., Princeton Univ., Princeton, New Jersey.
Goodman, S. E., and Hedetniemi, S. T.
  [1977] "Introduction to the Design and Analysis of Algorithms." McGraw-Hill, New York.
Gotleib, Calvin, C., and Gotlieb, Leo R.
  [1978] "Data types and structures." Prentice-Hall, Englewood Cliffs, New Jersey.
Hamacher, H.
  [1979] Numerical investigations on the maximal flow algorithm of Karzanov, *Computing* **22**, 17–29.
Hopcroft, John E., and Tarjan, Robert Endre
  [1972] Planarity testing in $V \log V$ steps: Extended abstract, *in* "Information Processing 71," Vol. 1, "Foundations and Systems," pp. 85–90. North-Holland, Amsterdam.
  [1974] Efficient planarity testing, *J. ACM* **21**, 549–568.
Horowitz, E., and Sahni, S.
  [1976] "Fundamentals of Data Structures." Computer Science Press, Potomac, Maryland.
Karp, Richard
  [1972] Reducibility among combinatorial problems, *in* "Complexity of Computer Computations" (R. E. Miller and J. W. Thatcher, eds.), pp. 85–103. Plenum, New York.
Karzanov, A. V.
  [1974] Determining the maximum flow in a network by the method of preflows, *Soviet Math. Dokl.* **15**, 434–437.
          See Hamacher [1979] for an implementation of this algorithm.
Knuth, Donald E.
  [1969] "The Art of Computer Programming," Vol. 1. Addison-Wesley, Reading, Massachusetts.
  [1973] "The Art of Computer Programming," Vol. 3. Addison-Wesley, Reading, Massachusetts.
Kuratowski, C.
  [1930] Sur le problème des corbes gauches en topologie, *Fund. Math.* **15**, 271–283.
Laderman, Julian D.
  [1976] A noncommutative algorithm for multiplying 3×3 matrices using 23 multiplications, *Bull. Amer. Math. Soc.* **82**, 126–128.
Lempel, A., Even, S., and Cederbaum, I.
  [1967] An algorithm for planarity testing of graphs, *in* "Theory of Graphs: Int. Symp., Rome, July 1966" (P. Rosentiehl, ed.), pp. 215–232. Gordon and Breach, New York.
Lewis, T. G., and Smith, M. Z.
  [1976] "Applying Data Structures." Houghton Mifflin, Boston, Massachusetts.

Malhotra, V. M., Kumar, M. Pramodh, and Maheshwari, S. N.

[1978]   An $O(n^3)$ algorithm for finding the maximum flows in networks, *Inf. Processing Lett.* **7**, 277–278.

Pan, Viktor Ya.

[1978]   Strassen's algorithm is not optimal; trilinear techniques of aggregating, unifying and cancelling for constructing fast algorithms for matrix operations, *Proc. 19th IEEE Annu. Symp. on Foundations of Computer Science, Ann Arbor, Michigan, 16–18 October*, pp. 166–176.

[1979a] New fast algorithms for matrix operations, *SIAM J. Comput.*, to be published.

[1979b] Field extension and trilinear aggregating, uniting and canceling for the acceleration of matrix multiplications, *Proc. 20th IEEE Annu. Symp. on Foundations of Computer Science, San Juan, Puerto Rico (29–31 October)*, pp. 28–38.

Poljak, S.

[1974]   A note on stable sets and colorings of graphs, *Commun. Math. Univ. Carolinae* **15**, 307–309.

Reingold, Edward M., Nievergelt, Jurg, and Deo, Narsingh

[1977]   "Combinatorial Algorithms: Theory and Practice." Prentice-Hall, Englewood Cliffs, New Jersey.

Shirey, R. W.

[1969]   Implementation and analysis of efficient graph planarity testing algorithms, Ph. D. thesis, Univ. of Wisconsin.

Strassen, V.

[1969]   Gaussian elimination is not optimal, *Numer. Math.* **13**, 354–356. MR40 # 2223.

Tarjan, Robert Endre

[1978]   Complexity of combinatorial algorithms, *SIAM Rev.* **20**, 457–491.

Weide, Bruce

[1977]   A survey of analysis techniques for discrete algorithms. *Comput. Surveys* **9**, 291–313.

Wirth, N.

[1976]   "Algorithms + Data Structures = Programs." Prentice-Hall, Englewood Cliffs, New Jersey.

CHAPTER 3

Perfect Graphs

1. The Star of the Show

In this section we introduce the main character of the book—*the perfect graph*. He was "discovered" by Claude Berge, who has been his agent since the early 1960s. P.G. has appeared in such memorable works as "Färbung von Graphen, deren sämtliche bzw. deren ungerade Kreise starr sind" and "Caractérisation des graphes non orientés dont on peut orienter les arrêtes de manière à obtenir le graphe d'une relation d'ordre." Despite his seemingly assuming name, P.G. has mixed the highbrow glamorous life with an intense dedication to improving the plight of mankind. His feature role in "Perfect graphs and an application to optimizing municipal services" has won him admiration and respect around the globe. Traveling incognito, a further sign of his modesty, he has been spotted by fans disguised as a graph parfait or as a (banana) split graph in a local ice cream parlor. So, ladies and gentlemen, without further ado, the management proudly presents

THE PERFECT GRAPH

Let us recall the following parameters of an undirected graph, which were defined in Section 1.1.

$\omega(G)$, the *clique number* of G: the size of the largest complete subgraph of G.

$\chi(G)$, the *chromatic number* of G: the fewest number of colors needed to properly color the vertices of G, or equivalently, the fewest number of stable sets needed to cover the vertices of G.

$\alpha(G)$, the *stability number* of G: the size of the largest stable set of G.

$k(G)$, the *clique cover number* of G: the fewest number of complete subgraphs needed to cover the vertices of G.

The intersection of a clique and a stable set of a graph G can be at most one vertex. Thus, for any graph G,

$$\omega(G) \leq \chi(G)$$

and

$$\alpha(G) \leq k(G).$$

These equalities are dual to one another since $\alpha(G) = \omega(\bar{G})$ and $k(G) = \chi(\bar{G})$.

Let $G = (V, E)$ be an undirected graph. The main purpose of this book is to study those graphs satisfying the properties

$$(P_1) \qquad \omega(G_A) = \chi(G_A) \qquad \text{(for all } A \subseteq V)$$

and

$$(P_2) \qquad \alpha(G_A) = k(G_A) \qquad \text{(for all } A \subseteq V).$$

Such a graph is called *perfect*. It is clear by duality that a graph G satisfies (P_1) if and only if its complement $\bar{G}$ satisfies (P_2). A much stronger result was conjectured by Berge [1961], cultivated by Fulkerson [1969, 1971, 1972], and finally proven by Lovász [1972a], namely, that (P_1) and (P_2) are equivalent. This has become known as the Perfect Graph theorem, which will be proved in the next section along with a third equivalent condition, due to Lovász [1972b],

$$(P_3) \qquad \omega(G_A)\alpha(G_A) \geq |A| \qquad \text{(for all } A \subseteq V).$$

In subsequent chapters it will be sufficient to show that a graph satisfies any (P_i) in order to conclude that it is perfect, and a perfect graph will satisfy all properties (P_i).

A fourth characterization of perfect graphs, due to Chvátal [1975], will be discussed in Section 3.3, and we shall encounter still another formulation in the chapter on superperfect graphs.

It is traditional to call a graph χ-*perfect* if it satisfies (P_1) and α-*perfect* if it satisfies (P_2). The Perfect Graph theorem then states that a graph is χ-perfect if and only if it is α-perfect. However, *the equivalence of (P_1) and (P_2) fails for uncountable graphs*. The interested reader may consult the following references on infinite perfect graphs: Hajnal and Surányi [1958], Perles [1963], and Nash-Williams [1967], Baumgartner, Malitz, and Reinhardt [1970], Trotter [1971], and Wagon [1978].

2. The Perfect Graph Theorem

In this section we shall show the equivalence of properties (P_1)–(P_3). A key to the proof is that multiplication of the vertices of a graph, as defined below, preserves each of the properties (P_i).

Let G be an undirected graph with vertex x. The graph $G \circ x$ is obtained from G by adding a new vertex x' which is connected to all the neighbors of x. We leave it to the reader to prove the elementary property

$$(G \circ x) - y = (G - y) \circ x \qquad \text{for distinct vertices } x \text{ and } y.$$

More generally, if $x_1, x_2, \ldots, x_n$ are the vertices of G and $\mathbf{h} = (h_1, h_2, \ldots, h_n)$ is a vector of non-negative integers, then $H = G \circ \mathbf{h}$ is constructed by substituting for each x_i a stable set of h_i vertices $x_i^1, \ldots, x_i^{h_i}$ and joining x_i^s with x_j^t iff x_i and x_j are adjacent in G. We say that H is obtained from G by *multiplication of vertices*.

Remark. The definition allows $h_i = 0$, in which case H includes no copy of x_i. Thus, every induced subgraph of G can be obtained by multiplication of the appropriate $(0, 1)$-valued vector.

Lemma 3.1 (Berge [1961]).   Let H be obtained from G by multiplication of vertices.

(i)   If G satisfies (P_1), then H satisfies (P_1).
(ii)   If G satisfies (P_2), then H satisfies (P_2).

Proof.   The lemma is true if G has only one vertex. We shall assume that (i) and (ii) are true for all graphs with fewer vertices than G. Let $H = G \circ \mathbf{h}$. If one of the coordinates of $\mathbf{h}$ equals zero, say $h_i = 0$, then H can be obtained from $G - x_i$ by multiplication of vertices. But, if G satisfies (P_1) [resp. (P_2)], then $G - x_i$ also satisfies (P_1) [resp. (P_2)]. In this case the induction hypothesis implies (i) and (ii).

Thus, we may assume that each coordinate $h_i \geq 1$, and since H can be built up from a sequence of smaller multiplications (Exercise 2), it is sufficient to prove the result for $H = G \circ x$. Let x' denote the added "copy" of x.

Assume that G satisfies (P_1). Since x and x' are nonadjacent, $\omega(G \circ x) = \omega(G)$. Let G be colored using $\omega(G)$ colors. Color x' the same color as x. This will be a coloring of $G \circ x$ in $\omega(G \circ x)$ colors. Hence, $G \circ x$ satisfies (i).

Next assume that G satisfies (P_2). We must show that $\alpha(G \circ x) = k(G \circ x)$. Let $\mathscr{K}$ be a clique cover of G with $|\mathscr{K}| = k(G) = \alpha(G)$, and let K_x be the clique of $\mathscr{K}$ containing x. There are two cases.

Case 1: x is contained in a maximum stable set S of G, i.e., $|S| = \alpha(G)$. In this case $S \cup \{x'\}$ is a stable set of $G \circ x$, so

$$\alpha(G \circ x) = \alpha(G) + 1.$$

Since $\mathcal{K} \cup \{\{x'\}\}$ covers $G \circ x$, we have that

$$k(G \circ x) \leq k(G) + 1 = \alpha(G) + 1 = \alpha(G \circ x) \leq k(G \circ x).$$

Thus, $\alpha(G \circ x) = k(G \circ x)$.

Case 2: No maximum stable set of G contains x. In this case,

$$\alpha(G \circ x) = \alpha(G).$$

Since each clique of $\mathcal{K}$ intersects a maximum stable set exactly once, this is true in particular for K_x. But x is not a member of any maximum stable set. Therefore, $D = K_x - \{x\}$ intersects each maximum stable set of G exactly once, so

$$\alpha(G_{V-D}) = \alpha(G) - 1.$$

This implies that

$$k(G_{V-D}) = \alpha(G_{V-D}) = \alpha(G) - 1 = \alpha(G \circ x) - 1.$$

Taking a clique cover of G_{V-D} of cardinality $\alpha(G \circ x) - 1$ along with the extra clique $D \cup \{x'\}$, we obtain a cover of $G \circ x$. Therefore,

$$k(G \circ x) = \alpha(G \circ x). \qquad \blacksquare$$

Lemma 3.2 (Fulkerson [1971], Lovász [1972b]). Let G be an undirected graph each of whose proper induced subgraphs satisfies (P_2), and let H be obtained from G by multiplication of vertices. If G satisfies (P_3), then H satisfies (P_3).

Proof. Let G satisfy (P_3) and choose H to be a graph having the smallest possible number of vertices which can be obtained from G by multiplication of vertices but which fails to satisfy (P_3) itself. Thus,

$$\omega(H)\alpha(H) < |X|, \qquad (1)$$

where X denotes the vertex set of H, yet (P_3) does hold for each proper induced subgraph of H.

As in the proof of the preceding lemma, we may assume that each vertex of G was multiplied by at least 1 and that some vertex u was multiplied by $h \geq 2$. Let $U = \{u^1, u^2, \ldots, u^h\}$ be the vertices of H corresponding to u. The vertex u^1 plays a distinguished role in the proof. By the minimality of H, (P$_3$) is satisfied by H_{X-u^1}, which gives

$$|X| - 1 = |X - u^1| \le \omega(H_{X-u^1})\alpha(H_{X-u^1}) \qquad \text{[by (P}_3\text{)]}$$

$$\le \omega(H)\alpha(H)$$

$$\le |X| - 1 \qquad \text{[by (1)]}.$$

Thus, equality holds throughout, and we can define

$$p = \omega(H_{X-u^1}) = \omega(H),$$

$$q = \alpha(H_{X-u^1}) = \alpha(H),$$

and

$$pq = |X| - 1. \qquad (2)$$

Since H_{X-U} is obtained from $G - u$ by multiplication of vertices, Lemma 3.1 implies that H_{X-U} satisfies (P$_2$). Thus, H_{X-U} can be covered by a set of q complete subgraphs of H, say $K_1, K_2, \ldots, K_q$. We may assume that the K_i are pairwise disjoint and that $|K_1| \ge |K_2| \ge \cdots \ge |K_q|$. Obviously,

$$\sum_{i=1}^{q} |K_i| = |X - U| = |X| - h = pq - (h - 1) \qquad \text{[by (2)]}.$$

Since $|K_i| \le p$, at most $h - 1$ of the K_i fail to contribute p to the sum. Hence,

$$|K_1| = |K_2| = \cdots = |K_{q-h+1}| = p.$$

Let H' be the subgraph of H induced by $X' = K_1 \cup \cdots \cup K_{q-h+1} \cup \{u^1\}$. Thus

$$|X'| = p(q - h + 1) + 1 < pq + 1 = |X| \qquad \text{[by (2)]}, \qquad (3)$$

so by the minimality of H,

$$\omega(H')\alpha(H') \ge |X'| \qquad \text{[by (P}_3\text{)]}. \qquad (4)$$

But $p = \omega(H) \ge \omega(H')$, so

$$\alpha(H') \ge |X'|/p \qquad \text{[by (4)]}$$

$$> q - h + 1 \qquad \text{[by (3)]}.$$

Let S' be a stable set of H' of cardinality $q - h + 2$. Certainly $u^1 \in S'$, for otherwise S' would contain two vertices of a clique (by the definition of H'). Therefore, $S = S' \cup U$ is a stable set of H with $q + 1$ vertices, contradicting the definition of q. ∎

Theorem 3.3 The Perfect Graph Theorem (Lovász [1972b]). For an undirected graph $G = (V, E)$, the following statements are equivalent:

(P_1) $\omega(G_A) = \chi(G_A)$ (for all $A \subseteq V$),

(P_2) $\alpha(G_A) = k(G_A)$ (for all $A \subseteq V$),

(P_3) $\omega(G_A)\alpha(G_A) \geq |A|$ (for all $A \subseteq V$).

Proof. We may assume that the theorem is true for all graphs with fewer vertices than G.

$(P_1) \Rightarrow (P_3)$. Suppose we can color G_A in $\omega(G_A)$ colors. Since there are at most $\alpha(G_A)$ vertices of a given color it follows that $\omega(G_A)\alpha(G_A) \geq |A|$.

$(P_3) \Rightarrow (P_1)$. Let $G = (V, E)$ satisfy (P_3); then by induction each proper induced subgraph of G satisfies (P_1)–(P_3). It is sufficient to show that $\omega(G) = \chi(G)$.

If we had a stable set S of G such that $\omega(G_{V-S}) < \omega(G)$, we could then paint S orange and paint G_{V-S} in $\omega(G) - 1$ other colors, and we would have $\omega(G) = \chi(G)$.

Suppose G_{V-S} has an $\omega(G)$-clique $K(S)$ for every stable set S of G. Let $\mathscr{S}$ be the collection of all stable sets of G, and keep in mind that $S \cap K(S) = \varnothing$. For each $x_i \in V$, let h_i denote the number of cliques $K(S)$ which contain x_i. Let $H = (X, F)$ be obtained from G by multiplying each x_i by h_i. On the one hand, by Lemma 3.2,

$$\omega(H)\alpha(H) \geq |X|.$$

On the other hand, using some simple counting arguments we can easily show that

$$|X| = \sum_{x_i \in V} h_i$$

$$= \sum_{S \in \mathscr{S}} |K(S)| = \omega(G)|\mathscr{S}|, \tag{5}$$

$$\omega(H) \leq \omega(G), \tag{6}$$

$$\alpha(H) = \max_{T \in \mathscr{S}} \sum_{x_i \in T} h_i \tag{7}$$

$$= \max_{T \in \mathscr{S}} \left[\sum_{S \in \mathscr{S}} |T \cap K(S)| \right] \tag{8}$$

$$\leq |\mathscr{S}| - 1, \tag{9}$$

which together imply that

$$\omega(H)\alpha(H) \le \omega(G)(|\mathscr{S}| - 1) < |X|,$$

a contradiction.*

$(P_2) \Leftrightarrow (P_3)$. By what we have already proved, we have the following implications:

$$G \text{ satisfies } (P_2) \Leftrightarrow \bar{G} \text{ satisfies } (P_1)$$
$$\Leftrightarrow \bar{G} \text{ satisfies } (P_3) \Leftrightarrow G \text{ satisfies } (P_3).\qquad \blacksquare$$

Corollary 3.4. A graph G is perfect if and only if its complement $\bar{G}$ is perfect.

Corollary 3.5. A graph G is perfect if and only if every graph H obtained from G by multiplication of vertices is perfect.

Historical note. The equivalence of (P_1) and (P_2) was almost proved by Fulkerson. He heard the news of the success of Lovász, who was not aware of Fulkerson's work at that time, from a postcard sent by Berge. Fulkerson immediately returned to his previous results on pluperfection and, within a few hours, obtained his own proof. Such are the joys and sorrows of research. His consolation, to our benefit, was that in the process of his investigations, Fulkerson invented and developed the notion of antiblocking pairs of polyhedra, an idea which has become an important topic in the rapidly growing field of polyhedral combinatorics.†

Briefly, and in our terminology, Fulkerson had proved the following:

Let $\mathscr{M}(G)$ be the collection of all graphs H which can be constructed from a graph G by multiplication of vertices. Then, H satisfies (P_1) for all $H \in \mathscr{M}(G)$ if and only if H satisfies (P_2) for all $H \in \mathscr{M}(G)$.

---

* Equations (5)–(9) are justified as follows:

(5) Consider the incidence matrix whose rows are indexed by the vertices $x_1, x_2, \ldots, x_n$ and whose columns correspond to the cliques $K(S)$ for $S \in \mathscr{S}$. Then, h_i equals the number of non-zeros in row i, and $|K(S)|$ equals the number of nonzeros in its corresponding column, which is by definition equal to $\omega(G)$.

(6) At most one "copy" of any vertex of G could be in a clique of H.

(7) If a maximum stable set of H contains some of the "copies" of x_i, then it will contain all of the "copies."

(8) Restrict attention to those rows of the matrix pertinent to (5) which belong to elements of T.

(9) $|T \cap K(S)| \le 1$ and $|T \cap K(T)| = 0$.

†Polyhedral combinatorics deals with the interplay between concepts from combinatorics and mathematical programming.

Clearly, this result together with Lemma 3.1 would give a proof of the equivalence of (P_1) and (P_2) for G.

3.   p-Critical and Partitionable Graphs*

An undirected graph G is called *p-critical* if it is minimally imperfect, that is, G is *not* perfect but every proper induced subgraph of G is a perfect graph. Such a graph, in particular, satisfies the inequalities

$$\alpha(G - x) = k(G - x) \qquad \text{and} \qquad \omega(G - x) = \chi(G - x)$$

for all vertices x, where $G - x$ denotes the resulting graph after deleting x. The following properties of p-critical graphs are easy consequences of the Perfect Graph theorem.

Theorem 3.6.   If G is a p-critical graph on n vertices, then

$$n = \alpha(G)\omega(G) + 1,$$

and for all vertices x of G,

$$\alpha(G) = k(G - x) \qquad \text{and} \qquad \omega(G) = \chi(G - x).$$

Proof.   By Theorem 3.3, since G is p-critical we have $n > \alpha(G)\omega(G)$ and $n - 1 \le \alpha(G - x)\omega(G - x)$ for all vertices x. Thus,

$$n - 1 \le \alpha(G - x)\omega(G - x) \le \alpha(G)\omega(G) < n.$$

Hence, $n - 1 = \alpha(G)\omega(G)$, $\alpha(G) = \alpha(G - x) = k(G - x)$, and

$$\omega(G) = \omega(G - x) = \chi(G - x). \qquad \blacksquare$$

Let $\alpha, \omega \ge 2$ be arbitrary integers. An undirected graph G on n vertices is called (α, ω)-*partitionable* if $n = \alpha\omega + 1$ and for all vertices x of G

$$\alpha = k(G - x), \qquad \omega = \chi(G - x).$$

We have shown in Theorem 3.6 that every p-critical graph is (α, ω)-partitionable with $\alpha = \alpha(G)$ and $\omega = \omega(G)$. A more general result holds.

Remark 3.7.   After removing any vertex x of an (α, ω)-partitionable graph, the remaining graph has $\alpha\omega$ vertices, chromatic number ω, and clique cover number α. So an ω-coloring of $G - x$ will partition the vertices into ω stable sets, one of which must be at least of size α. Similarly, a minimum clique

---

* Sections 3.3–3.5 were written jointly with Mark Buckingham.

cover of $G - x$ will partition the vertices into α cliques, one of which must be at least of size ω.

Theorem 3.8. If G is an (α, ω)-partitionable graph, then $\alpha = \alpha(G)$ and $\omega = \omega(G)$.

Proof. Let $G = (V, E)$ be (α, ω)-partitionable. By Remark 3.7, $\alpha \le \alpha(G)$ and $\omega \le \omega(G)$. Conversely, take a maximum stable set S of G and let $y \in V - S$. Then S is also a maximum stable set of $G - y$, so

$$\alpha(G) = |S| = \alpha(G - y) \le k(G - y) = \alpha.$$

Thus, $\alpha(G) \le \alpha$. Similarly, $\omega(G) \le \omega$. Therefore, $\alpha = \alpha(G)$ and $\omega = \omega(G)$. ∎

Theorem 3.8 shows that the integers α and ω for a partitionable graph are unique. Therefore, we shall simply use the term partitionable graph and assume that $\alpha = \alpha(G)$ and $\omega = \omega(G)$. The class of p-critical graphs is properly contained in the class of partitionable graphs which, in turn, is properly contained in the class of imperfect graphs (Exercise 10).

Lemma 3.9. If G is a partitionable graph on n vertices, then the following conditions hold:

   (i) G contains a set of n maximum cliques $K_1, K_2, \ldots, K_n$ that cover each vertex of G exactly $\omega(G)$ times;
   (ii) G contains a set of n maximum stable sets $S_1, S_2, \ldots, S_n$ that cover each vertex of G exactly $\alpha(G)$ times; and
   (iii) $K_i \cap S_j = \emptyset$ if and only if $i = j$.

Proof. Choose a maximum clique K of G and, for each $x \in K$, choose a minimum clique cover $\mathcal{K}_x$ of $G - x$. By Remark 3.7, all of the members of $\mathcal{K}_x$ must be cliques of size ω. Finally, let $\mathbf{A}$ be the $n \times n$ matrix whose first row is the characteristic vector of K and whose subsequent rows are the characteristic vectors of each of the cliques in $\mathcal{K}_x$ for all $x \in K$. (Note that the number of rows is $1 + \alpha\omega = n$.)

Each vertex $y \notin K$ is covered once by $\mathcal{K}_x$ for all $x \in K$. Each vertex $z \in K$ is covered once by K and once by $\mathcal{K}_x$ for all $z \ne x \in K$. Therefore, every vector is covered ω times. For each row $\mathbf{a}_i$ of $\mathbf{A}$ we let K_i be the clique whose characteristic vector is $\mathbf{a}_i$. We may express (i) by the matrix equation $\mathbf{1A} = \omega\mathbf{1}$, where $\mathbf{1}$ is the row vector containing all ones. Condition (i) will be satisfied once we show that the K_i are distinct.

For each i, pick a vertex $v \in K_i$ and let $\mathcal{S}$ denote a minimum stable set covering (coloring) of $G - v$. By Remark 3.7 and an easy counting exercise, there must be some stable set $S_i \in \mathcal{S}$ such that $K_i \cap S_i = \emptyset$. Let $\mathbf{b}_i$ be the characteristic vector of S_i, and let **B** denote the $n \times n$ matrix having rows $\mathbf{b}_i$ for $i = 1, \ldots, n$. Since $\mathbf{1} \cdot \mathbf{b}_i^\mathsf{T} = \alpha$, we have

$$\mathbf{1AB^T} = \omega\mathbf{1B^T} = \omega\alpha\mathbf{1} = (n - 1)\mathbf{1}.$$

But $\mathbf{a}_i \cdot \mathbf{b}_i^\mathsf{T} = 0$, so $\mathbf{AB^T} = \mathbf{J} - \mathbf{I}$, where $\mathbf{J}$ is the matrix containing all ones and $\mathbf{I}$ is the identity matrix. This proves (iii).

Finally, both **A** and **B** are nonsingular matrices since $\mathbf{J} - \mathbf{I}$ is nonsingular. Thus, the K_i are distinct and the S_i are distinct. Furthermore,

$$\begin{aligned} \mathbf{1B} &= \mathbf{1BA^T(A^T)}^{-1} = \mathbf{1(J - I)(A^T)}^{-1} = (n - 1)\mathbf{1(A^T)}^{-1} \\ &= [(n - 1)/\omega]\mathbf{1} = \alpha\mathbf{1}, \end{aligned}$$

which proves (ii).   ∎

The next result shows that *all* the maximum cliques and stable sets of G are among those in Lemma 3.9.

Lemma 3.10. A partitionable graph G contains exactly n maximum cliques and n maximum stable sets.

Proof. Let **A** and **B** be the matrices whose rows are the characteristic vectors of the cliques and stable sets, respectively, satisfying $\mathbf{AB^T} = \mathbf{J} - \mathbf{I}$ as specified in Lemma 3.9. Suppose that **c** is the characteristic vector of some maximum clique of G. We will show that **c** is a row of **A**.

We first observe that $\mathbf{A}^{-1} = \omega^{-1}\mathbf{J} - \mathbf{B^T}$ since

$$\mathbf{A}(\omega^{-1}\mathbf{J} - \mathbf{B^T}) = \omega^{-1}\mathbf{AJ} - \mathbf{AB^T} = \mathbf{J} - \mathbf{AB^T} = \mathbf{I}.$$

A solution **t** to the equation $\mathbf{tA} = \mathbf{c}$ will satisfy

$$\mathbf{t} = \mathbf{cA}^{-1} = \omega^{-1}\mathbf{cJ} - \mathbf{cB^T} = \omega^{-1}(\omega\mathbf{1}) - \mathbf{cB^T} = \mathbf{1} - \mathbf{cB^T}.$$

Therefore, **t** is a (0, 1)-valued vector. Also,

$$\mathbf{t} \cdot \mathbf{1^T} = (\mathbf{1} - \mathbf{cB^T}) \cdot \mathbf{1^T} = n - \alpha\mathbf{c} \cdot \mathbf{1^T} = n - \alpha\omega = 1.$$

Therefore, **t** is a unit vector. This implies that **c** is a row of **A**.

Similarly, every characteristic vector of a maximum stable set is a row of **B**.   ∎

Theorem 3.11. Let G be an undirected graph on n vertices, and let $\alpha = \alpha(G)$ and $\omega = \omega(G)$. Then G is partitionable if and only if the following conditions hold:

(i) $n = \alpha\omega + 1$;

(ii) G has exactly n maximum cliques and n maximum stable sets;

(iii)   every vertex of G is contained in exactly ω maximum cliques and in exactly α maximum stable sets;

(iv)   each maximum clique intersects all but one maximum stable set and vice versa.

Proof. ($\Rightarrow$)   This implication follows from Lemmas 3.9 and 3.10.

($\Leftarrow$)   Following our previous notation, conditions (ii)–(iv) imply that

$$\mathbf{AJ} = \mathbf{JA} = \omega\mathbf{J}, \qquad \mathbf{BJ} = \mathbf{JB} = \alpha\mathbf{J}, \qquad \mathbf{AB}^\mathsf{T} = \mathbf{J} - \mathbf{I},$$

where $\mathbf{A}$ and $\mathbf{B}$ are $n \times n$ matrices whose rows are the characteristic vectors of the maximum cliques and maximum stable sets, respectively. Let x_i be a vertex of G and let $\mathbf{h}_i^\mathsf{T}$ be its corresponding column in $\mathbf{A}$. Since

$$\mathbf{A}^\mathsf{T}\mathbf{B} = \mathbf{B}^{-1}\mathbf{B}\mathbf{A}^\mathsf{T}\mathbf{B} = \mathbf{B}^{-1}(\mathbf{J} - \mathbf{I})\mathbf{B} = \mathbf{B}^{-1}(\alpha\mathbf{J} - \mathbf{B})$$
$$= \alpha\alpha^{-1}\mathbf{J} - \mathbf{I} = \mathbf{J} - \mathbf{I},$$

we obtain $\mathbf{h}_i\mathbf{B} = \mathbf{1} - \mathbf{e}_i$, where $\mathbf{e}_i$ is the ith unit vector. Thus, $\mathbf{h}_i$ designates ω rows of $\mathbf{B}$ (i.e., stable sets of G) which cover $G - x_i$. Thus, $\chi(G - x_i) \leq \omega$. By a similar argument, $k(G - x_i) \leq \alpha$ for all x_i. But since $n - 1 = \alpha\omega$, we must have $\chi(G - x_i) = \omega$ and $k(G - x_i) = \alpha$. Therefore, G is partitionable. ∎

Corollary 3.12   (Padberg [1974]).   If G is a p-critical graph, then conditions (i)–(iv) of Theorem 3.11 hold.

Padberg's investigation of the facial structure of polyhedra associated with $(0, 1)$-valued matrices first led him to a proof of Corollary 3.12. (We shall discuss some of Padberg's work in Section 3.5.) The proof presented here, using only elementary linear algebra, is due to Bland, Huang, and Trotter [1979]. Additional results on p-critical graphs can be found in Section 3.6.

The only p-critical graphs known are the chordless cycles of odd length and their complements. Figures 3.1 and 3.2 illustrate the conditions of Theorem 3.11 for the graphs C_5 and $\bar{C}_7$.

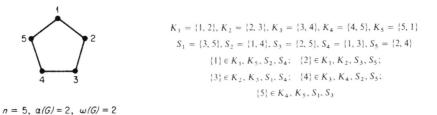

$K_1 = \{1, 2\}, K_2 = \{2, 3\}, K_3 = \{3, 4\}, K_4 = \{4, 5\}, K_5 = \{5, 1\}$

$S_1 = \{3, 5\}, S_2 = \{1, 4\}, S_3 = \{2, 5\}, S_4 = \{1, 3\}, S_5 = \{2, 4\}$

$\{1\} \in K_1, K_5, S_2, S_4; \quad \{2\} \in K_1, K_2, S_3, S_5;$

$\{3\} \in K_2, K_3, S_1, S_4; \quad \{4\} \in K_3, K_4, S_2, S_5;$

$\{5\} \in K_4, K_5, S_1, S_3$

$n = 5, \ \alpha(G) = 2, \ \omega(G) = 2$

Figure 3.1.   The graph C_5 and its maximum clique and stable set structure as specified in Theorem 3.11.

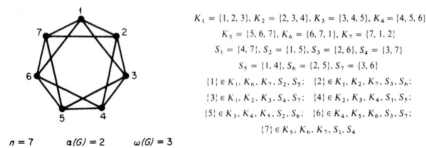

$K_1 = \{1, 2, 3\}, K_2 = \{2, 3, 4\}, K_3 = \{3, 4, 5\}, K_4 = \{4, 5, 6\}$

$K_5 = \{5, 6, 7\}, K_6 = \{6, 7, 1\}, K_7 = \{7, 1, 2\}$

$S_1 = \{4, 7\}, S_2 = \{1, 5\}, S_3 = \{2, 6\}, S_4 = \{3, 7\}$

$S_5 = \{1, 4\}, S_6 = \{2, 5\}, S_7 = \{3, 6\}$

$\{1\} \in K_1, K_6, K_7, S_2, S_5;\quad \{2\} \in K_1, K_2, K_7, S_3, S_6;$

$\{3\} \in K_1, K_2, K_3, S_4, S_7;\quad \{4\} \in K_2, K_3, K_4, S_1, S_5;$

$\{5\} \in K_3, K_4, K_5, S_2, S_6;\quad \{6\} \in K_4, K_5, K_6, S_3, S_7;$

$\{7\} \in K_5, K_6, K_7, S_1, S_4$

$n = 7 \qquad \alpha(G) = 2 \qquad \omega(G) = 3$

Figure 3.2. The graph $\bar{C}_7$ and its maximum clique and stable set structure as specified in Theorem 3.11.

4. A Polyhedral Characterization of Perfect Graphs

Let **A** be an $m \times n$ matrix. We consider the two polyhedra

$$P(\mathbf{A}) = \{\mathbf{x} \mid \mathbf{Ax} \le \mathbf{1}, \mathbf{x} \ge \mathbf{0}\}$$

and

$$P_I(\mathbf{A}) = \text{convex hull}(\{\mathbf{x} \mid \mathbf{x} \in P(\mathbf{A}), \mathbf{x} \text{ integral}\}),$$

where **x** is an n-vector and **1** is the m-vector of all ones. Clearly $P_I(\mathbf{A}) \subseteq P(\mathbf{A})$, and for (0, 1)-valued matrices **A** having no zero column, $P(\mathbf{A})$ and $P_I(\mathbf{A})$ are bounded and are within the unit hypercube in $\mathbb{R}^n$. An important example of such a matrix is the maximal cliques-versus-vertices incidence matrix of an undirected graph G. This is called the *clique matrix* if all the maximal cliques are included. The clique matrix of G is unique up to permutations of the rows and columns (see Figure 3.3).

Let **A** be any $m \times n$ (0, 1)-valued matrix having no zero columns. The *derived graph* of **A** has n vertices $v_1, v_2, \ldots, v_n$ corresponding to the columns of **A**, and an edge connecting v_i and v_j whenever the ith and jth columns of **A** have a 1 in some row $\mathbf{a}_k$. Clearly every row of **A** forms a (not necessarily

$$\begin{pmatrix} 1 & 1 & 1 & 0 & 0 \\ 0 & 1 & 0 & 1 & 0 \\ 0 & 0 & 1 & 0 & 1 \end{pmatrix}$$

Figure 3.3. A graph and its clique matrix.

maximal) clique in its derived graph. Many matrices have the same derived graph. For example, if $\mathbf{A}$ is either the clique matrix or the edge incidence matrix of G, then the derived graph of $\mathbf{A}$ will be G.

Lemma 3.13.   Let G be an undirected graph, and let $\mathbf{A}$ be any $(0, 1)$-valued matrix having no zero column whose derived graph equals G. Then $\mathbf{x}$ is an extremum of $P_I(\mathbf{A})$ if and only if $\mathbf{x}$ is the characteristic vector of some stable set of G.

Proof.   If $\mathbf{x}$ is an extremum of $P_I(\mathbf{A})$, then $\mathbf{x}$ must be integral, and since $\mathbf{A}$ is $(0, 1)$-valued without a zero column, $\mathbf{x} \leq \mathbf{1}$. Thus, $\mathbf{x}$ is the characteristic vector of some set of vertices S. Suppose there exist vertices u and v of S that are connected in G; hence some row $\mathbf{a}_k$ of $\mathbf{A}$ has a 1 in columns u and v. This yields $\mathbf{a}_k \cdot \mathbf{x} \geq 2$, yet $\mathbf{Ax} \leq \mathbf{1}$. Therefore, S must be a stable set.

Conversely, given that $\mathbf{x}$ is a characteristic vector of a stable set of G, certainly $\mathbf{x} \in P_I(\mathbf{A})$. Let $\mathbf{x}$ be expressed as a convex combination of some set of extrema $\{\mathbf{b}^{(1)}, \mathbf{b}^{(2)}, \ldots, \mathbf{b}^{(s)}\}$ of $P_I(\mathbf{A})$; that is,

$$x_k = \sum_i c^{(i)} b_k^{(i)}, \qquad 1 = \sum_i c^{(i)}, \qquad 0 \leq c^{(i)} \leq 1.$$

Thus, if $x_k = 1$, then $b_k^{(i)} = 1$ for all i, and if $x_k = 0$, then $b_k^{(i)} = 0$ for all i. Therefore, $\mathbf{x} = \mathbf{b}^{(i)}$ and $\mathbf{x}$ is an extremum of $P_I(\mathbf{A})$.   ∎

Theorem 3.14(Chvátal [1975]).   Let $\mathbf{A}$ be the clique matrix of an undirected graph G. Then G is perfect if and only if $P_I(\mathbf{A}) = P(\mathbf{A})$.

To prove the theorem we shall use a result from linear programming used by Edmonds [1965] and others:

Lemma 3.15.   Given bounded polyhedra S and T, where S has a finite number of extrema,

$$S = T \qquad \text{iff} \qquad \max_{\text{subj } \mathbf{x} \in S} \mathbf{c} \cdot \mathbf{x} = \max_{\text{subj } \mathbf{x} \in T} \mathbf{c} \cdot \mathbf{x} \quad (\forall \mathbf{c}, \text{ integral}).$$

Proof of Theorem 3.14.   Assume that $P_I(\mathbf{A}) = P(\mathbf{A})$. Let G_U be an induced subgraph of G, and let $\mathbf{u}$ denote the characteristic vector of U. We have,

$$\alpha(G_U) = \max_{\text{subj } \mathbf{x} \in P_I(\mathbf{A})} \mathbf{ux} = \max_{\text{subj } \mathbf{Ax} \leq \mathbf{1}, \, \mathbf{x} \geq 0} \mathbf{ux} = \min_{\text{subj } \mathbf{yA} \geq \mathbf{u}, \, \mathbf{y} \geq 0} \mathbf{y} \cdot \mathbf{1}.$$

The first equality follows from the fact that maximums are always achievable at some extremum and the extrema of $P_I(\mathbf{A})$ correspond to stable sets (Lemma 3.13). The second equality follows from Lemma 3.15 setting $\mathbf{c} = \mathbf{u}$, and the third equality comes from the duality theorem of linear programming.

Therefore, choose $\mathbf{y} \geq \mathbf{0}$ such that $\sum y_i = \alpha(G_U)$ and $\mathbf{u} \leq \mathbf{y}A$. Denoting the jth column of A by $\mathbf{a}^j$, we obtain

$$|U| = \sum_{j \in U} u_j \leq \sum_{j \in U} \mathbf{y} \cdot \mathbf{a}^j = \mathbf{y} \cdot \sum_{j \in U} \mathbf{a}^j \leq \mathbf{y} \cdot (\omega(G_U)\mathbf{1}) = \alpha(G_U)\omega(G_U).$$

Thus, by Theorem 3.3, G is perfect.

Conversely, assume that G is perfect. For any integer vector $\mathbf{c}$, form the graph H by multiplying the ith vertex of G by max $(0, c_i)$ for each i. By Lemma 3.1, H is perfect. We have the following:

$\alpha(H) = \alpha_c(G)$	The maximum weighted stable set of G given by $\mathbf{c}$.
$= \max \mathbf{c} \cdot \mathbf{x}$ $\text{subj } x \in P_I(A)$	The maximum can always be found at an extremum, which corresponds to a stable set (Lemma 3.13).
$\leq \max \mathbf{c} \cdot \mathbf{x}$ $\text{subj } x \in P(A)$	$P_I(A) \subseteq P(A)$.
$= \min \mathbf{y} \cdot \mathbf{1}$ $\text{subj } yA \geq c,\, y \geq 0$	Duality theorem.
$\leq \min \mathbf{y} \cdot \mathbf{1}$ $\text{subj } yA \geq c,\, \text{non-negative integral } y$	The constraint set is smaller.
$= k_c(G)$	The minimum clique covering of G such that vertex i is covered c_i times. The constraint $\mathbf{y}A \geq \mathbf{c}$, non-negative integral $\mathbf{y}$, specifies such a covering.
$= k(H).$	Any clique of H corresponds to a clique of G, thus $k(H) \geq k_c(G)$; if vertex i of G is covered by c_i cliques, then there are c_i cliques in H, each covering a different copy of i, so $k_c(G) \geq k(H)$.

But $\alpha(H) = k(H)$. Thus,

$$\max_{\text{subj } x \in P_I(A)} \mathbf{c} \cdot \mathbf{x} = \max_{\text{subj } x \in P(A)} \mathbf{c} \cdot \mathbf{x}$$

and, by Lemma 3.15, $P_I(A) = P(A)$.   ∎

Remark.   The first half of the proof of Theorem 3.14 still holds under a weakened hypothesis on A:

If A is a $(0, 1)$-valued matrix having no zero column whose derived graph equals G, then $P_I(A) = P(A)$ implies that G is perfect.

5. A Polyhedral Characterization of p-Critical Graphs

Manfred Padberg first discovered the properties shown in Section 3.3 of p-critical graphs while investigating the facial structure of the polyhedra $P(\mathbf{A})$ for general $(0, 1)$-valued matrices $\mathbf{A}$. In doing so, he also discovered a polyhedral characterization of p-critical graphs. In Padberg [1973, 1974], he used the results of Lovász and Chvátal to produce these results. In a later work, Padberg [1976b], he developed a more general approach, which enabled him to prove the same results directly and to prove the theorems of Lovász and Chvátal in a different manner.

The matrix $\mathbf{A}$ is said to be *perfect* if $P(\mathbf{A})$ is *integral*, that is, $P(\mathbf{A})$ has only integer extrema: $P_I(\mathbf{A}) = P(\mathbf{A})$. $\mathbf{A}$ is said to be *almost perfect* if $P(\mathbf{A})$ is *almost integral*, that is, (i) $P_I(\mathbf{A}) \neq P(\mathbf{A})$ ($P(\mathbf{A})$ has at least one nonintegral extremum), and (ii) the polyhedra $P_j(\mathbf{A}) = P(\mathbf{A}) \cap \{\mathbf{x} \in \mathbb{R}^n \,|\, x_j = 0\}$ are all integral, $j = 1, 2, \ldots, n$.

For the remainder of this section, $\mathbf{A}$ will always denote an $m \times n$ $(0, 1)$-valued matrix having no zero column, and P, P_I, and P_j will denote $P(\mathbf{A})$, $P_I(\mathbf{A})$, and $P_j(\mathbf{A})$, respectively.

Padberg's results, although not stated in the following manner, include the following six theorems.

Theorem 3.16. If $\mathbf{A}$ is perfect, then $\mathbf{A}$ is an augmented clique matrix of its derived graph, that is, $\mathbf{A}$ is the clique matrix possibly augmented with some redundant rows corresponding to nonmaximal cliques.

Let $\mathbf{J}$ denote the matrix of all ones and $\mathbf{I}$ the identity matrix. We say that $\mathbf{A}$ *contains* the $n \times n$ submatrix $\mathbf{J} - \mathbf{I}$ if some permutation of $\mathbf{J} - \mathbf{I}$ occurs as an $n \times n$ submatrix of $\mathbf{A}$.

Theorem 3.17. If $\mathbf{A}$ is almost perfect, then either (i) $\mathbf{A}$ is an augmented clique matrix of its derived graph or (ii) $\mathbf{A}$ contains the $n \times n$ submatrix $\mathbf{J} - \mathbf{I}$.

Theorem 3.18. Let G be the derived graph of $\mathbf{A}$. If $\mathbf{A}$ is almost perfect and does not contain the $n \times n$ submatrix $\mathbf{J} - \mathbf{I}$, then

(i) $n = \alpha(G)\omega(G) + 1$;
(ii) every vertex of G is in exactly ω cliques of size ω and in exactly α stable sets of size α;
(iii) G has exactly n maximum cliques and n maximum stable sets;
(iv) there is a numbering of the maximum cliques $K_1, K_2, \ldots, K_n$ and maximum stable sets $S_1, S_2, \ldots, S_n$ of G such that $K_i \cap S_j = \varnothing$ if and only if $i = j$.

$$A = \begin{pmatrix} 1 & 1 & 0 & 1 & 0 & 0 \\ 0 & 1 & 1 & 0 & 1 & 0 \\ 1 & 0 & 1 & 0 & 0 & 1 \end{pmatrix} \qquad G = \qquad A' = \begin{pmatrix} 1 & 1 & 0 & 1 & 0 & 0 \\ 0 & 1 & 1 & 0 & 1 & 0 \\ 1 & 0 & 1 & 0 & 0 & 1 \\ 1 & 1 & 1 & 0 & 0 & 0 \end{pmatrix}$$

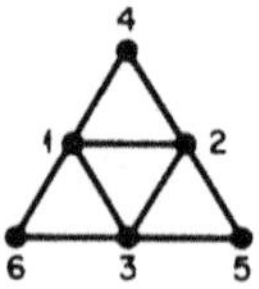

Figure 3.4. The derived graph G of the matrix A is a perfect graph, yet $P(A)$ has $(\tfrac{1}{2}, \tfrac{1}{2}, \tfrac{1}{2}, 0, 0, 0)$ as an extremum; thus A is an imperfect matrix. A' is the clique matrix of G and is perfect.

Theorem 3.19. A is perfect if and only if A is an augmented clique matrix of its derived graph and the derived graph is perfect.

Corollary 3.20. A is almost perfect if and only if either (i) A is an augmented clique matrix of its derived graph and the derived graph is almost perfect (p-critical) or (ii) A has no row of all ones and contains the $n \times n$ submatrix $J - I$ for $n \geq 3$. Furthermore, in (ii) the derived graph is complete.

Corollary 3.21. Every p-critical graph has the four properties of Theorem 3.18.

Note carefully the wording of Theorem 3.19. It is very possible that A is not a perfect matrix and yet its derived graph G is perfect and every row of A corresponds to a maximal clique of G. Of course, in this case, by Theorem 3.19, the matrix is missing a row corresponding to some other maximal clique (see Figure 3.4).

Theorems 3.16, 3.17, 3.19, and Corollary 3.20 are very useful when considering graphs as incidence matrices. Corollary 3.21 is a restatement of Corollary 3.12.

To show Theorems 3.16 and 3.17 we will turn to the concept of *antiblocking polyhedra* (Fulkerson [1971, 1972]). Two polyhedra P_1 and P_2 are an *antiblocking pair* if $P_1 = \{x \,|\, xP_2 \leq 1, \; x \geq 0\}$ or $P_2 = \{y \,|\, yP_1 \leq 1, \; y \geq 0\}$, the conditions being equivalent. If P_2 is generated from a $(0, 1)$-valued matrix A_2 having no zero column, then we have the property, among many others, that every extremum of P_1 is a projection of some row of A_2 and every nonredundant row of A_2 is an extremem of P_1 (Fulkerson [1972]). The same result holds if we interchange the indices.

Let $b^{(1)}, b^{(2)}, \ldots, b^{(r)}$ be the extrema of P_I and denote the matrix having rows $b^{(1)}, b^{(2)}, \ldots, b^{(r)}$ by B. Define $Q = P(B)$, $Q_I = P_I(B)$ and $Q_j = P_j(B)$ for $j = 1, 2, \ldots, n$. The polyhedra P_I and Q are an antiblocking pair (Fulkerson [1972]). By Lemma 3.13, the rows of B correspond to all of the stable sets of G, the derived graph of A. Thus, B is an augmented clique matrix of the complement $\bar{G}$. See also Monma and Trotter [1979].

Proof of Theorem 3.16.   Let $\mathbf{A}$ be perfect ($P_I = P$) and G be its derived graph. Since P_I and Q are an antiblocking pair, P and Q are also an antiblocking pair. By the properties of antiblocking pairs, the extrema of Q must all be projections of the rows of $\mathbf{A}$, so Q is integral. By Lemma 3.13, all the stable sets of $\bar{G}$, in other words all the cliques of G, are extrema of Q, since $Q = Q_I$. Thus, every clique of G must be a projection of some row of $\mathbf{A}$. Therefore, $\mathbf{A}$ is an augmented clique matrix of its derived graph.   ∎

Proof of Theorem 3.17.   Assuming that $\mathbf{A}$ is almost perfect, P_j is integral for $j = 1, 2, \ldots, n$. By a similar argument to that for Theorem 3.16, each Q_j is also integral. This follows since $P(\mathbf{A}) \cap \{\mathbf{x} \in \mathbb{R}^n | x_j = 0\}$ is the same as removing the jth column from $\mathbf{A}$ and forming its polyhedron.

Case 1: Q is not integral and thus is almost integral. In this case Padberg was able to show by a direct analysis of the facets of P that P and Q_I are an antiblocking pair. As in the proof of Theorem 3.16, we have that $\mathbf{A}$ is an augmented clique matrix of its derived graph.

Case 2: Q is integral. In this case Padberg was able to show by the non-integrality of P that $\mathbf{1}$ is an extremum of Q. This means that $\mathbf{B}$ must be the identity matrix (or a permutation of it). This in turn implies that the derived graph of $\mathbf{A}$ is complete. Therefore, for P_j to be integral, some row k of the matrix formed by deleting the jth column of $\mathbf{A}$ must be all ones (Theorem 3.16). Yet no row in $\mathbf{A}$ can have all ones since $\mathbf{A}$ is only almost perfect. Thus, row k in $\mathbf{A}$ must be all ones except for a zero in column j. Since this is true for all $j = 1, 2, \ldots, n$, $\mathbf{A}$ contains the $n \times n$ submatrix $\mathbf{J} - \mathbf{I}$.   ∎

Although Theorem 3.18 is essentially contained among the results of Section 3.3, Padberg's proof does not use the Perfect Graph theorem and his technique is valuable in its own right. Before proving Theorem 3.18 we state Padberg's cornerstone lemma.

Lemma 3.22.   If $\mathbf{x}$ is a nonintegral extremum of an almost integral polyhedron P, then for every $n \times n$ nonsingular submatrix $\mathbf{A}_1$ of $\mathbf{A}$ such that $\mathbf{A}_1 \mathbf{x} = \mathbf{1}$, there exists an $n \times n$ nonsingular submatrix $\mathbf{B}_1$ of $\mathbf{B}$ satisfying the matrix equation

$$\mathbf{B}_1 \mathbf{A}_1^\mathrm{T} = \mathbf{J} - \mathbf{I}.$$

Furthermore,

$$\mathbf{x} = (1/(n-1))\mathbf{B}_1^\mathrm{T}\mathbf{1}.$$

As a quick observation, we note that for any noninteger extremum $\mathbf{x}$ of P, $\mathbf{x} > \mathbf{0}$. If for some k, $x_k = 0$, then $\mathbf{x} \in P_k$ and thus is an extremum of P_k. But then $\mathbf{x}$ would have to be integral. The only way $\mathbf{x}$ could be an extremum of P

is to satisfy n linearly independent constraints of $\mathbf{Ax} \le \mathbf{1}$. Let $\mathbf{A}_1$ be the $n \times n$ nonsingular submatrix of $\mathbf{A}$. Thus, for each $\mathbf{x}$ there does exist such an $\mathbf{A}_1$ as specified in Lemma 3.22.

Padberg was also able to show that $\mathbf{x}$, a noninteger extremum of P, is the unique noninteger extremum; that $\mathbf{y} = (1/(n-1))\mathbf{A}_1^{\mathrm{T}}\mathbf{1}$, for any $\mathbf{A}_1$ of Lemma 3.22, is an extremum of Q; and that for any $\mathbf{A}_1$ and corresponding $\mathbf{B}_1$ of Lemma 3.22, $\mathbf{x} = |\det \mathbf{A}_1^{-1}|\mathbf{1}$ and $\mathbf{y} = |\det \mathbf{B}_1^{-1}|\mathbf{1}$. Armed with these matrix equations, the proof of Theorem 3.18 is a straightforward exercise in linear algebra.

Proof of Theorem 3.18. Let G be the derived graph of $\mathbf{A}$, where $\mathbf{A}$ is almost perfect and does not contain the $n \times n$ submatrix $\mathbf{J} - \mathbf{I}$. By the definition of almost perfect we have the existence of a noninteger extremum $\mathbf{x}$ of P. By Lemma 3.22 and the previous discussion, $\mathbf{x}$ is unique and there exist $n \times n$ nonsingular submatrices $\mathbf{A}_1$ of $\mathbf{A}$ and $\mathbf{B}_1$ of $\mathbf{B}$ such that $\mathbf{A}_1\mathbf{x} = \mathbf{1}$ and $\mathbf{B}_1\mathbf{A}_1^{\mathrm{T}} = \mathbf{J} - \mathbf{I}$. Moreover, for all such $\mathbf{A}_1$ and $\mathbf{B}_1$, $\mathbf{x} = (1/(n-1))\mathbf{B}_1^{\mathrm{T}}\mathbf{1} = |\det \mathbf{A}_1^{-1}|\mathbf{1}$, and $\mathbf{y}$, defined by $\mathbf{y} = (1/(n-1))\mathbf{A}_1^{\mathrm{T}}\mathbf{1} = |\det \mathbf{B}_1^{-1}|\mathbf{1}$, is an extremum of Q.

We shall first show that $\mathbf{A}_1$ is unique, in that any row $\mathbf{a}_k$ of $\mathbf{A}$ satisfying $\mathbf{a}_k \cdot \mathbf{x} = 1$ is in $\mathbf{A}_1$. We have the following implications:

$$\mathbf{B}_1\mathbf{A}_1^{\mathrm{T}} = \mathbf{J} - \mathbf{I} \Rightarrow \mathbf{A}_1\mathbf{B}_1^{\mathrm{T}} = \mathbf{J} - \mathbf{I} \Rightarrow \mathbf{B}_1^{\mathrm{T}} = \mathbf{X} - \mathbf{A}_1^{-1} \Rightarrow \mathbf{A}_1^{-1} = \mathbf{X} - \mathbf{B}_1^{\mathrm{T}},$$

where $\mathbf{X}$ is the $n \times n$ matrix having n columns of $\mathbf{x}$. Thus, if $\mathbf{a}_k \cdot \mathbf{x} = 1$, then $\mathbf{a}_k\mathbf{A}_1^{-1} = \mathbf{a}_k\mathbf{X} - \mathbf{a}_k\mathbf{B}_1^{\mathrm{T}}$ is 0 or 1, yet $\mathbf{a}_k\mathbf{A}_1^{-1} \cdot \mathbf{1} = \mathbf{a}_k \cdot \mathbf{x} = 1$. Therefore, $\mathbf{a}_k\mathbf{A}_1^{-1} = \mathbf{e}_j$, the jth unit vector, for some $j \in \{1, 2, \ldots, n\}$. This implies that $\mathbf{a}_k$ is equal to the jth row of $\mathbf{A}_1$, that is, $\mathbf{a}_k$ is in $\mathbf{A}_1$. Finally, since $\mathbf{x} = |\det \mathbf{A}_1^{-1}|\mathbf{1}$, we have that $\mathbf{A}_1$ contains exactly all the rows of $\mathbf{A}$ having the maximum number of ones. By Theorem 3.17, $\mathbf{A}$ is an augmented clique matrix of G. Therefore, $\mathbf{A}_1$ must contain exactly all the maximum cliques of G.

A similar argument holds for $\mathbf{y}$, $\mathbf{B}_1$, $\mathbf{B}$ and $\bar{G}$. Since

$$\mathbf{B}_1\mathbf{y} = \mathbf{B}_1((1/(n-1))\mathbf{A}_1^{\mathrm{T}}\mathbf{1}) = \mathbf{1},$$

we have $\mathbf{B}_1^{-1} = \mathbf{Y} - \mathbf{A}_1^{\mathrm{T}}$. Thus for any row $\mathbf{b}_k$ of $\mathbf{B}$ satisfying $\mathbf{b}_k \cdot \mathbf{y} = 1$, we have $\mathbf{b}_k\mathbf{B}_1^{-1} = \mathbf{b}_k\mathbf{Y} - \mathbf{b}_k\mathbf{A}_1^{\mathrm{T}}$, and yet $\mathbf{b}_k\mathbf{B}_1^{-1} \cdot \mathbf{1} = \mathbf{b}_k \cdot \mathbf{y} = 1$. So $\mathbf{b}_k$ is in $\mathbf{B}_1$. Since $\mathbf{y} = |\det \mathbf{B}_1^{-1}|\mathbf{1}$, and since by construction $\mathbf{B}$ is an augmented clique matrix of $\bar{G}$, we have that $\mathbf{B}_1$ must contain exactly all the maximum cliques of $\bar{G}$.

(i) The row sum of $\mathbf{A}_1$ is $\omega(G)$, yet $\mathbf{A}_1\mathbf{1} = \mathbf{A}_1|\det \mathbf{A}_1|\mathbf{x} = |\det \mathbf{A}_1|\mathbf{1}$; thus $|\det \mathbf{A}_1| = \omega(G)$. Similarly for $\mathbf{B}_1$, the row sum is $\alpha(G)$, yet

$$\mathbf{B}_1\mathbf{1} = \mathbf{B}_1|\det \mathbf{B}_1|\mathbf{y} = |\det \mathbf{B}_1|\mathbf{1};$$

so $|\det \mathbf{B}_1| = \alpha(G)$. Therefore, $\alpha(G)\omega(G) = |\det \mathbf{B}_1\mathbf{A}_1^{\mathrm{T}}| = |\det (\mathbf{E} - \mathbf{I})| = |(-1)^{n-1}(n-1)| = n - 1$. Thus, $n = \alpha(G)\omega(G) + 1$.

(ii)   Since $y = (1/(n - 1))A_1^T1$, we have $(1/\alpha)1 = (1/\alpha\omega)$, A_1^T1, and thus $\omega1 = A_1^T1$. That is, all the column sums of A_1 are ω. Therefore, every vertex is in exactly ω cliques of size ω.

Similarly for x, $x = (1/(n - 1))B_1^T1$ implies $(1/\omega)1 = (1/\alpha\omega)B_1^T1$, and hence $\alpha1 = B_1^T1$. Therefore, every vertex is in exactly α stable sets of size α.

(iii) A_1 is an $n \times n$ nonsingular matrix containing exactly all the maximum cliques of G; therefore, G has exactly n maximum cliques. By a similar argument on B_1, G has exactly n maximum stable sets.

(iv)   Let K_i correspond to the ith row of A_1 for $i = 1, 2, \ldots, n$, and S_j correspond to the jth row of B_1 for $j = 1, 2, \ldots, n$. Since $B_1A_1^T = J - I$, the maximum cliques $K_1, K_2, \ldots, K_n$ and maximum stable sets $S_1, S_2, \ldots, S_n$ of G are numbered such that $K_i \cap S_j = \varnothing$ if and only if $i = j$.   ∎

The "only if" condition of Theorem 3.19 is a stronger statement than Theorem 3.16; it states that the derived graph itself is perfect, which also turns out to be a sufficient condition for A to be perfect. In fact, Theorem 3.19 is precisely Chvátal's Theorems 3.14 and 3.16 put together. A more direct proof here will be instructive. Again we need an intermediate result of Padberg's.

Lemma 3.23. P is integral if and only if $\max_{\text{subj } x \in P} q \cdot x \equiv 0$ mod 1 for all $(0, 1)$-valued q.

It is well known that for a general matrix A with non-negative entries and no zero column, satisfying $\max_{\text{subj } x \in P} c \cdot x \equiv 0$ mod 1 for all non-negative c is equivalent to P being integral. But for our matrix A, considering only $(0, 1)$-valued q is sufficient.

Proof of Theorem 3.19.   ($\Leftarrow$)   Let A be an augmented clique matrix of its derived graph G, where G is perfect. Let q be a $(0, 1)$-valued vector and G' its corresponding induced subgraph of G. Then

$$\alpha(G') = \max_{\text{subj } x \in P_I} q \cdot x \leq \max_{\text{subj } x \in P} q \cdot x = \min_{\text{subj } yA \geq q, \ y \geq 0} y \cdot 1$$

$$\leq \min_{\text{subj } yA \geq q, \ y \geq 0, \text{ integral}} y \cdot 1 = k(G').$$

The first equality is clear because of Lemma 3.13 and the fact that an optimal solution can always be found at an extremum. The last equality is true since A is an augmented clique matrix and any optimal y is $(0, 1)$-valued. The inequalities have been seen before in Section 3.4.

Now since G is perfect, we must have equality everywhere. Thus,

$$\max_{\text{subj } x \in P} q \cdot x \equiv 0 \text{ mod 1}.$$

Finally, since $\mathbf{q}$ was arbitrary, Lemma 3.23 implies that P is integral, and thus $\mathbf{A}$ is perfect.

($\Rightarrow$)   Let $\mathbf{A}$ be perfect. By Theorem 3.16, $\mathbf{A}$ is an augmented clique matrix of its derived graph. To show that G is perfect we shall use induction on the size of the induced subgraphs.

For $|G'| = 0$ it is clear that $\alpha(G') = k(G')$. Assume that every k-vertex induced subgraph is perfect. Given $|G'| = k + 1$, let $\mathbf{q}$ be the characteristic vector of G'. Since P is integral and $\mathbf{A}$ is an augmented clique matrix of G,

$$\alpha(G') = \max_{\text{subj } \mathbf{x} \in P} \mathbf{q} \cdot \mathbf{x} = \min_{\text{subj } \mathbf{y}\mathbf{A} \geq \mathbf{q},\ \mathbf{y} \geq 0} \mathbf{y} \cdot \mathbf{1} \leq \min_{\text{subj } \mathbf{y}\mathbf{A} \geq \mathbf{q},\ \mathbf{y} \geq 0,\ \text{integral}} \mathbf{y} \cdot \mathbf{1}$$

$$= k(G'). \tag{10}$$

We claim that there is an integer optimal solution for $\min_{\text{subj } \mathbf{y}\mathbf{A} \geq \mathbf{q},\ \mathbf{y} \geq 0} \mathbf{y} \cdot \mathbf{1}$. We know that an optimal solution $\bar{\mathbf{y}}$ exists. If $\bar{\mathbf{y}}$ is integral we are done; otherwise there is a k such that $0 < \bar{y}_k < 1$. Clearly the kth row $\mathbf{a}_k$ of $\mathbf{A}$ has the property $\mathbf{a}_k \cdot \mathbf{q} > 0$, for otherwise $\bar{\mathbf{y}}$ would not be optimal. Define $\bar{\bar{q}}_i = q_i$ for $a_{ki} = 0$ and $\bar{\bar{q}}_i = 0$ for $a_{ki} = 1$. Since $\bar{\bar{\mathbf{q}}}$ is the characteristic vector of a smaller induced subgraph, and since (10) still holds, there is an integer optimal solution $\bar{\bar{\mathbf{y}}}$ for $\bar{\bar{\mathbf{q}}}$. Clearly any optimal solution for $\bar{\bar{\mathbf{q}}}$ has its kth component zero; thus $\bar{\mathbf{y}}$ is feasible but not optimal for $\bar{\bar{\mathbf{q}}}$. Yet $\mathbf{y}^*$, where $\mathbf{y}^* = \bar{\bar{\mathbf{y}}}$ except for $y_k^* = 1$, is feasible for $\mathbf{q}$. That is,

$$\min_{\text{subj } \mathbf{y}\mathbf{A} \geq \bar{\bar{\mathbf{q}}},\ \mathbf{y} \geq 0} \mathbf{y} \cdot \mathbf{1} < \min_{\text{subj } \mathbf{y}\mathbf{A} \geq \mathbf{q},\ \mathbf{y} \geq 0} \mathbf{y} \cdot \mathbf{1} \leq \mathbf{y}^* \cdot \mathbf{1} = \min_{\text{subj } \mathbf{y}\mathbf{A} \geq \bar{\bar{\mathbf{q}}},\ \mathbf{y} \geq 0} \mathbf{y} \cdot \mathbf{1} + 1.$$

Therefore, $\mathbf{y}^*$ is an integer optimal solution for $\min_{\text{subj } \mathbf{y}\mathbf{A} \geq \mathbf{q},\ \mathbf{y} \geq 0} \mathbf{y} \cdot \mathbf{1}$ and thus $\alpha(G') = k(G')$.   ∎

The observant reader will notice that the same "only if" proof could have been used in Theorem 3.14.

The proofs of Corollaries 3.20 and 3.21 are now easy.

Proof of Corollary 3.20.   ($\Leftarrow$)   *Case 1*: Let $\mathbf{A}$ be an augmented clique matrix of its derived graph G, where G is p-critical. Since deleting any vertex j of G results in a perfect graph, all the P_j are integral. Yet by Theorem 3.19, $\mathbf{A}$ is imperfect because G is imperfect; therefore $\mathbf{A}$ is almost perfect.

Case 2: Let $\mathbf{A}$ have no row of all ones and contain the $n \times n$ submatrix $\mathbf{J} - \mathbf{I}$ for $n \geq 3$. Since each P_j is obtained from the matrix $\mathbf{A}$ with its jth column deleted, and since this submatrix has a row containing all ones, all P_j are integral. Yet $(1/(n - 1))\mathbf{1}$ is an extremum of P, since every row has at most $n - 1$ ones and $\mathbf{J} - \mathbf{I}$ is an $n \times n$ submatrix. Therefore $\mathbf{A}$ is almost perfect.

($\Rightarrow$)   Given that $\mathbf{A}$ is almost perfect, we apply Theorem 3.17 to obtain two cases.

Case 1: **A** *is an augmented clique matrix of its derived graph G*. By Theorem 3.19 each P_j is integral, the submatrix of **A** obtained by deleting the *j*th column is perfect, and thus the deletion of any vertex *j* of *G* results in a perfect graph. Yet by Theorem 3.19 again, *G* itself is not perfect since **A** is not perfect. Therefore, *G* is p-critical.

Case 2: **A** *contains the* $n \times n$ *submatrix* **J** − **I**. Clearly **A** does not contain a row of all ones, for otherwise **A** would be perfect. Finally, we must certainly have $n \geq 3$, thus *G* is complete.   ∎

Proof of Corollary 3.21.   Given a p-critical graph *G*, form **A**, its clique matrix. By Corollary 3.20, case 1, **A** is almost perfect. Certainly *G* is the derived graph of **A**, and thus the hypothesis of Theorem 3.18 is satisfied.   ∎

6.   The Strong Perfect Graph Conjecture

The odd cycle C_{2k+1} (for $k \geq 2$) is not a perfect graph since $\alpha(C_{2k+1}) = k$ and $k(C_{2k+1}) = k + 1$ (or, alternatively, since $\omega(C_{2k+1}) = 2$ and $\chi(C_{2k+1}) = 3$). However, every proper subgraph of C_{2k+1} is perfect. Thus, C_{2k+1} is a p-critical graph (i.e., minimally imperfect) and by the Perfect Graph theorem its complement $\bar{C}_{2k+1}$ is also p-critical. To date, these are the only known p-critical graphs.

During the second international meeting on graph theory, held at Halle-on-Saal in March 1960, Claude Berge raised the question of whether or not other p-critical graphs besides the odd cycles and their complements exist. He conjectured that there are none, and this has come to be known as the *strong perfect graph conjecture (SPGC)*. (Actually, the word "conjecture" first appeared in Berge [1962].)

The strong perfect graph conjecture may be stated in several equivalent forms:

$SPGC_1$.   An undirected graph is perfect if and only if it contains no induced subgraph isomorphic to C_{2k+1} or $\bar{C}_{2k+1}$ (for $k \geq 2$).

$SPGC_2$.   An undirected graph *G* is perfect if and only if in *G* and in $\bar{G}$ every odd cycle of length ≥ 5 has a chord.

$SPGC_3$.   The only p-critical graphs that exist are C_{2k+1} and $\bar{C}_{2k+1}$ (for $k \geq 2$).

The graphs C_{2k+1} and $\bar{C}_{2k+1}$ are commonly referred to as the *odd hole* and the *odd antihole*, respectively.

We have seen in Sections 3.3 and 3.5 that p-critical graphs reflect an extraordinary amount of symmetry (as indeed they should if the SPGC turns out to be true). Let G be a p-critical graph on n vertices, and let $\alpha = \alpha(G)$ and $\omega = \omega(G)$. Then the following conditions hold for G.

Lovász condition

$$n = \alpha\omega + 1$$

Padberg conditions

Every vertex is in exactly ω maximum cliques (of size ω).
Every vertex is in exactly α maximum stable sets (of size α).
G has exactly n maximum cliques (of size ω).
G has exactly n maximum stable sets (of size α).
The maximum cliques and maximum stable sets can be indexed K_1, $K_2, \ldots, K_n$ and $S_1, S_2, \ldots, S_n$, respectively, so that $|K_i \cap S_j| = 1 - \delta_{ij}$, where δ_{ij} is the Kronecker delta.

Clearly, any p-critical graph must be connected. But C_n is the only connected graph on n vertices for which $\omega = 2$ and having exactly n undirected edges such that each vertex is an endpoint of exactly two of these edges. So, by Padberg's conditions we obtain another equivalent form of the strong perfect graph conjecture:

$SPGC_4$. There is no p-critical graph with $\alpha > 2$ and $\omega > 2$.

Recall from Section 3.3 that a partitionable graph on n vertices satisfies the Lovász and Padberg conditions.

Figures 3.5 and 3.9 give two examples of $(3, 3)$-partitionable graphs which fail to be p-critical. For this reason, the Lovász and Padberg conditions alone are not sufficient to prove the SPGC. Nevertheless, partitionable graphs do give us further reductions of the SPGC.

One special type of partitionable graph is easy to describe. The undirected graph C_n^d has vertices $v_1, v_2, v_3, \ldots, v_n$ with v_i and v_j joined by an edge if and only if i and j differ by at most d. (Here and in the next theorem all subscript arithmetic is taken modulo n.) It is easy to see that the graph $C_{\alpha\omega+1}^{\omega-1}$ is an (α, ω)-partitionable graph. When $\omega = 2$, then $C_{\alpha\omega+1}^{\omega-1}$ is simply the odd hole $C_{2\alpha+1}$; when $\alpha = 2$, then $C_{\alpha\omega+1}^{\omega-1}$ is the odd antihole $\bar{C}_{2\omega+1}$.

Theorem 3.24 (Chvátal [1976]). For any integers $\alpha \geq 3$ and $\omega \geq 3$, the partitionable graph $C_{\alpha\omega+1}^{\omega-1}$ is not p-critical.

Proof. Let $\alpha \geq 3$ and $\omega \geq 3$ be given. We will show that $C_{\alpha\omega+1}^{\omega-1}$ contains a proper induced subgraph H which is not perfect.

If we index the $n = \alpha\omega - 1$ maximal cliques $\{K_i\}$ of $C_{\alpha\omega+1}^{\omega-1}$ so that $K_i = \{v_i, v_{i+1}, \ldots, v_{i+\omega-1}\}$ for each $1 \leq i \leq n$, then the clique matrix of the graph has the familiar cyclical pattern, as shown in Figure 3.6. Let H denote the

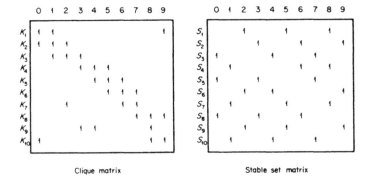

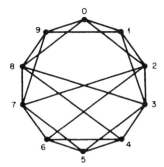

Figure 3.5. A graph satisfying the Lovász and Padberg conditions which fails to be p-critical. The clique matrix and stable set matrix indicate the required indexing of the maximum cliques and maximum stable sets. This example was discovered independently by Huang [1976] and by Chvátal, Graham, Perold, and Whitesides [1979].

subgraph remaining after deleting the $\alpha + 2$ vertices v_n, v_2, $v_{\omega+1}$, $v_{\omega+3}$, and all $v_{t\omega+2}$ for $t = 2, 3, \ldots, \alpha - 1$. In the deleting process every maximum clique has lost at least one of its members, so $\omega(H) \leq \omega - 1$. Therefore, it suffices to show that H cannot be colored using $\omega - 1$ colors.

Suppose that H is $\omega - 1$ colorable. Let v_1 be colored *black* and let the $\omega - 2$ additional colors be called the *rainbow*. We have the following series of implications:

$$\{v_1, v_3, \ldots, v_\omega\} \subset K_1 \Rightarrow \{v_3, \ldots, v_\omega\} \text{ requires the entire rainbow};$$

$$\{v_3, \ldots, v_\omega, v_{\omega+2}\} \subset K_3 \Rightarrow v_{\omega+2} \text{ is black};$$

$$\{v_{\omega+2}, v_{\omega+4}, \ldots, v_{2\omega+1}\} \subset K_{\omega+2}$$
$$\Rightarrow \{v_{\omega+4}, \ldots, v_{2\omega+1}\} \text{ requires the entire rainbow};$$

$$\{v_{\omega+4}, \ldots, v_{2\omega+1}, v_{2\omega+3}\} \subset K_{\omega+4} \Rightarrow v_{2\omega+3} \text{ is black};$$

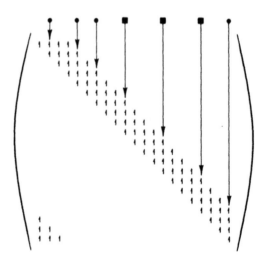

Figure 3.6.   The clique matrix of $C_{2\omega+1}^{\omega-1}$, where $\omega = 4$ and $\alpha = 5$. The markers designate which vertices are to be deleted to obtain an imperfect subgraph.

and finally, by induction on t,

$v_{t\omega+3}$ is black

$\Rightarrow \{v_{t\omega+4}, \dots, v_{(t+1)\omega+1}\}$ requires the entire rainbow
$\Rightarrow \{v_{(t+1)\omega+3}\}$ is black,

for $t = 2, \dots, \alpha - 2$.

Therefore, both v_1 and $v_{(\alpha-1)\omega+3}$ are black, but they are both contained in the clique $K_{(\alpha-1)\omega+3}$, a contradiction. Hence, $\chi(H) > \omega - 1 \geq \omega(H)$ and H is imperfect as required.   ∎

As a corollary of Theorem 3.24 we obtain another equivalent version of the strong perfect graph conjecture:

SPGC$_5$.   If G is p-critical with $\alpha(G) = \alpha$ and $\omega(G) = \omega$, then G contains an induced subgraph isomorphic to $C_{\alpha\omega+1}^{\omega-1}$.

Chvátal, Graham, Perold, and Whitesides [1979] have presented two procedures for constructing (α, ω)-partitionable graphs other than $C_{\alpha\omega+1}^{\omega-1}$.

If we we restrict the universe of graphs being considered by making an extra assumption about their structure, then, in certain cases, the SPGC can be shown to hold. Table 3.1 lists some successful restrictions. For the most part the original proofs cited do not make use of the Padberg conditions. Tucker [1979] has incorporated the Padberg conditions into new proofs of the SPGC for $K_{1,3}$-free graphs and 3-chromatic graphs.

Table 3.1

Classes of graphs for which the strong perfect graph conjecture
is known to hold

Planar graphs	Tucker [1973a]
$K_{1,3}$-free graphs	Parthasarathy and Ravindra [1976]
Circular-arc graphs	Tucker [1975]
⋈-free graphs	Parthasarathy and Ravindra [1979]
3-chromatic graphs	
(actually, any graph with $\omega \leq 3$)	Tucker [1977]
Toroidal graphs; graphs having	
maximum vertex degree ≤ 6	Grinstead [1978]

The strong perfect graph conjecture remains a formidable challenge to us. Its solution has eluded researchers for two decades. Perhaps in the third decade a reader of this book will settle the problem.

EXERCISES

1. Let x and y be distinct vertices of a graph G. Prove that $(G \circ x) - y = (G - y) \circ x$.

2. Let $x_1, x_2, \ldots, x_n$ be the vertices of a graph G and let $H = G \circ h$ where $h = (h_1, h_2, \ldots, h_n)$ is a vector of non-negative integers.

Verify that H can be constructed by the following procedure:

```
begin
  H ← G;
  for i ← 1 to n do
    if h_i = 0 then H ← H - x_i;
    else while h_i > 0 do
      begin
        H ← H ∘ x_i;
        h_i ← h_i - 1;
      end
end
```

3. Give an example of a graph G for which $\alpha(G) = k(G)$ and $\omega(G) < \chi(G)$. Why does this not contradict the Perfect Graph theorem?

4. Suppose G satisfies $\alpha(G) = k(G)$. Let $\mathscr{K}$ be a clique cover of G where $|\mathscr{K}| = k(G)$, and let $\mathscr{S}$ be the collection of all stable sets of cardinality $\alpha(G)$. Show that

$$|S \cap K| = 1 \qquad \text{for all } S \in \mathscr{S} \text{ and } K \in \mathscr{K}.$$

Give a dual statement for a graph satisfying $\omega(G) = \chi(G)$.

5.   Prove the following: For any integer k, there exists a graph G such that $\omega(G) = 2$ and $\chi(G) = k$. Thus, the gap between the clique number and the chromatic number can be arbitrarily large (Tutte [1954], Kelly and Kelly [1954], Zykov [1952]; see also Sachs [1969]).

6.   Prove that an n-vertex graph G is an odd chordless cycle if and only if $n = 2k + 1$, $\alpha(G) = k$, and $\alpha(G - v - w) = k$ for all vertices v and w of G (Melnikov and Vising [1971], Greenwell [1978]).

7.   An undirected graph G is *unimodular* if its clique matrix $\mathbf{A}$ has the property that every square submatrix of $\mathbf{A}$ has determinant equal to 0, $+1$, or -1. Prove the following:

   (i)   The graph in Figure 3.7 is unimodular;
   (ii)   unimodularity is a hereditary property;
   (iii) a bipartite graph is unimodular;
   (iv)   a unimodular graph is perfect (if necessary, for (iv) see Berge [1975]).

Figure 3.7

8.   Show that the five versions of the strong perfect graph conjecture given in this chapter are equivalent.

9.   Prove that G is p-critical if and only if G is partitionable but no proper induced subgraph of G is partitionable.

10.   Show that the graph in Figure 3.8 is partitionable but not p-critical. Show that the graph in Figure 3.9 is imperfect but not partitionable.

Figure 3.8

Figure 3.9

11. Let **A** and **B** be $n \times n$ matrices and let α and ω be integers. Using matrix operations give a short proof of the following: If $\mathbf{AJ} = \mathbf{JA} = \omega\mathbf{J}$, $\mathbf{BJ} = \mathbf{JB} = \alpha\mathbf{J}$, and $\mathbf{AB}^\mathrm{T} = \mathbf{J} - \mathbf{I}$, then $\alpha\omega = n - 1$.

12. Let $G = (X, E)$ and $H = (Y, F)$ be undirected graphs. Their normal product is defined to be the graph $G \cdot H$ whose vertex set is the Cartesian product $X \times Y$ with vertices (x, y) and (x', y') adjacent if and only if

$$x = x' \quad \text{and} \quad yy' \in F \qquad \text{or} \qquad xx' \in E \quad \text{and} \quad y = y'$$

or

$$xx' \in E \quad \text{and} \quad yy' \in F.$$

Prove the following:

  (i) $\chi(G \cdot h) \geq \max\{\chi(G), \chi(H)\}$;
  (ii) $\omega(G \cdot H) = \omega(G)\omega(H)$;
  (iii) $\alpha(G \cdot H) \geq \alpha(G)\alpha(H)$;
  (iv) $k(G \cdot H) \leq k(G)k(H)$.

13. Let G^r denote the normal product of G with itself $r - 1$ times, i.e., $G^1 = G$ and $G^r = G \cdot G^{r-1}$. Let

$$c(G) = \sup \sqrt[r]{\alpha(G^r)}.$$

Prove that $\alpha(G) = k(G)$ implies $c(G) = \alpha(G)$. For an application of this to zero-capacity codes, see Berge [1973, p. 382; 1975, p. 13].

Bibliography

Balinski, M. L., and Hoffman, A. J., eds.
  [1978] "*Polyhedral Combinatorics*," Math. Programming Studies, Vol. 8. North-Holland, Amsterdam.
Baumgartner, J. E., Malitz, J., and Reinhardt, W.
  [1970] Embedding trees in the rationals. *Proc. Nat. Acad. Sci. U.S.A.* **67**, 1748–1753.
Berge, Claude
  [1961] Farbung von Graphen, deren sämtliche bzw. deren ungerade Kreise starr sind, *Wiss. Z. Martin-Luther-Univ., Halle-Wittenberg Math.-Natur, Reihe*, 114–115.

[1962] Sur une conjecture relative au problème des codes optimaux, *Comm. 13ieme Assemblee Gen. URSI*, Tokyo, 1962.

[1967] Some classes of perfect graphs, "Graph Theory and Theoretical Physics," pp. 155–165. Academic Press, London–New York. MR38 #1017.

[1969] Some classes of perfect graphs, "Combinatorial Mathematics and its Applications," pp. 539–552. Univ. North Carolina Press, Chapel Hill. MR42 #100.

[1973] "Graphs and Hypergraphs," Chapter 16. North-Holland, Amsterdam. MR50 #9640.

[1975] Perfect graphs, *in* "Studies in Graph Theory," Part I (D. R. Fulkerson, ed.), M.A.A. Studies in Mathematics Vol. 11, pp. 1–22. Math. Assoc. Amer., Washington, D.C. MR53 #10585.

[1976] Short note about the history of the perfect graph conjecture. (mimeographed).

Bland, Robert, G., Huang, H.-C., and Trotter, Leslie E., Jr.

[1979] Graphical properties related to minimal imperfection, *Discrete Math.* **27**, 11–22.

Bollobás, Béla

[1978] "Extremal Graph Theory," pp. 263–270. Academic Press, London.

Chvátal, Václav

[1973] Edmonds polytopes and a hierarchy of combinatorial problems, *Discrete Math.* **4**, 305–337. MR47 #1635.

[1975] On certain polytopes associated with graphs, *J. Combin. Theory B* **18**, 138–154. MR51 #7949.

[1976] On the strong perfect graph conjecture, *J. Combin. Theory B* **20**, 139–141. MR54 #129.

Chvátal, Václav, Graham, R. L., Perold, A. F., and Whitesides, S. H.

[1979] Combinatorial designs related to the strong perfect graph conjecture, *Discrete Math.* **26**, 83–92.

Commoner, F. G.

[1973] A sufficient condition for a matrix to be totally unimodular, *Networks* **3**, 351–365. MR49 #331.

de Werra, D.

[1978] On line perfect graphs, *Math. Programming* **15**, 236–238.
Gives alternate proof of result of L. E. Trotter [1977].

Edmonds, Jack

[1965] Maximum matching and a polyhedron with 0,1-vertices, *J. Res. Nat. Bur. Standards Sect. B* **69**, 125–130. MR32 #1012.

Fulkerson, Delbert Ray

[1969] The perfect graph conjecture and pluperfect graph theorem, *2nd Chapel Hill Conf. on Combin. Math. and its Appl.*, 171–175.

[1971] Blocking and anti-blocking pairs of polyhedra, *Math. Programming* **1**, 168–194. MR45 #3222.

[1972] Anti-blocking polyhedra, *J. Combin. Theory* **12**, 50–71. MR44 #2629.

[1973] On the perfect graph theorem, *in* "Mathematical Programming," (T. C. Hu and S. Robinson, eds.), pp. 68–76. Academic Press, New York. MR51 #10147.

Greenwell, Don

[1978] Odd cycles and perfect graphs, "Theory and Applications of Graphs," Lecture Notes in Math. Vol. 642, pp. 191–193. Springer-Verlag, Berlin.

Grinstead, Charles M.

[1978] Toroidal graphs and the strong perfect graph conjecture, Ph.D. thesis, UCLA.

Hajnal, Andras, and Surányi, Janos
[1958] Über die Auflösung von Graphen in vollständige Teilgraphen, *Ann. Univ. Sci. Budapest Eötvös. Sect. Math.* **1**, 113–121. MR21 #1944.
Huang, H.-C.
[1976] Investigations on combinatorial optimization, Ph.D. thesis, Yale Univ.
Jolivet, J. L.
[1975] Graphes parfaits pour une propriété, *P. Cahiers Centre Etudes Rech. Opér.* **17**, 253–256. MR53 #7841.
Karpetjan, I. A.
[1976] Critical and essential edges in perfect graphs (Russian; Armenian summary), *Akad. Nauk Armjan. SSR Dokl.* **63**, 65–70.
Kelly, John B., and Kelly, L. M.
[1954] Path and circuits in critical graphs, *Amer. J. Math.* **76**, 786–792. MR16, p. 387.
Lovász, László
[1972a] Normal hypergraphs and the perfect graph conjecture, *Discrete Math.* **2**, 253–267. MR46 #1624.
[1972b] A characterization of perfect graphs, *J. Combin. Theory B* **13**, 95–98. MR46 #8885.
Markosjan, S. E.
[1975] Perfect and critical graphs (Russian; Armenian summary), *Akad. Nauk Armjan. SSR Dokl.* **60**, 218–223. MR53 #10659.
Markosjan, S. E., and Karpetjan, I. A.
[1976] Perfect graphs (Russian; Armenian summary). *Akad. Nauk Armjan. SSR Dokl.* **63**, (1976), 292–296. MR56 #8427.
Melnikov, L. S., and Vising, V. G.
[1971] Solution to Toft's problem (Russian), *Diskret. Analiz.* **19**, 11–14. MR46 #3379. See Greenwell [1978].
Monma, C. L., and Trotter, Leslie E., Jr.
[1979] On perfect graphs and polyhedra with (0,1)-valued extreme points. *Math. Programming* **17**, 239–242.
Nash-Williams, C. St. J. A.
[1967] Infinite graphs—a survey, *J. Combin. Theory* **3**, 286–301. MR35 #5351.
Olaru, Elefterie
[1969] Über die Überdeckung von Graphen mit Cliquen, *Wiss. Z. Tech. Hochsch. Ilmenau* **15**, 115–121. MR43 #3162.
[1972] Beiträge zur Theorie der perfekten Graphen, *Elektron. Informationsverarb. Kybernet.* **8**, (1972), 147–172. MR47 #8338.
[1973a] Über perfekte und kritisch imperfekte Graphen, *Ann. St. Univ. Iasi* **19**, 477–486. MR54 #5053.
[1973b] Zur Charakterisierung perfekter Graphen, *Electron. Informationsverarb. Kybernet.* **9**, 543–548. MR51 #10167.
[1977] Zur Theorie der perfekten Graphen, *J. Combin. Theory B* **23**, 94–105.
Padberg, Manfred W.
[1973] On the facial structure of set packing polyhedra, *Math. Programming* **5**, 199–215. MR51 #4990.
[1974] Perfect zero-one matrices, *Math. Programming* **6**, 180–196. MR49 #4809.
[1975] Characterizations of totally unimodular, balanced and perfect matrices, *in* "Combinatorial Programming: Methods and Applications," Proc. NATO Advanced Study Inst., Versailles, France, 1974, pp. 275–284. Reidel, Dordrecht, MR53 #10291. An excellent survey article with good bibliography and clearly written.
[1976a] A note on the total unimodularity of matrices, *Discrete Math.* **14**, 273–278. MR54 #2685.

[1976b] Almost integral polyhedra related to certain combinatorial optimization problems, *Linear Algebra and Appl.* **15**, 69–88.

Parthasarathy, K. R., and Ravindra, G.

[1976] The strong perfect-graph conjecture is true for $K_{1,3}$-free graphs, *J. Combin. Theory B* **21**, 212–223. MR55 #10308.

[1979] The validity of the strong perfect-graph conjecture for $(K_4 - e)$-free graphs, *J. Combin. Theory B* **26**, 98–100.

Perles, M. A.

[1963] On Dilworth's theorem in the infinite case, *Israel J. Math.* **1**, 108–109. MR29 #5758.

Pretzel, Oliver

[1979] Another proof of Dilworth's decomposition theorem, *Discrete Math.* **25**, 91–92.

Ravindra, G.

[1975] On Berge's conjecture concerning perfect graphs, *Proc. Indian Nat. Sci. Acad.* **41A**, 294–296.

Sachs, Horst

[1969] Finite graphs (Investigations and generalizations concerning the construction of finite graphs having given chromatic number and no triangles), "Recent Progress in Combinatorics," pp. 175–184. Academic Press, New York. MR42 #2980.

[1970] On the Berge conjecture concerning perfect graphs, "Combinatorial Structures and their Applications," pp. 377–384. Gordon & Breach, New York. MR42 #7549.

Tamir, A.

[1976] On totally unimodular matrices, *Networks* **6**, 373–382.

Tomescu, Ioan

[1971] Sur le nombre des cliques maximales d'un graphe et quelques problèmes sur les graphes parfaits, *Rev. Roumaine Math. Pures Appl.* **16**, 1115–1126. MR45 #103.

Trotter, Leslie E., Jr.

[1977] Line perfect graphs, *Math. Programming* **12**, 255–259. MR56 #15501.
Characterizes those graphs G whose line graphs $L(G)$ are perfect.

Trotter, William Thomas, Jr.

[1971] A note on triangulated graphs, *Notices Amer. Math. Soc.* **18**, 1045 (A).

Tucker, Alan C.

[1972] The strong perfect graph conjecture and an application to a municipal routing problem, *in* "Graph Theory and Applications," Proc. Conf. Western Michigan Univ., Kalamazoo, Lecture Notes in Math., Vol. 303, pp. 297–303. Springer-Verlag, Berlin. MR49 #7181.

[1973a] The strong perfect graph conjecture for planar graphs, *Canad. J. Math.* **25**, 103–114. MR47 #4868.

[1973b] Perfect graphs and an application to optimizing municipal services, *SIAM Rev.* **15**, 585–590. MR48 #3817.

[1977] Critical perfect graphs and perfect 3-chromatic graphs, *J. Combin. Theory B* **23**, 143–149.

[1978] Circular arc graphs: new uses and a new algorithm, *in* "Theory and Application of Graphs," Lecture Notes in Math. 642; pp. 580–589. Springer-Verlag, Berlin.

[1979] On Berge's strong perfect graph conjecture, *Ann. N.Y. Acad. Sci.* **319**, 530–535.

Tutte, W. (alias B. Descartes)

[1954] Solution to advanced problem No. 4526, *Amer. Math. Monthly* **61**, 352.

Wagon, Stanley

[1978] Infinite triangulated graphs, *Discrete Math.* **22**, 183–189.

Zykov, A. A.

[1952] On some properties of linear complexes (Russian), *Math. Sbornik N.S.* **24** (1949), 163–188, English transl. *Amer. Math. Soc. Transl.* No. 79. MR11, p. 733.

CHAPTER 4

Triangulated Graphs

1. Introduction

One of the first classes of graphs to be recognized as being perfect was the class of triangulated graphs. Hajnal and Surányi [1958] showed that triangulated graphs satisfy the *perfect property* P_2 (α-perfection), and Berge [1960] proved that they satisfy P_1 (χ-perfection). These two results, in large measure, inspired the conjecture that P_1 and P_2 were equivalent, a statement that we now know to be true (Theorem 3.3). Thus, the study of triangulated graphs can well be thought of as the beginning of the theory of perfect graphs.

We briefly looked at the triangulated graph property in the sneak preview Section 1.3. For completeness' sake, we shall repeat the definition here and mention a few basic properties.

An undirected graph G is called *triangulated* if every cycle of length strictly greater than 3 possesses a chord, that is, an edge joining two nonconsecutive vertices of the cycle. Equivalently, G does not contain an induced subgraph isomorphic to C_n for $n > 3$. Being triangulated is a hereditary property inherited by all the induced subgraphs of G. You may recall from Section 1.3 that the interval graphs constitute a special type of triangulated graph. Thus we have our first example of triangulated graphs.

In the literature, triangulated graphs have also been called *chordal, rigid-circuit, monotone transitive,* and *perfect elimination* graphs.

2. Characterizing Triangulated Graphs

A vertex x of G is called *simplicial* if its adjacency set $\text{Adj}(x)$ induces a complete subgraph of G, i.e., $\text{Adj}(x)$ is a clique (not necessarily maximal).

Dirac [1961], and later Lekkerkerker and Boland [1962], proved that a triangulated graph always has a simplicial vertex (in fact at least two of them), and using this fact Fulkerson and Gross [1965] suggested an iterative procedure to recognize triangulated graphs based on this and the hereditary property. Namely, *repeatedly locate a simplicial vertex and eliminate it from the graph, until either no vertices remain and the graph is triangulated or at some stage no simplicial vertex exists and the graph is not triangulated.* The correctness of this procedure is proved in Theorem 4.1. Let us state things more algebraically.

Let $G = (V, E)$ be an undirected graph and let $\sigma = [v_1, v_2, \ldots, v_n]$ be an ordering of the vertices. We say that σ is a *perfect vertex elimination scheme* (or *perfect scheme*) if each v_i is a simplicial vertex of the induced subgraph $G_{\{v_i, \ldots, v_n\}}$. In other words, each set

$$X_i = \{v_j \in \text{Adj}(v_i) \mid j > i\}$$

is complete. For example, the graph G_1 in Figure 4.1 has a perfect vertex elimination scheme $\sigma = [a, g, b, f, c, e, d]$. It is not unique; in fact G_1 has 96 different perfect elimination schemes. In contrast to this, the graph G_2 has no simplicial vertex, so we cannot even start constructing a perfect scheme — it has none.

A subset $S \subset V$ is a *vertex separator* for nonadjacent vertices a and b (or an *a–b separator*) if the removal of S from the graph separates a and b into distinct connected components. If no proper subset of S is an a–b separator, then S is a *minimal vertex separator* for a and b. Consider again the graphs of Figure 4.1. In G_2, the set $\{y, z\}$ is a minimal vertex separator for p and q, whereas $\{x, y, z\}$ is a minimal vertex separator for p and r. (How is it possible that both are *minimal* vertex separators, yet one is contained in the other?) In G_1, every minimal vertex separator has cardinality 2. This is an unusual phenomenon. However, notice also that the two vertices of such a separator of G_1 are adjacent, in every case. This latter phenomenon actually occurs for all triangulated graphs, as you will see in Theorem 4.1.

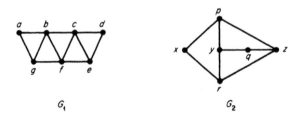

Figure 4.1.   Two graphs, one triangulated and one not triangulated.

We now give two characterizations of triangulated graphs, one algorithmic (Fulkerson and Gross [1965]) and the other graph theoretic (Dirac [1961]).

Theorem 4.1. Let G be an undirected graph. The following statements are equivalent:

   (i) G is triangulated.
   (ii) G has a perfect vertex elimination scheme. Moreover, *any* simplicial vertex can start a perfect scheme.
   (iii) Every minimal vertex separator induces a complete subgraph of G.

Proof. (iii) $\Rightarrow$ (i)   Let $[a, x, b, y_1, y_2, \ldots, y_k, a]$ ($k \geq 1$) be a simple cycle of $G = (V, E)$. Any minimal a–b separator must contain vertices x and y_i for some i, so $xy_i \in E$, which is a chord of the cycle.

(i) $\Rightarrow$ (iii)   Suppose S is a minimal a–b separator with G_A and G_B being the connected components of G_{V-S} containing a and b, respectively. Since S is minimal, each $x \in S$ is adjacent to some vertex in A and some vertex in B. Therefore, for any pair $x, y \in S$ there exist paths $[x, a_1, \ldots, a_r, y]$ and $[y, b_1, \ldots, b_t, x]$, where each $a_i \in A$ and $b_i \in B$, such that these paths are chosen to be of smallest possible length. It follows that $[x, a_1, \ldots, a_r, y, b_1, \ldots, b_t, x]$ is a simple cycle whose length is at least 4, implying that it must have a chord. But $a_i b_j \notin E$ by the definition of vertex separator, and $a_i a_j \notin E$ and $b_i b_j \notin E$ by the minimality of r and t. Thus, the only possible chord is $xy \in E$. ∎

Remark.   It also follows that $r = t = 1$, implying that for all $x, y \in S$ there exist vertices in A and B which are adjacent to both x and y. A stronger result is given in Exercise 12.

Before continuing with the remaining implications, we pause for a message from our lemma department.

Lemma 4.2 (Dirac [1961]).   Every triangulated graph $G = (V, E)$ has a simplicial vertex. Moreover, if G is not a clique, then it has two nonadjacent simplicial vertices.

Proof.   The lemma is trivial if G is complete. Assume that G has two non-adjacent vertices a and b and that the lemma is true for all graphs with fewer vertices than G. Let S be a minimal vertex separator for a and b with G_A and G_B being the connected components of G_{V-S} containing a and b, respectively.

By induction, either the subgraph G_{A+S} has two nonadjacent simplicial vertices one of which must be in A (since S induces a complete subgraph) or G_{A+S} is itself complete and any vertex of A is simplicial in G_{A+S}. Furthermore, since $\text{Adj}(A) \subseteq A + S$, a simplicial vertex of G_{A+S} in A is simplicial in all of G. Similarly B contains a simplicial vertex of G. This proves the lemma.

We now rejoin the proof of the theorem which is still in progress.

(i) $\Rightarrow$ (ii)   According to the lemma, if G is triangulated, then it has a simplicial vertex, say x. Since $G_{V-\{x\}}$ is triangulated and smaller than G, it has, by induction, a perfect scheme which, when adjoined as a suffix of x, forms a perfect scheme for G.

(ii) $\Rightarrow$ (i)   Let C be a simple cycle of G and let x be the vertex of C with the smallest index in a perfect scheme. Since $|\mathrm{Adj}(x) \cap C| \geq 2$, the eventual simpliciality of x guarantees a chord in C. ∎

3. Recognizing Triangulated Graphs by Lexicographic Breadth-First Search

From Lemma 4.2 we learned that the Fulkerson–Gross recognition procedure affords us a choice of at least two vertices for each position in constructing a perfect scheme for a triangulated graph. Therefore, we can freely choose a vertex v_n to *avoid* during the whole process, saving it for the last position in a scheme. Similarly, we can pick any vertex v_{n-1} adjacent to v_n to save for the $(n-1)$st position. If we continued in this manner, we would be constructing a scheme *backwards*! This is exactly what Leuker [1974] and Rose and Tarjan [1975] have done in order to give a linear-time algorithm for recognizing triangulated graphs. The version presented in Rose, Tarjan, and Leuker [1976] uses a *lexicographic* breadth-first search in which the usual queue of vertices is replaced by a queue of (unordered) subsets of the vertices which is sometimes refined but never reordered. The method (Figure 4.2) is as follows:

```
      begin
1.    assign the label ∅ to each vertex;
2.    for i ← n to 1 step − 1 do
3.        select: pick an unnumbered vertex v with largest label;
4.            σ(i) ← v; comment This assigns to v the number i.
5.            update: for each unnumbered vertex w∈ Adj(v) do add i to label(w);
      end
```

Figure 4.2.   Algorithm 4.1: Lex BFS.

Algorithm 4.1.   Lexicographic breadth-first search.

Input: The adjacency sets of an undirected graph $G = (V, E)$.
Output: An ordering σ of the vertices.
Method: The vertices are numbered from n to 1 in the order that they are selected in line 3. This numbering fixes the positions of an elimination scheme

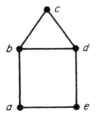

Figure 4.3.

σ. For each vertex x, the *label* of x will consist of a set of numbers listed in decreasing order. The vertices can then be lexicographically ordered according to their labels. (Lexicographic order is just dictionary order, so that $9761 < 985$ and $643 < 6432$.) Ties are broken arbitrarily.

Example.   We shall apply Algorithm 4.1 to the graph in Figure 4.3. The vertex a is selected arbitrarily in line 3 during the first pass. The evolution of the labeling and the numbering are illustrated in Figure 4.4. Notice that the final numbering $\sigma = [c, d, e, b, a]$ is a perfect vertex elimination scheme. This is no accident.

For each value of i, let $L_i(x)$ denote the label of x when statement 4 is executed, i.e., when the ith vertex is numbered. Remember, the index is *decremented* at each successive iteration. For example, $L_n(x) = \emptyset$ for all x and $L_{n-1}(x) = \{n\}$ iff $x \in \text{Adj}(\sigma(n))$. The following properties are of prime importance:

(L1) $L_i(x) \leq L_j(x)$ $(j \leq i)$;
(L2) $L_i(x) < L_i(y) \Rightarrow L_j(x) < L_j(y)$ $(j < i)$;
(L3) if $\sigma^{-1}(a) < \sigma^{-1}(b) < \sigma^{-1}(c)$ and $c \in \text{Adj}(a) - \text{Adj}(b)$, then there exists a vertex $d \in \text{Adj}(b) - \text{Adj}(a)$ with $\sigma^{-1}(c) < \sigma^{-1}(d)$.

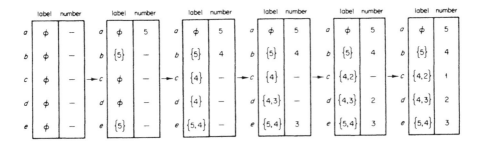

Figure 4.4.

Property (L1) says that the label of a vertex may get larger but never smaller as the algorithm proceeds. Property (L2) states that once a vertex gets ahead of another vertex, they stay in that order. Finally, (L3) gives a condition under which there must be a suitable vertex d which was numbered before c (in time) and hence received a larger number.

Lexicographic breadth-first search can be used to recognize triangulated graphs as demonstrated by the next theorem.

Theorem 4.3. An undirected graph $G = (V, E)$ is triangulated if and only if the ordering σ produced by Algorithm 4.1 is a perfect vertex elimination scheme.

Proof. If $|V| = n = 1$, then the proof is trivial. Assume that the theorem is true for all graphs with fewer than n vertices and let σ be the ordering produced by Algorithm 4.1 when applied to a triangulated graph G. By induction, it is sufficient to show that $x = \sigma(1)$ is a simplicial vertex of G.

Suppose x is not simplicial. Choose vertices $x_1, x_2 \in \text{Adj}(x)$ with $x_1 x_2 \notin E$ so that x_2 is as large as possible (with respect to the ordering σ). (Remember, σ increases as you approach the root of the search tree.) Consider the following inductive procedure. Assume we are given vertices $x_1, x_2, \ldots, x_m$ with these properties: for all $i, j > 0$,

(1) $x, x_i \in E \Leftrightarrow i \leq 2$,
(2) $x_i x_j \in E \Leftrightarrow |i - j| = 2$,
(3) $\sigma^{-1}(x_1) < \sigma^{-1}(x_2) < \cdots < \sigma^{-1}(x_m)$,
(4) x_j is the largest vertex (with respect to σ) such that

$$x_{j-2} x_j \in E \qquad \text{but} \qquad x_{j-3} x_j \notin E.$$

(For notational reasons let $x_0 = x$ and $x_{-1} = x_1$.) The situation for $m = 2$ was constructed initially.

The vertices x_{m-2}, x_{m-1}, and x_m satisfy the hypothesis of property (L3) as a, b, and c, respectively. Hence, choose x_{m+1} to be the largest vertex (with respect to σ) larger than x_m which is adjacent to x_{m-1} but not adjacent to x_{m-2}. Now, if x_{m+1} were adjacent to x_{m-3}, then (L3) applied to the vertices $x_{m-3}, x_{m-2}, x_{m+1}$ would imply the existence of a vertex larger than x_{m+1} (hence larger than x_m) which is adjacent to x_{m-2} but not to x_{m-3}, contradicting the maximality of x_m in (4). Therefore x_{m+1} is not adjacent to x_{m-3}. Finally, it follows from (1), (2), and chordality that $x_i x_{m+1} \notin E$ for $i = 0, 1, \ldots, m - 4, m$.

Clearly this inductive procedure continues indefinitely, but the graph is finite, a contradiction. Therefore, the vertex x must be simplicial, and the theorem is proved in one direction. The converse follows from Theorem 4.1. ∎

In an unpublished work, Tarjan [1976] has shown another method of searching a graph that can be used to recognize triangulated graphs. It is called *maximum cardinality search* (MCS), and it is defined as follows:

MCS: The vertices are to be numbered from n to 1.
The next vertex to be numbered is always one which is adjacent to the most numbered vertices, ties being broken arbitrarily.

Using an argument similar to the proof of Theorem 4.3, one can show that G is triangulated if and only if every MCS ordering of the vertices is a perfect ellimination scheme. It should be pointed out that there are MCS orderings which cannot be obtained by Lex BFS, there are Lex BFS orderings which are not MCS, and there exist perfect elimination schemes which are neither MCS nor Lex BFS. Exercises 27 and 28 develop some of the results on MCS. Both Lex BFS and MCS are special cases of a general method for finding perfect elimination schemes recently developed by Alan Hoffman and Michel Sakarovich.

4.   The Complexity of Recognizing Triangulated Graphs

Having proved the correctness of Algorithm 4.1, let us now analyze its complexity. We first describe an implementation of Lex BFS, then show that it requires $O(|V| + |E|)$ time. We do not actually calculate the labels, but rather we keep the unnumbered vertices in lexicographic order.

Data Structure

We use a queue Q of sets

$$S_l = \{v \in V \,|\, \text{label}(v) = l \text{ and } \sigma^{-1}(v) \text{ undefined}\}$$

ordered lexicographically from smallest to largest; each set S_l is represented by a doubly linked list. Initially there is but one set, $S_\phi = V$. Each set S_l has a FLAG initially set at 0. For a vertex w, the array element SET(w) points to $S_{\text{label}(w)}$ and another array gives the address of w in SET(w) for deletion purposes. A list FIX LIST, initially empty, is also used, and simple arrays represent σ and σ^{-1}.

Implementation

Select as v in line 3 any vertex in the last set of Q and delete v from SET(v). Create a new set $S_{l \cdot i}$ for each old set S_l containing an unnumbered vertex

$w \in \text{Adj}(v)$. We delete from S_l all such vertices w and place them in the new set $S_{l \cdot i}$, which is inserted into the queue of sets immediately following S_l. Clearly this method maintains the proper lexicographic ordering without our actually having to calculate the labels. More specifically, *update* can be implemented as follows:

```
    ┌  for all unnumbered w ∈ Adj(v) do
    │     begin
    │        if FLAG(SET(w)) = 0 then
    │           begin
    │              Create new set S′ and insert it in Q immediately in back of SET(w);
    │              FLAG(SET(w)) ← 1; FLAG(S′) ← 0; put a pointer to SET (w) on FIX LIST;
    │           end
    │        let S′ be the set immediately in back of SET(w) in Q; delete w from SET(w); add w to S′;
 5. ┤        SET(w) ← S′;
    │     end
    │  for each set S on FIX LIST do
    │     begin
    │        FLAG(S) ← 0;
    │        if S is empty then
    │           delete S from Q;
    └     end
```

It is easy to verify that, as presented, statement 5 requires $O(|\text{Adj}(v)|)$ time. Consequently, the **for** loop between statements 2 and 5 uses $O(|V| + |E|)$ time. Initializing the data structure including statement 1 takes $O(|V|)$ time. This proves the following result.

Theorem 4.4. Algorithm 4.1 can be implemented to carry out lexicographic breadth-first search on an undirected graph $G = (V, E)$ in $O(|V| + |E|)$ time and space.

Example. Let Q_i denote the queue of sets of unnumbered vertices just before $\sigma(i)$ is defined in Algorithm 4.1. Figure 4.5b gives Q_9, Q_8, and Q_7 for the graph in Figure 4.5a. For convenience, the vertices are identified with their eventual position in σ. Figure 4.5c shows the data structure for Q_7 before the FIX LIST has been emptied and with the implicit labels in parentheses.

In order to use Lex BFS to recognize triangulated graphs, we need an efficient method to test whether or not a given ordering σ of the vertices is a perfect vertex elimination scheme. This is proved by the next algorithm.

Algorithm 4.2. Testing a perfect elimination scheme.
Input. The adjacency sets of an undirected graph $G = (V, E)$ and an ordering σ of V.

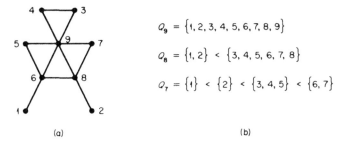

$$Q_9 = \{1, 2, 3, 4, 5, 6, 7, 8, 9\}$$

$$Q_8 = \{1, 2\} < \{3, 4, 5, 6, 7, 8\}$$

$$Q_7 = \{1\} < \{2\} < \{3, 4, 5\} < \{6, 7\}$$

(a)   (b)

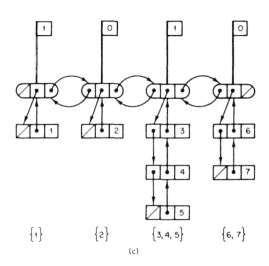

$\{1\}$ $\{2\}$ $\{3, 4, 5\}$ $\{6, 7\}$

(c)

Figure 4.5.

Output. "True" if σ is a perfect vertex elimination scheme and "false" otherwise.

Method. A single call to the procedure PERFECT(σ), given in Figure 4.6. The list $A(u)$ collects all the vertices which will eventually have to be checked for adjacency with u. The actual checking is delayed until the iteration when $u = \sigma(i)$ in lines 8 and 9. This technique is used so that in the $\sigma^{-1}(v)$-th iteration there is no search of Adj(u).

Complexity. Arrays are used for σ and σ^{-1} and lists hold Adj(v) and $A(v)$. Lines 4–7 can be implemented simultaneously in one scan of Adj(v). The **go to** in line 5 will be executed exactly $j - 1$ times, where j is the number of connected components of G. The list $A(u)$ will represent a *set with repetitions*. The test in line 8 simply checks for a vertex w on the list $A(v)$ which is *not*

```
boolean procedure PERFECT (σ):
begin
1.    for all vertices v do A(v) ← ∅;;
2.    for i ← 1 to n − 1 do
         begin
3.          v ← σ(i);
4.          X ← {x ∈ Adj(v) | σ⁻¹(v) < σ⁻¹(x)};
5.          if X = ∅ then go to 8;;
6.          u ← σ (min {σ⁻¹(x) | x ∈ X});
7.          concatenate X − {u} to A(u);
8.          if A(v) − Adj(v) ≠ ∅ then
9.             return "false";
         end
10.   return "true";
end
```

Figure 4.6.   Procedure to test a perfect vertex elimination scheme.

adjacent to v, can be done in $O(|\text{Adj}(v)| + |A(v)|)$ time by using an array TEST of size n initially set to all zeros as follows:

```
      begin
        for w ∈ Adj(v) do TEST(w) ← 1;;
        for w ∈ A(v) do
8.         if TEST(w) = 0 then
             return "nonempty";
        for w ∈ Adj(v) do TEST(w) ← 0;;
        return "empty";
      end
```

Thus, the entire algorithm can be performed in time and space proportional to

$$|V| + \sum_{v \in V} |\text{Adj}(v)| + \sum_{u \in V} |A(u)|,$$

where has $A(u)$ is its final value. Now, the middle summand is larger than the last since a given $\text{Adj}(v)$ appears as part of at most one of the lists $A(u)$. Hence, both summands can be replaced by $O(|E|)$. This proves the complexity part of the next theorem.

Theorem 4.5.   Algorithm 4.2 correctly tests whether or not an ordering σ of the vertices is a perfect vertex elimination scheme. It can be implemented to run in time and space proportional to $|V| + |E|$.

Proof.   The algorithm returns "false" during the $\sigma^{-1}(u)$-th iteration if and only if there exist vertices v, u, w $(\sigma^{-1}(v) < \sigma^{-1}(u) < \sigma^{-1}(w))$, where u is defined in line 4 during the $\sigma^{-1}(v)$-th iteration, and

$$u, w \in \text{Adj}(v) \qquad \text{but } u \text{ is not adjacent to } w.$$

Clearly, if we get "false," then σ is not a perfect elimination scheme.

Conversely, suppose σ is not perfect elimination and the algorithm returns "true." Let v be the vertex with $\sigma^{-1}(v)$ largest possible such that $X = \{w \mid w \in \mathrm{Adj}(v) \text{ and } \sigma^{-1}(v) < \sigma^{-1}(w)\}$ is *not* complete. Let u be the vertex of X defined in line 6 during the $\sigma^{-1}(v)$-th iteration, after which (in line 7) $X - \{u\}$ is added to $A(u)$. Since during the $\sigma^{-1}(u)$-th iteration line 9 is not executed,

$$\text{every } x \in X - \{u\} \text{ is adjacent to } u,$$

and

$$\text{every pair } x, y \in X - \{u\} \text{ is adjacent.}$$

The latter statement follows from the maximality of $\sigma^{-1}(v)$. Thus, X is complete, a contradiction.   ∎

Corollary 4.6.   Triangulated graphs can be recognized in linear time.

Proof.   The proof follows from Theorems 4.3–4.5.   ∎

5.   Triangulated Graphs as Intersection Graphs

We have seen in Chapter 1 that the interval graphs are a proper subclass of the triangulated graphs. This leads naturally to the problem of characterizing triangulated graphs as the intersection graphs of some topological family slightly more general than intervals on a line. In this section we shall show that a graph is triangulated if and only if it is the intersection graph of a family of subtrees of a tree. (See Figure 4.7.)

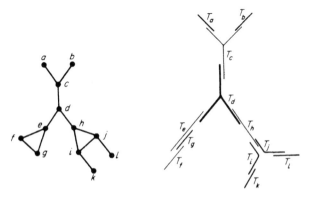

Figure 4.7.   A triangulated graph and a subtree representation for it.

A family $\{T_i\}_{i \in I}$ of subsets of a set T is said to satisfy the *Helly property* if $J \subseteq I$ and $T_i \cap T_j \neq \varnothing$ for all $i, j \in J$ implies that $\bigcap_{j \in J} T_j \neq \varnothing$.

If we let T be a tree and let each T_i be a subtree of T, then we can prove the following result.

Proposition 4.7. A family of subtrees of a tree satisfies the Helly property.

Proof. Suppose $T_i \cap T_j \neq \varnothing$ for all $i, j \in J$. Consider three points a, b, c on T. Let S be the set of indices s such that T_s contains at least two of these three points, and let P_1, P_2, P_3 be the simple paths in T connecting a with b, b with c, and a with c, respectively. Since T is a tree, it follows that $P_1 \cap P_2 \cap P_3 \neq \varnothing$, but each $T_s (s \in S)$ contains one of these paths P_i. Therefore,

$$\bigcap_{s \in S} T_s \supseteq P_1 \cap P_2 \cap P_3 \neq \varnothing.$$

The lemma is proved by induction. Let us assume that

$$[T_i \cap T_j \neq \varnothing \quad \text{for all} \quad i, j \in J] \Rightarrow \bigcap_{j \in J} T_j \neq \varnothing \qquad (1)$$

for all index sets J of size $\leq k$. This is certainly true for $k = 2$. Consider a family of subtrees $\{T_{i_1}, \ldots, T_{i_{k+1}}\}$. By the induction hypothesis there exist points a, b, c on T such that

$$a \in \bigcap_{j=1}^{k} T_{i_j}, \qquad b \in \bigcap_{j=2}^{k+1} T_{i_j}, \qquad c \in T_{i_1} \cap T_{i_{k+1}}.$$

Moreover, every T_{i_j} contains at least two of the points a, b, c. Hence, by the preceeding paragraph, $\bigcap_{j=1}^{k+1} T_{i_j} \neq \varnothing$. ∎

Theorem 4.8 (Walter [1972], Gavril [1974a], and Buneman [1974]). Let $G = (V, E)$ be an undirected graph. The following statements are equivalent:

(i) G is a triangulated graph.

(ii) G is the intersection graph of a family of subtrees of a tree.

(iii) There exists a tree $T = (\mathcal{K}, \mathcal{E})$ whose vertex set $\mathcal{K}$ is the set of maximal cliques of G such that each of the induced subgraphs $T_{\mathcal{K}_v} (v \in V)$ is connected (and hence a subtree), where $\mathcal{K}_v$ consists of those maximal cliques which contain v.

Proof. (iii) $\Rightarrow$ (ii) Assume that there exists a tree $T = (\mathcal{K}, \mathcal{E})$ satisfying statement (iii). Let $v, w \in V$. Now

$$vw \in E, \quad v, w \in A \quad \text{for some clique} \quad A \in \mathcal{K},$$

$$\mathcal{K}_v \cap \mathcal{K}_w \neq \varnothing, \qquad T_{\mathcal{K}_v} \cap T_{\mathcal{K}_w} \neq \varnothing.$$

Thus G is the intersection graph of the family of subtrees $\{T_{\mathcal{K}_v} | v \in V\}$ of T.

(ii) $\Rightarrow$ (i)   Let $\{T_v\}_{v \in V}$ be a family of subtrees of a tree T such that $vw \in E$ iff $T_v \cap T_w \neq \emptyset$.

Suppose G contains a chordless cycle $[v_0, v_1, \ldots, v_{k-1}, v_0]$ with $k > 3$ corresponding to the sequence of subtrees $T_0, T_1, \ldots, T_{k-1}, T_0$ of the tree T; that is, $T_i \cap T_j \neq \emptyset$ if and only if i and j differ by at most one modulo k. All arithmetic will be done mod k.

Choose a point a_i from $T_i \cap T_{i+1}$ $(i = 0, \ldots, k - 1)$. Let b_i be the last common point on the (unique) simple paths from a_i to a_{i-1} and a_i to a_{i+1}. These paths lie in T_i and T_{i+1}, respectively, so that b_i also lies in $T_i \cap T_{i+1}$. Let P_{i+1} be the simple path connecting b_i and b_{i+1}. Clearly $P_i \subseteq T_i$, so $P_i \cap P_j = \emptyset$ for i and j differing by more than 1 mod k. Moreover, $P_i \cap P_{i+1} = \{b_i\}$ for $i = 0, \ldots, k - 1$. Thus, $\bigcup_i P_i$ is a simple cycle in T, contradicting the definition of a tree.

(i) $\Rightarrow$ (iii)   We prove the implication by induction on the size of G. Assume that the theorem is true for all graphs having fewer vertices than G. If G is complete, then T is a single vertex and the result is trivial. If G is disconnected with components $G_1, \ldots, G_k$, then by induction there exists a corresponding tree T_i satisfying (iii) for each G_i. We connect a point of T_i with a point of T_{i+1} $(i = 1, \ldots, k - 1)$ to obtain a tree satisfying (iii) for G.

Let us assume that G is connected but not complete. Choose a simplicial vertex a of G and let $A = \{a\} \cup \mathrm{Adj}(a)$. Clearly, A is a maximal clique of G. Let

$$U = \{u \in A \mid \mathrm{Adj}(u) \subset A\}$$

and

$$Y = A - U.$$

Note that the sets U, Y, and $V - A$ are nonempty since G is connected but not complete. Consider the induced subgraph $G' = G_{V-U}$, which is triangulated and has fewer vertices than G. By induction, let T' be a tree whose vertex set K' is the set of maximal cliques of G' such that for each vertex $v \in V - U$ the set $K'_v = \{X \in K' \mid v \in X\}$ induces a connected subgraph (subtree) of T'.

Remark.   Either $K = K' + \{A\} - \{Y\}$ or $K = K' + \{A\}$ depending upon whether or not Y is a maximal clique of G'.

Let B be a maximal clique of G' containing Y.

Case 1.   If $B = Y$, then we obtain T from T' by renaming B, A.
Case 2.   If $B \neq Y$, then we obtain T from T' by connecting the new vertex A to B.

In either case, $K_u = \{A\}$ for all u in U and $K_v = K'_v$ for all v in $V - A$, each of which induces a subtree of T. We need only worry about the sets K_y $(y \in Y)$.

In case 1, $K_y = K'_y + \{A\} - \{B\}$, which induces the same subtree as K'_y since only names were changed. In case 2, $K_y = K'_y + \{A\}$, which clearly induces a subtree.

Thus, we have constructed the required tree T and the proof of the theorem is complete. ∎

Buneman [1972, 1974] discusses the application of the subtree intersection model in constructing evolutionary trees and in certain other classificatory problems.

An undirected graph $G = (V, E)$ is called a *path graph* if it is the intersection graph of a family of paths in a tree. Renz [1970] showed that G is a path graph if and only if G is triangulated and G is the intersection graph of a family $\mathscr{F}$ of paths in an undirected graph such that $\mathscr{F}$ satisfies the Helly property. Gavril [1978] presented an efficient algorithm for recognizing path graphs; he also proved a theorem for path graphs analogous to the equivalence of (ii) and (iii) in Theorem 4.8 (see Exercise 26).

6.   Triangulated Graphs Are Perfect

Occasionally, the minimum graph coloring problem and the maximum clique problem can be simplified using the *principle of separation into pieces* (Berge [1973, p. 329]). This method is described in the following theorem and its proof. In particular, it is applicable to triangulated graphs.

Theorem 4.9.   Let S be a vertex separator of a connected undirected graph $G = (V, E)$, and let $G_{A_1}, G_{A_2}, \ldots, G_{A_t}$ be the connected components of G_{V-S}. If S is a clique (not necessarily maximal), then

$$\chi(G) = \max_i \chi(G_{S+A_i})$$

and

$$\omega(G) = \max_i \omega(G_{S+A_i}).$$

Proof.   Clearly $\chi(G) \geq \chi(G_{S+A_i})$ for each i, so $\chi(G) \geq k = \max_i \chi(G_{S+A_i})$. In fact, G can be colored using exactly k colors. First color G_S, then independently extend the coloring to each *piece* G_{S+A_i}. This composite will be a coloring of G. Thus, $\chi(G) = k$.

Next, certainly $\omega(G) \geq \omega(G_{S+A_i})$ for each i, so $\omega(G) \geq \max_i \omega(G_{S+A_i})$ $= m$. Let X be a maximum clique of G, i.e., $|X| = \omega(G)$. It is impossible that two vertices of X lie in G_{A_i} and G_{A_j} $(i \neq j)$ since the vertices are connected. Thus, X lies wholly in one of the pieces, say G_{S+A_r}. Hence, $m \geq \omega(G_{S+A_r}) \geq |X| = \omega(G)$. Therefore, $\omega(G) = m$.   ∎

Corollary 4.10.   Let S be a separating set of a connected undirected graph $G = (V, E)$, and let $G_{A_1}, G_{A_2}, \ldots, G_{A_t}$ be the connected components of G_{V-S}. If S is a clique, and if each subgraph G_{S+A_i} is perfect, then G is perfect.

Proof.   Assume that the result is true for all graphs with fewer vertices than G. It suffices to show that $\chi(G) = \omega(G)$. Using Theorem 4.9 and the fact that each graph G_{S+A_i} is perfect, we have

$$\chi(G) = \max_i \chi(G_{S+A_i}) = \max_i \omega(G_{S+A_i}) = \omega(G). \qquad \blacksquare$$

We are now ready to state the main result.

Theorem 4.11 (Berge [1960], Hajnal and Surányi [1958]).   Every triangulated graph is perfect.

Proof.   Let G be a triangulated graph, and assume that the theorem is true for all graphs having fewer vertices than G. We may assume that G is connected, for otherwise we consider each component individually. If G is complete, then G is certainly perfect. If G is not complete, then let S be a minimal vertex separator for some pair of nonadjacent vertices. By Theorem 4.1, S is a clique. Moreover, by the induction hypothesis, each of the (triangulated) subgraphs G_{S+A_i}, as defined in Corollary 4.10, is perfect. Thus, by Corollary 4.10, G is perfect.   ∎

Remark.   The proofs in this section used only the perfect graph property (P_1) (Berge [1960]). Historically, however, until Theorem 3.3 was proved, the arguments had to be carried out for property (P_2) as well (Hajnal and Surányi [1958]).

Let $\mathscr{G}$ denote the class of all undirected graphs satisfying the property that every odd cycle of length greater than or equal to 5 has at least two chords. Clearly, every triangulated graph is in $\mathscr{G}$. Our ultimate goal in the remainder of this section is to prove that the graphs in $\mathscr{G}$ are perfect. The technique used to show this will be constructive in the following sense: Given a k-coloring of a graph $G \in \mathscr{G}$, we will show how to reduce it into an ω-coloring of G, where $k \geq \omega = \omega(G)$, by performing a sequence of color interchanges called *switchings*.

Let G be an undirected graph which has been properly colored. An (α, β)-*chain* in G is a chain whose vertices alternate between the colors α and β. Let

$G_{\alpha\beta}$ denote the subgraph induced by the vertices of G which are colored α or β. An $\langle \alpha, \beta \rangle$ *switch with respect to* G consists of the following operation:

Either interchange the colors in a nontrivial connected component of $G_{\alpha\beta}$ and leave all other colors unchanged, or recolor all isolated vertices of $G_{\alpha\beta}$ using β and leave all other colors unchanged.

Note that the result of an $\langle \alpha, \beta \rangle$ switch with respect to G is again a proper coloring of G.

Lemma 4.12. Let $G \in \mathscr{G}$ be properly colored, and let x be any vertex of G. Let vertices $y, z \in \text{Adj}(x)$ be colored α and β, respectively, with $\alpha \neq \beta$. If y and z are linked by an (α, β)-chain in G, then they are linked by an (α, β)-chain in $G_{\text{Adj}(x)}$.

Proof. Let $\mu = [y = x_0, x_1, x_2, \ldots, x_l = z]$ be an $[\alpha, \beta]$ chain in G of minimum length between y and z. Clearly, l must be odd. We claim that $\{x_0, x_1, x_2, \ldots, x_l\} \subseteq \text{Adj}(x)$.

The claim is certainly true if $l = 1$. Let us assume that $l \geq 3$ and that the claim is true for all minimum (α, β)-chains of odd length strictly less than l. Now, the cycle $\bar{\mu} = [x, x_0, x_1, \ldots, x_l, x]$ has odd length $l + 2 \geq 5$, and all of its chords must have x as an endpoint since a chord between an α vertex and a β vertex of μ would give a shorter chain. Therefore, every subchain $\mu[x_s, x_t] = [x_s, \ldots, x_t]$ of μ is a minimum (α, β)-chain, and since $G \in \mathscr{G}$ the cycle $\bar{\mu}$ has at least two chords, xx_i and xx_j $(i < j)$.

If $\mu[x_0, x_i]$, $\mu[x_i, x_j]$, and $\mu[x_j, x_l]$ all have odd length, then applying the induction hypothesis to each of them we obtain $\{x_0, x_1, \ldots, x_l\} \subseteq \text{Adj}(x)$. Otherwise, at least one of $\mu[x_0, x_i]$ or $\mu[x_j, x_l]$ has even length. Without loss of generality, assume that $\mu[x_0, x_i]$ has even length so that $\mu[x_i, x_l]$ has odd length. By induction, $\{x_i, x_{i+1}, \ldots, x_l\} \subseteq \text{Adj}(x)$. In particular, $x_{i+1} \in \text{Adj}(x)$, so $\mu[x_0, x_{i+1}]$ has odd length and by induction $\{x_0, x_1, \ldots, x_{i+1}\} \subseteq \text{Adj}(x)$. This proves the claim. ∎

Let $G' = G_{\text{Adj}(x)}$. Lemma 4.12 says that a nontrivial connected component of $G_{\alpha\beta}$ contains only one nontrivial connected component of $G'_{\alpha\beta}$ or only isolated α vertices of $G'_{\alpha\beta}$ or only isolated β vertices of $G'_{\alpha\beta}$.

Lemma 4.13. Let f be a proper coloring of a graph $G \in \mathscr{G}$, and let x be a vertex of G colored γ. Let $f_{G'}$ be the restriction of f to the subgraph G' induced by those vertices adjacent to x whose colors are from some arbitrary subset Q of colors with $\gamma \notin Q$. If $f_{G'}$ can be transformed into a coloring g' of G' by a sequence of switchings with respect to G' (using colors from Q), then f can be transformed into a coloring f' of G by a sequence of switchings with respect to G such that $f'_{G'} = g'$.

Proof. It is sufficient to consider the case of a single $\langle \alpha, \beta \rangle$ switch with respect to G', where $\alpha, \beta \neq \gamma$. Suppose that a connected component $H'_{\alpha\beta}$ of $G'_{\alpha\beta}$ was switched. If $H'_{\alpha\beta}$ is nontrivial, then by Lemma 4.12 the same result could be obtained by switching the component of $G_{\alpha\beta}$, containing $H'_{\alpha\beta}$. If $H'_{\alpha\beta}$ has only one vertex, then all isolated vertices of $G'_{\alpha\beta}$ were switched to β. In this case the same result could be obtained by switching all nontrivial components of $G_{\alpha\beta}$ which contain isolated α vertices of $G'_{\alpha\beta}$ plus switching all isolated vertices of $G_{\alpha\beta}$ to β.   ∎

Theorem 4.14 (Meyniel [1976]).   Let $G \in \mathcal{G}$ and let f be a k-coloring of G. Then there exists a q-coloring g of G with $q = \chi(G)$ which is obtainable from f by a sequence of switchings with respect to G.

Proof.   The theorem is obviously true for graphs with one vertex. Assume that the theorem is true for all graphs with fewer vertices than G.

Consider a k-coloring f of G using the colors $\{\alpha_1, \alpha_2, \ldots, \alpha_k\}$ with $k > q = \chi(G)$. Choose a vertex x with color $\alpha \neq \alpha_1, \alpha_2, \ldots, \alpha_q$; if there is none, the proof is finished. Let G' be the subgraph induced by the vertices colored $\alpha_1, \alpha_2, \ldots, \alpha_q$ and adjacent to x. Clearly,

$$q' = \chi(G') \leq \chi(G_{\text{Adj}(x)}) \leq q - 1.$$

Since $G' \in \mathcal{G}$, the induction hypothesis implies that there exists a q'-coloring g' of G' which is obtainable from $f_{G'}$ by a sequence of switchings with respect to G'. By Lemma 4.13, g' can also be obtained from f by a sequence of switchings with respect to G. After performing this sequence of switchings, we can recolor x with one of the colors $\alpha_1, \alpha_2, \ldots, \alpha_q$ which is unused by g' (since $q' \leq q - 1$). Thus, we have enlarged the set of vertices colored $\alpha_1, \alpha_2, \ldots, \alpha_q$. Repeating this process until all vertices of G are colored $\alpha_1, \alpha_2, \ldots, \alpha_q$ will yield a minimum coloring.   ∎

We are now ready to show that the graphs in $\mathcal{G}$ are perfect. Gallai [1962] originally proved the case where each odd cycle has two noncrossing chords; a shorter proof appeared in Surányi [1968]. The case where each odd cycle has two crossing chords was proved by Olaru [1969] (see Sachs [1970]). The general case, as presented here, is due to Meyniel [1976].

Theorem 4.15.   If G is an undirected graph such that every odd cycle has two chords, then G is perfect.

Proof.   Let $G \in \mathcal{G}$ with $\chi(G) = q$, and let H be an induced subgraph of G satisfying

$$\chi(H) = q,$$

$$\chi(H - x) = q - 1 \qquad \text{for every vertex } x \text{ of } H.$$

Choose a vertex x of H and a $(q - 1)$-coloring f of $H - x$, and let H' be the subgraph induced by $\text{Adj}_H(x)$. If H' were $(q - 2)$-colorable, then by Theorem 4.14 f restricted to H' could be transformed into a $(q - 2)$-coloring of H' by a sequence of switchings with respect to H'. Then by Lemma 4.13 there would exist a $(q - 1)$-coloring of $H - x$ using $q - 2$ colors for $\text{Adj}_H(x)$. But this would imply that $\chi(H) = q - 1$, a contradiction.

Therefore, $\{x\} \cup \text{Adj}_H(x)$ is not $(q - 1)$-colorable, and hence it must be the entire vertex set of H. Since this argument holds for all x, it follows that H is a q-clique. Thus, $\chi(G) = \omega(G) = q$. In like manner, $\chi(G') = \omega(G')$ for all induced subgraphs G' of G since being in $\mathscr{G}$ is a hereditary property. Thus G is perfect.   ∎

7. Fast Algorithms for the COLORING, CLIQUE, STABLE SET, and CLIQUE-COVER Problems on Triangulated Graphs

Let $G = (V, E)$ be a triangulated graph, and let σ be a perfect elimination scheme for G. It was first pointed out by Fulkerson and Gross [1965] that every maximal clique was of the form $\{v\} \cup X_v$ where

$$X_v = \{x \in \text{Adj}(v) \mid \sigma^{-1}(v) < \sigma^{-1}(x)\}.$$

This elementary fact is easily shown. By the definition of σ, each $\{v\} \cup X_v$ is complete. Let w be the first vertex in σ contained in an arbitrary maximal clique A; then $A = \{w\} \cup X_w$. Therefore, we have the following result.

Proposition 4.16 (Fulkerson and Gross [1965]).   A triangulated graph on n vertices has at most n maximal cliques, with equality if and only if the graph has no edges.

It is easy enough to modify Algorithm 4.2 to print out each set $\{v\} \cup X_v$. However, some of these will not be maximal, and we would like to filter them out. The mechanism that we employ is the observation that $\{u\} \cup X_u$ is *not* maximal iff for some i, in line 7 of Algorithm 4.2, X_u is concatenated to $A(u)$ (Exercise 13). The modified algorithm is as follows:

Algorithm 4.3.   Chromatic number and maximal cliques of a trangulated graph.

Input: The adjacency sets of a triangulated graph G and a perfect elimination scheme σ.

Output: All maximal cliques of G and the chromatic number $\chi(G)$.

Method: A single call to the procedure CLIQUES(σ) given in Figure 4.8. The number $S(v)$ indicates the size of the largest set that would have been con-catenated to $A(v)$ in Algorithm 4.2. A careful comparison will reveal that Algorithm 4.3 is a modification of Algorithm 4.2.

Theorem 4.17.   Algorithm 4.3 correctly calculates the chromatic number and all maximal cliques of a triangulated graph $G = (V, E)$ in $O(|V| + |E|)$ time.

The proof is similar to that of Theorem 4.5.

Next we tackle the problem of finding the stability number $\alpha(G)$ of a tri-angulated graph. Better yet, since G is perfect, let us demand that we produce both a stable set and clique cover of size $\alpha(G)$. A solution is given by Gavril.

Let σ be a perfect elimination scheme for $G = (V, E)$. We define inductively a sequence of vertices $y_1, y_2, \ldots, y_t$ in the following manner: $y_1 = \sigma(1)$; y_i is the first vertex in σ which follows y_{i-1} and which is not in $X_{y_1} \cup X_{y_2} \cup \cdots \cup X_{y_{i-1}}$; all vertices following y_t are in $X_{y_1} \cup \cdots \cup X_{y_t}$. Hence

$$V = \{y_1, y_2, \ldots, y_t\} \cup X_{y_1} \cup \cdots \cup X_{y_t}.$$

The following theorem applies.

Theorem 4.18 (Gavril [1972]).   The set $\{y_1, y_2, \ldots, y_t\}$ is a maximum stable set of G, and the collection of sets $Y_i = \{y_i\} \cup X_{y_i}$ $(i = 1, 2, \ldots, t)$ comprises a minimum clique cover of G.

```
              procedure CLIQUES (σ):
              begin
   1.            χ ← 1;
   2.            for all vertices v do S(v) ← 0;;
   3.            for i ← 1 to n do
                 begin
   4.                v ← σ(i);
   5.                X ← {x ∈ Adj(v) | σ⁻¹(v) < σ⁻¹(x)};
   6.                if Adj(v) = Ø then print {v};;
   7.                if X = Ø then go to 13;;
   8.                u ← σ(min{σ⁻¹(x) | x ∈ X});
   9.                S(u) ← max{S(u), |X| − 1};
  10.                if S(v) < |X| then do
                     begin
  11.                    print {v} ∪ X;
  12.                    χ = max{χ, 1 + |X|};
                     end
  13.             end
  14.          print "The chromatic number is", χ;
              end
```

Figure 4.8.   Procedure to list all maximal cliques of a triangulated graph, given a perfect elimination scheme.

Proof.   The set $\{y_1, y_2, \ldots, y_t\}$ is stable since if $y_j y_i \in E$ for $j < i$, then $y_i \in X_{y_j}$, which cannot be. Thus $\alpha(G) \geq t$. On the other hand, each of the sets $Y_i = \{y_i\} \cup X_{y_i}$ is a clique, and so $\{Y_1, \ldots, Y_t\}$ is a clique cover of G. Thus, $\alpha(G) = k(G) = t$, and we have produced the desired maximum stable set and minimum clique cover.   ∎

Implementing this procedure to run efficiently is a straightforward exercise and is left for the reader (Exercise 25). For a treatment of the maximum weighted stable set problem, see Frank [1976].

EXERCISES

1.   Show that for $n \geq 5$ the graph $\bar{C}_n$ is not triangulated.

2.   Using Theorem 4.1, condition (iii), prove that every interval graph is triangulated. What is the interpretation of a separator in an interval representation of a graph?

3.   Prove properties (L1)–(L3) of lexicographic breadth-first search (Section 4.3).

4.   Apply Algorithm 4.1 to the graph in Figure 3.3 by arbitrarily selecting the vertex of degree 2 in line 3 during the first pass of the algorithm. (i) What is the perfect scheme you get? (ii) Find a perfect scheme of G which cannot possibly arise from Algorithm 4.1.

The class of undirected graphs known as *k-trees* is defined recursively as follows: A k-tree on k vertices consists of a clique on k vertices (k-clique); given any k-tree T_n on n vertices, we construct a k-tree on $n + 1$ vertices by adjoining a new vertex x_{n+1} to T_n, which is made adjacent to each vertex of some k-clique of T_n and nonadjacent to the remaining $n - k$ vertices. Notice that a 1-tree is just a tree in the usual sense, and that a k-tree has at least k vertices. Exercises 5–7 below are due to Rose [1974]. Harary and Palmer [1968] discuss 2-trees.

5.   Show that a k-tree has a perfect vertex elimination scheme and is therefore triangulated. Give an example of a triangulated graph which is not a k-tree for any k.

6.   Prove the following result: An undirected graph $G = (V, E)$ is a k tree if and only if

   (i) G is connected,
   (ii) G has a k-clique but no $(k + 2)$-clique, and
   (iii) every minimal vertex separator of G is a k-clique.

7.   Let $G = (V, E)$ be a triangulated graph which has a k-clique but no $(k + 2)$-clique. Prove that $\|E\| \leq k|V| - \frac{1}{2}k(k + 1)$ with equality holding if and only if G is a k-tree.

8. Show that every 3-tree is planar.

9. Let G be an undirected graph and let H be constructed as follows. The vertices of H correspond to the edges of G, and two vertices of H are adjacent if their corresponding edges form two sides of a triangle in G. Prove that G is a 2-tree if and only if H is a cactus of triangles.

10. Show that every vertex of a minimal $x-y$ separator is adjacent to some vertex in each of the connected components containing x and y, respectively.

11. Let S be a minimal $x-y$ separator of a connected graph G. Show that every path in G from x to y contains a member of S and that every $s \in S$ is contained in some path μ from x to y which involves no other element of S, that is, $\mu \cap S = \{s\}$.

12. Prove the following: For any minimal vertex separator S of a triangulated graph $G = (V, E)$, there exists a vertex c in each connected component of G_{V-S} such that $S \subseteq \text{Adj}(c)$. (Hint: Prove the inclusion for each subset $X \subseteq S$ using induction.)

13. Program Algorithms 4.1 and 4.2 using the data structures suggested and test some graphs for the triangulated graph property.

14. Give a representation of the graph in Figure 4.5a as intersecting subtrees of a tree.

15. Prove that G is triangulated if and only if G is the intersection graph of a family $\mathcal{T}$ of subtrees of a tree where no member of $\mathcal{T}$ contains another member of $\mathcal{T}$ (Gavril [1974a]).

16. Give an algorithm which constructs for any triangulated graph G a collection of subtrees of a tree whose intersection graph is isomorphic to G.

17. Prove the following: H is a tree if and only if every family of paths in H satisfies the Helly property.

18. Prove the following theorem of Renz [1970]: G is the intersection graph of a family of paths in a tree iff G is triangulated and is the intersection graph of a family of arcs of a graph satisfying the Helly property.

19. Using the Helly property for subtrees of a tree, show directly that (ii) implies (iii) in Theorem 4.8. (Hint: for each clique A of the intersection graph, paint the subtree corresponding to the intersection of all members of A red and paint the remainder of the tree green. What does it look like when you collapse each red piece to a point?)

20. Prove Corollary 4.10 using the perfect graph property (P_2) instead of (P_1).

21. The line graph $L(G)$ of G is defined to be the undirected graph whose vertices correspond to the edges of G, and two vertices of $L(G)$ are joined by an edge if and only if they correspond to adjacent edges in G. Prove that G is triangulated if and only if $L(G)$ is triangulated.

22.   Prove that Algorithm 4.3 correctly calculates the chromatic number and all maximal cliques of a triangulated graph.

23.   Let σ be a perfect vertex elimination scheme for a triangulated graph $G = (V, E)$. Let $H = (V, F)$ be an orientation of G, where $xy \in F$ iff $\sigma^{-1}(x) < \sigma^{-1}(y)$. Show that H is acyclic. Let τ be any topological sorting of H. Show that τ is also a perfect elimination scheme for G.

24.   Prove that a height function h (see Chapter 2, Exercise 8) of the acyclic oriented graph H defined in the preceding exercise is a minimum coloring of the triangulated graph G. Thus, a triangulated graph can be colored with a minimum number of colors in time proportional to its size.

25.   Modify Algorithm 4.3 so that, in addition, it prints out a maximum stable set and prints an asterisk next to those cliques which together comprise a minimum clique cover.

26.   Prove the following: $G = (V, E)$ is a path graph if and only if there exists a tree T whose vertex set is $\mathscr{K}$ (the maximal cliques of G) such that for all $v \in V$, the induced subgraph $T_{\mathscr{K}_v}$ is a path in T. ($\mathscr{K}_v$ denotes the set of maximal cliques which contain v.) (Gavril [1978].)

27.   Let $G = (V, E)$ be an undirected graph, and let $\sigma = [v_1, v_2, \ldots, v_n]$ be an ordering of V. Consider the following property:

   (T):   If $\sigma^{-1}(u) < \sigma^{-1}(v) < \sigma^{-1}(w)$ and $w \in \mathrm{Adj}(u) - \mathrm{Adj}(v)$, then there exists an x such that $\sigma^{-1}(v) < \sigma^{-1}(x)$ and $x \in \mathrm{Adj}(v) - \mathrm{Adj}(w)$.

   Prove that if G is a triangulated graph and σ satisfies (T), then σ is a perfect elimination scheme for G (Tarjan [1976]).

28.   (i)   Prove that any MCS order, as defined at the end of Section 4.3, satisfies property (T) from the preceding exercise.

   (ii)   Give an implementation of MCS to recognize triangulated graphs in $O(n + e)$ time. (Hint. To achieve linearity you may wish to link together all unnumbered vertices which are currently adjacent to the same number of numbered vertices (Tarjan [1976]).)

29.   An undirected graph is called *i-triangulated* if every odd cycle with more than three vertices has a set of chords which form with the cycle a planar graph whose unbounded face is the exterior of the cycle and whose bounded faces are all triangles. Prove that a graph is *i*-triangulated if and only if every cycle of odd length k has $k - 3$ chords that do not cross one another (Gallai [1962]).

Bibliography

Berge, Claude
   [1960]   Les problèmes de colorations en théorie des graphs, *Publ. Inst. Statist. Univ. Paris* **9**, 123–160.

[1973] "Graphs and Hypergraphs," Chapter 16. North-Holland, Amsterdam. MR50 #9640.
Buneman, Peter
[1972] The recovery of trees from measures of dissimilarity, "Mathematics in the Archaeological and Historical Sciences," pp. 387–395. Edinburgh Univ. Press, Edinburgh.
[1974] A characterization of rigid circuit graphs, *Discrete Math.* **9**, 205–212. MR50 #9686.
Cantalupi, Gabriella Tazzi, and Zucchetti, Bianca Ricetti
[1972] Singrammi triangolati sferici, *Inst. Lombarado Accad. Sci. Lett. Rend. A* **106**, 697–703. MR48 #8287.
Which connected triangulated graphs can be drawn on the sphere?
Dirac, G. A.
[1961] On rigid circuit graphs, *Abh. Math. Sem. Univ. Hamburg* **25**, 71–76. MR24 #57.
Frank, Andras
[1976] Some polynomial algorithms for certain graphs and hypergraphs, *Proc. 5th British Combin. Conf.*, Congressus Numerantium No. XV, Utilitas Math., Winnipeg, MR53 #13500.
Fulkerson, D. R., and Gross, O. A.
[1965] Incidence matrices and interval graphs, *Pacific J. Math.* **15**, 835–855. MR32 #3881.
Gallai, Tibor
[1962] Graphen mit triangulierbaren ungeraden Vielecken, *Magyar Tud. Akad. Mat. Kutató Int. Közl.* **7**, 3–36. MR26 #3039.
Gavril, Fanica
[1972] Algorithms for minimum coloring, maximum clique. minimum covering by cliques, and maximum independent set of a chordal graph, *SIAM J. Comput.* **1**, 180–187. MR48 #5922.
[1974a] The intersection graphs of subtrees in trees are exactly the chordal graphs, *J. Combin. Theory B* **16**, 47–56. MR48 #10868.
[1974b] An algorithm for testing chordality of graphs, *Inform. Process. Lett.* **3**, 110–112. MR52 #9671.
[1977] Algorithms on clique separable graphs, *Discrete Math.* **19**, 159–165.
[1978] A recognition algorithm for the intersection graphs of paths in trees, *Discrete Math.* **23**, 211–227.
Hajnal, Andras, and Surányi, Janos
[1958] Über die Auflösung von Graphen in vollständige Teilgraphen, *Ann. Univ. Sci. Budapest Eötvös. Sect. Math.* **1**, 113–121. MR21 #1944.
Harary, Frank, and Palmer, Edgar M.
[1968] On acyclic simplicial complexes. *Mathematika* **15**, 115–122. MR37 #3936.
Lekkerkerker, C. G., and Boland, J. Ch.
[1962] Representation of a finite graph by a set of intervals on the real line, *Fund. Math.* **51**, 45–64. MR25 #2596.
Leuker, George S.
[1974] Structured breadth first search and chordal graphs, Princeton Univ. Tech. Rep. TR-158.
Meyniel, H.
[1976] On the perfect graph conjecture, *Discrete Math.* **16**, 339–342. MR55 #12568.
Ohtsuki, Tatsuo
[1976] A fast algorithm for finding an optimal ordering for vertex elimination on a graph, *SIAM J. Comput.* **5**, 133–145. MR52 #13515.
Ohtsuki, Tatsuo, Cheung, L. K., and Fujisawa, T.
[1976] Minimal triangulation of a graph and optimal pivoting order in a sparse matrix. *J. Math. Anal. Appl.* **54**, 622–633.

Olaru, Elefterie
   [1969] Über die Überdeckung von Graphen mit Cliquen, *Wiss. Z. Tech. Hochsch. Ilmenau*
         15, 115–121. MR43 #3162.

Renz, P. L.
   [1970] Intersection representations of graphs by arcs, *Pacific J. Math.* **34**, 501–510.
         MR42 #5839.

Rose, Donald J.
   [1970] Triangulated graphs and the elimination process, *J. Math. Anal. Appl.* **32**, 597–609.
         MR42 #5840.

   [1972] A graph-theoretic study of the numerical solution of sparse positive definite systems
         of linear equations, *in* "Graph Theory and Computing," (Ronald C. Read, ed.),
         pp. 183–217. Academic Press, New York. MR49 #6579.

   [1974] On simple characterizations of *k*-trees, *Discrete Math* **7**, 317–322. MR49 #101.

Rose, Donald J., and Tarjan, Robert Endre
   [1975] Algorithmic aspects of vertex elimination. *Proc. 7th Annu. ACM Symp. Theory*
         Comput., 245–254. MR56 #7320.

   [1978] Algorithmic aspects of vertex elimination of directed graphs, *SIAM J. Appl. Math.*
         34, 176–197.

Rose, Donald J., Tarjan, Robert Endre, and Leuker, George S.
   [1976] Algorithmic aspects of vertex elimination on graphs, *SIAM J. Comput.* **5**, 266–283.
         MR53 #12077.

Sachs, Horst
   [1970] On the Berge conjecture concerning perfect graphs, *in* "Combinatorial Structures
         and their Applications," pp. 377–384. Gordon & Breach, New York. MR42 #7549.

Surányi, L.
   [1968] The covering of graphs by cliques, *Studia Sci. Math. Hungar.* **3**, 345–349. MR38 #76.

Tarjan, Robert Endre
   [1976] Maximum cardinality search and chordal graphs. Stanford Univ. Unpublished
         Lecture Notes CS 259.

Trotter, William Thomas, Jr.
   [1971] A note on triangulated graphs, *Notices Amer. Math. Soc.* **18**, 1045 (A).

Wagon, Stanley
   [1978] Infinite triangulated graphs, *Discrete Math* **22**, 183–189.

Walter, J. R.
   [1972] Representations of rigid cycle graphs, Ph.D. thesis, Wayne State Univ.

   [1978] Representations of chordal graphs as subtrees of a tree, *J. Graph Theory* **2**, 265–267.

CHAPTER 5

Comparability Graphs

1. Γ-Chains and Implication Classes

This chapter is devoted to the class of perfect graphs known as comparability graphs or transitively orientable graphs. These graphs were encountered in Section 1.3 in connection with interval graphs (Proposition 1.3), but our treatment here will be independent of that brief introduction.

An undirected graph $G = (V, E)$ is a *comparability graph* if there exists an orientation (V, F) of G satisfying

$$F \cap F^{-1} = \emptyset, \qquad F + F^{-1} = E, \qquad F^2 \subseteq F,$$

where $F^2 = \{ac \,|\, ab, \, bc \in F$ for some vertex $b\}$. The relation F is a strict partial ordering of V whose comparability relation is exactly E, and F is called a *transitive orientation* of G (or of E). Comparability graphs are also known as *transitively orientable* graphs and *partially orderable* graphs. Examples of some comparability graphs can be found in Section 1.3.

Let us see what happens when we try to assign a transitive orientation to the 4-cycle (Figure 5.1a). Arbitrarily choosing $ab \in F$ *forces* us to orient the bottom edge toward b and the top edge toward d (for otherwise transitivity would be violated). These in turn, force the remaining edge to be oriented toward d. Applying the same idea to the graph in Figure 5.1b, we find that a contradiction arises, namely, choosing $ab \in F$ forces successively the orientations $cb, cd, cf, ef, bf,$ and ba. This graph is not a comparability graph. We now make the notion of forcing more precise.

Define the binary relation Γ on the edges of an undirected graph $G = (V, E)$ as follows:

$$ab \, \Gamma \, a'b' \qquad \text{iff} \qquad \begin{cases} \text{either} & a = a' \text{ and } bb' \notin E \\ \text{or} & b = b' \text{ and } aa' \notin E \end{cases}$$

(a)   (b)

Figure 5.1.   Examples of forcing. The arbitrary choice of $ab \in F$ forces the other indicated orientations.

We say that *ab directly forces a'b'* whenever $ab \; \Gamma \; a'b'$. Since E is irreflexive, $ab \; \Gamma \; ab$; however, $ab \; \Gamma\hspace{-0.6em}/\; ba$. The reader should not continue until he is convinced of this fact.

The reflexive, transitive closure Γ^* of Γ is easily shown to be an equivalence relation on E and hence partitions E into what we shall call the *implication classes* of G. Thus edges ab and cd are in the same implication class if and only if there exists a sequence of edges

$$ab = a_0 b_0 \; \Gamma \; a_1 b_1 \; \Gamma \cdots \Gamma \; a_k b_k = cd, \qquad \text{with} \quad k \geq 0.$$

Such a sequence is called a *Γ-chain* from ab to cd, and we say that ab (eventually) *forces cd* whenever $ab \; \Gamma^* \; cd$.

The reader can easily verify the properties

$$ab \; \Gamma \; a'b' \Leftrightarrow ba \; \Gamma \; b'a',$$

$$ab \; \Gamma^* \; a'b' \Leftrightarrow ba \; \Gamma^* \; b'a',$$

which follow directly from the definitions.

Let $\mathscr{I}(G)$ denote the collection of implication classes of G. We define

$$\hat{\mathscr{I}}(G) = \{\hat{A} \mid A \in \mathscr{I}(G)\},$$

where $\hat{A} = A \cup A^{-1}$ is the symmetric closure of A. The members of $\hat{\mathscr{I}}(G)$ are called the *color classes* of G for reasons that will become evident later.

Examples.   The graph G in Figure 5.2 has eight implication classes:

$$A_1 = \{ab\}, \qquad A_2 = \{cd\}, \qquad A_3 = \{ac, ad, ae\}, \qquad A_4 = \{bc, bd, be\},$$
$$A_1^{-1} = \{ba\}, \qquad A_2^{-1} = \{dc\}, \qquad A_3^{-1} = \{ca, da, ea\}, \qquad A_4^{-1} = \{cb, db, eb\}.$$

So we have $\hat{\mathscr{I}}(G) = \{\hat{A}_1, \hat{A}_2, \hat{A}_3, \hat{A}_4\}$. On the other hand, the graph in Figure 5.1b has only one implication class:

$$A = \{ab, cb, cd, cf, ef, bf, ba, bc, dc, fc, fe, fb\}$$

and $A = \hat{A}$.

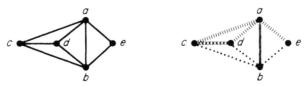

Figure 5.2. An undirected graph G and a coloring of its edges according to the classes of $\mathscr{I}(G)$.

Theorem 5.1. Let A be an implication class of an undirected graph G. If G has a transitive orientation F, then either $F \cap \hat{A} = A$ or $F \cap \hat{A} = A^{-1}$ and, in either case, $A \cap \hat{A} = \varnothing$.

Proof. We defined Γ in order to capture the fact that, for any transitive orientation F of G,

$$\text{if } ab \; \Gamma \; a'b' \quad \text{and} \quad ab \in F, \quad \text{then} \quad a'b' \in F.$$

Applying this property repeatedly, we obtain $F \cap A = \varnothing$ or $A \subseteq F$. Since (i) $A \subseteq F + F^{-1}$ and (ii) $F \cap F^{-1} = \varnothing$, we have the implications

$$F \cap A = \varnothing \Rightarrow A \subseteq F^{-1} \qquad [\text{by (i)}]$$
$$\Rightarrow A^{-1} \subseteq F \Rightarrow F \cap \hat{A} = A^{-1},$$

and

$$A \subseteq F \Rightarrow A^{-1} \subseteq F^{-1} \Rightarrow F \cap A^{-1} = \varnothing \qquad [\text{by (ii)}]$$
$$\Rightarrow F \cap \hat{A} = A.$$

In either case $A \cap A^{-1} = \varnothing$.   ∎

The converse of Theorem 5.1 is also valid, namely, *if $A \cap A^{-1} = \varnothing$ for every implication class A, then G has a transitive orientation.* This result will be proved as part of Theorem 5.27. Theorem 5.27 also provides the justification for an algorithm which assigns a transitive orientation to a comparability graph.

Remark. Many readers may wonder whether an arbitrary union of implication classes $F = \bigcup_i A_i$ satisfying $F \cap F^{-1} = \varnothing$ and $F + F^{-1} = E$ is necessarily a transitive orientation of G. The answer is no. As a counter-example, consider a triangle which has $8 = 2^3$ such orientations, two of which fail to be transitive.

Next we present two lemmas which will be useful throughout this chapter.

Let $ab = a_0 b_0 \; \Gamma \; a_1 b_1 \; \Gamma \cdots \Gamma \; a_k b_k = cd$ be given. For each $i = 1, \ldots, k$ we have

$$a_{i-1} b_{i-1} \; \Gamma \; a_i b_{i-1} \; \Gamma \; a_i b_i,$$

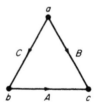

Figure 5.3.

since the added middle edge equals one of the other two. Hence we may state the following:

Lemma 5.2.   If $ab\ \Gamma^* cd$, then there exists a Γ-chain from ab to cd of the form

$$ab = a_0 b_0\ \Gamma\ a_1 b_0\ \Gamma\ a_1 b_1\ \Gamma\ a_2 b_1\ \Gamma \cdots \Gamma\ a_k b_k = cd.$$

Such a chain will be called a *canonical Γ-chain*.

Lemma 5.3 (The Triangle Lemma).   Let A, B, and C be implication classes of an undirected graph $G = (V, E)$ with $A \neq B$ and $A \neq C^{-1}$ and having edges $ab \in C$, $ac \in B$, and $bc \in A$ (see Figure 5.3).

(i)   If $b'c' \in A$, then $ab' \in C$ and $ac' \in B$.
(ii)   If $b'c' \in A$ and $a'b' \in C$, then $a'c' \in B$.
(iii)   No edge in A touches the vertex a.

Proof.   By Lemma 5.2 there exists a canonical Γ-chain

$$bc = b_0 c_0\ \Gamma\ b_1 c_0\ \Gamma\ b_1 c_1\ \Gamma \cdots \Gamma\ b_k c_k = b'c'.$$

By induction on i, we have the following implications:

$$[B \ni ac_i\ \not\Gamma\ b_{i+1} c_i \in A] \Rightarrow ab_{i+1} \in E,$$

$$b_{i+1} b_i \notin E \Rightarrow ab_{i+1}\ \Gamma\ ab_i \in C,$$

$$[C^{-1} \ni b_{i+1} a\ \not\Gamma\ b_{i+1} c_{i+1} \in A] \Rightarrow ac_{i+1} \in E,$$

$$c_{i+1} c_i \notin E \Rightarrow ac_{i+1}\ \Gamma\ ac_i \in B.$$

Therefore, in particular, $ab' = ab_k \in C$ and $ac' = ac_k \in B$. This proves (i).

Next, let us assume that $b'c' \in A$ and $a'b' \in C$. By part (i), $ac' \in B$. Consider a new canonical Γ-chain,

$$ab = a_0 b_0\ \Gamma\ a_1 b_0\ \Gamma\ a_1 b_1\ \Gamma \cdots \Gamma\ a_l b_l = a'b'.$$

This chain gives rise to the chain

$$ac' = a_0 c' \; \Gamma \; a_1 c' \; \Gamma \cdots \Gamma \; a_l c' = a'c'.$$

Thus, $ac' \; \Gamma^* \; a'c'$ and $a'c' \in B$, which proves (ii).
    Finally, part (i) immediately implies (iii).   ∎

Theorem 5.4.   Let A be an implication class of an undirected graph $G = (V, E)$. Exactly one of the following alternatives holds:

(i) $A = \hat{A} = A^{-1}$;
(ii) $A \cap A^{-1} = \varnothing$, A and A^{-1} are transitive, and they are the only transitive orientations of $\hat{A}$.

Proof.   (i)   Assume $A \cap A^{-1} \neq \varnothing$. Let $ab \in A \cap A^{-1}$, so $ab \; \Gamma^* \; ba$. For any $cd \in A$, $cd \; \Gamma^* \; ab$ and $dc \; \Gamma^* \; ba$. Since Γ^* is an equivalence relation, $cd \; \Gamma^* \; dc$ and $dc \in A$. Thus $A = \hat{A}$.
    (ii)   Assume $A \cap A^{-1} = \varnothing$ and let $ab, bc \in A$. Now $ac \notin E \Rightarrow ab \; \Gamma \; cb \Rightarrow cb \in A \Rightarrow bc \in A^{-1}$, a contradiction. Thus $ac \in E$.
    Let B be the implication class of G containing ac, and suppose $A \neq B$. Since $A \neq A^{-1}$ and $ab \in A$, the Triangle Lemma 5.3(i) implies that $ab \in B$, a contradiction. Thus $ac \in A$, and A is transitive. Moreover, A being transitive implies that A^{-1} is transitive.
    Finally, A is an implication class of $\hat{A}$, so by Theorem 5.1 A and A^{-1} are the only transitive orientations of $\hat{A}$.   ∎

Corollary 5.5.   Each color class of an undirected graph G either has exactly two transitive orientations, one being the reversal of the other, or has no transitive orientation. If in G there is a color class having no transitive orientation, then G fails to be a comparability graph.

2.   Uniquely Partially Orderable Graphs

Let H_0 be a graph with n vertices $v_1, v_2, \ldots, v_n$ and let $H_1, H_2, \ldots, H_n$ be n disjoint graphs.* The *composition* graph $H = H_0[H_1, H_2, \ldots, H_n]$ is formed as follows: For all $1 \leq i, j \leq n$, replace vertex v_i in H_0 with the graph H_i and make each vertex of H_i adjacent to each vertex of H_j whenever v_i is

---

* The graphs may be directed or undirected.

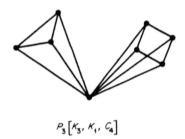

$$P_3[K_3, K_1, C_4]$$

Figure 5.4.   The composition of some undirected graphs.

adjacent to v_j in H_0. Formally, for $H_i = (V_i, E_i)$ we define $H = (V, E)$ as follows:

$$V = \bigcup_{i \geq 1} V_i;$$

$$E = \bigcup_{i \geq 1} E_i \cup \{xy \,|\, x \in V_i,\ y \in V_j \text{ and } v_i v_j \in E_0\}.$$

We may also denote $E = E_0[E_1, E_2, \ldots, E_n]$. We call H_0 the *outer factor* and $H_1, \ldots, H_n$ the *inner factors* (see Figures 5.4 and 5.5).

Theorem 5.6.   Let $G = G_0[G_1, G_2, \ldots, G_n]$, where the G_i are disjoint undirected graphs. Then G is a comparability graph if and only if each G_i $(0 \leq i \leq n)$ is a comparability graph.

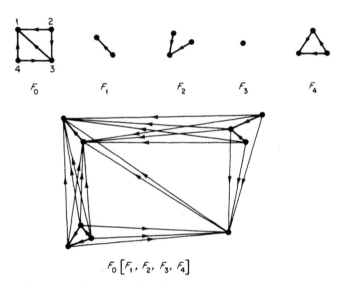

$$F_0[F_1, F_2, F_3, F_4]$$

Figure 5.5.   The composition of some transitively oriented graphs.

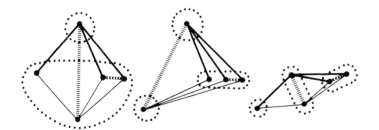

Figure 5.6. Three decompositions of the same graph. The edges are marked according to their color classes.

Proof. Let $F_0, F_1, \ldots, F_n$ be transitive orientations of $G_0, G_1, \ldots, G_n$, respectively. It is easy to show that $F_0[F_1, \ldots, F_n]$ is a transitive orientation of G. The converse follows from the hereditary property of comparability graphs. ∎

A graph is called *decomposable* if it can be expressed as a nontrivial composition of some of its induced subgraphs; otherwise, it is called *indecomposable*. Three decompositions of the same graph are illustrated in Figure 5.6. Of course, any graph G has the trivial decompositions $G = K_1[G]$ and $G = G[K_1, K_1, \ldots, K_1]$. Formally, $G = (V, E)$ is *decomposable* if there exists a partition $V = V_1 + V_2 + \cdots + V_r$ of the vertices into nonempty pairwise disjoint subsets with $1 < r < |V|$ such that

$$G = G_R[G_{V_1}, G_{V_2}, \ldots, G_{V_r}]$$

for any set of representatives $R = \{x_1, x_2, \ldots, x_r\}$, $x_i \in V_i$. Such a partition is said to *induce* a *proper decomposition* of G. Theorem 5.6 may be reinterpreted as follows.

Corollary 5.7. Let F be a transitive orientation of a comparability graph G. If $G = G_R[G_{V_1}, \ldots, G_{V_r}]$ is a proper decomposition of G, then $F = F_R[F_{V_1}, \ldots, F_{V_r}]$.

Let us examine the effect of this decomposition on the color classes. Notice in Figure 5.6 that each color class occurs either entirely within one internal factor or entirely within the external edges. This phenomenon is true in general.

Theorem 5.8. Let $G = G_0[G_1, \ldots, G_n]$ be the composition of disjoint undirected graphs $G_i = (V_i, E_i)$ $(i = 0, 1, \ldots, n)$. If $\hat{A}$ is a color class of G, then one of the following alternatives holds:

(i) $\hat{A} \subseteq E_j$ for exactly one index $j \geq 1$, or
(ii) $\hat{A} \cap E_j = \emptyset$ for all indices $j \geq 1$.

Proof.   By our original definition of forcing, every color class $\hat{A}$ is a connected (partial) subgraph of G. Suppose that $\hat{A} \cap E_j \neq \varnothing$ for some $j \geq 1$. Let $ab \in \hat{A} \cap E_j$ and consider an edge $a'b' \Gamma ab$. Clearly $a'b' \notin E_k$ for any $k \neq j$, $k \geq 1$, since edges in different internal components never share a vertex. Moreover, $a'b'$ cannot be an external edge because if it were then by the definition of composition the vertices a, a', b, b' would induce a triangle in G, implying that $a'b' \,\cancel{\Gamma}\, ab$. Hence, $a'b'$ must also be in E_j. Thus, by connectivity, $\hat{A} \subseteq E_j$.   ∎

Let $G = (V, E)$ be an undirected graph. A subset $Y \subseteq V$ is called *partitive* if for each $x \in V - Y$ either $Y \cap \mathrm{Adj}(x) = \varnothing$ or $Y \subseteq \mathrm{Adj}(x)$. A partitive set Y is *nontrivial* if $1 < |Y| < |V|$. On the one hand, any internal factor of a decomposition of G is partitive. On the other hand, a partitioning of the vertices $V = \{v_1\} + \cdots + \{v_k\} + Y$ where Y is partitive induces a proper decomposition of G. Therefore, we may conclude the following remark.

Remark 5.9.   *G has a nontrivial partitive set if and only if G is decomposable.*

Before continuing, we present two simple consequences of the Triangle lemma.

Proposition 5.10.   *If Y is the set of vertices spanned by a color class $\hat{A}$ of an undirected graph $G = (V, E)$, then Y is partitive.*

Proof.   If $Y = V$, then the result is trivial. Otherwise, let $a \in V - Y$, and suppose that $b \in Y \cap \mathrm{Adj}(a)$. Then, $ab \in E - \hat{A}$ and $bc \in \hat{A}$ for some $c \in Y$, which implies that $ac \in E - \hat{A}$. Applying Lemma 5.3(i), we obtain that $Y \subseteq \mathrm{Adj}(a)$.   ∎

Proposition 5.11.   *An undirected graph $G = (V, E)$ may have at most one color class which spans all of V.*

Proof.   Suppose that two distinct color classes $\hat{A}$ and $\hat{B}$ both span V. Then for every vertex b there exist edges $ab \in \hat{B}$ and $bc \in \hat{A}$. Since $\hat{A} \neq \hat{B}$, the edge ac is in E. What color is it? Let $\hat{C}$ denote the color class containing ac. If $\hat{C} \neq \hat{A}$, then Lemma 5.3(iii) implies that no edge from $\hat{A}$ may touch vertex a, a contradiction. Hence $\hat{C} = \hat{A} \neq \hat{B}$, and Lemma 5.3(iii) now implies that no edge from $\hat{B}$ may touch vertex c, another contradiction. Therefore, $\hat{A}$ and $\hat{B}$ cannot both span all of V.   ∎

A comparability graph G is called *uniquely partially orderable* (UPO) if it has exactly two transitive orientations, one being the reversal of the other. Clearly, a comparability graph is UPO if and only if it has exactly one color class (see Corollary 5.5).

Theorem 5.12 (Shevrin and Filippov [1970]; Trotter, Moore, and Sumner [1976]).   Let G be a connected comparability graph. The following conditions are equivalent.

(i) G is UPO.
(ii)   Every nontrivial partitive set of G is a stable set.
(iii)   For every proper decomposition of G, each internal factor is a stable set (i.e., all edges are external).

Proof.   The following proof is due to Arditti [1976a]. By the comments preceding Remark 5.9, (ii) and (iii) are equivalent. If G is UPO, then G has exactly one color class, and this class spans V. Therefore, by Theorem 5.8 any proper decomposition of G must make all edges external. Thus (i) implies (iii). Next, suppose G is not UPO; then by Proposition 5.11 G has a color class which only spans a proper subset Y of V. By Proposition 5.10, Y is a nontrivial partitive set which is not a stable set. Thus (ii) implies (i).   ∎

Corollary 5.13.   Let G be a comparability graph. If G is indecomposable, then G is UPO.

Proof.   If G is indecomposable, then G is connected and it satisfies condition (iii) of Theorem 5.12 vacuously. Hence G is UPO.   ∎

3.   The Number of Transitive Orientations

In this section we shall examine the interaction between implication classes. In the process we will obtain a formula for the number $t(G)$ of transitive orientations of a comparability graph G and a procedure for constructing them. Our treatment follows Golumbic [1977a], in which most of this theory was developed. An alternate method for calculating $t(G)$ appears in Shevrin and Filippov [1970].

Example.   A transitive orientation of any graph partially orders its vertices. Consider a transitive orientation F of the complete graph K_{r+1} on $r + 1$ vertices. Since in F each pair of distinct vertices is comparable, the partial ordering is actually a linear ordering (total ordering). Conversely, any linear ordering of the vertices of K_{r+1} yields a transitive orientation by directing each edge from smaller to larger. Therefore,

$$t(K_{r+1}) = \text{the number of linear orderings of } r + 1 \text{ elements}$$
$$= (r + 1)!$$

Let $G = (V, E)$ be an undirected graph. A complete subgraph (V_S, S) on $r + 1$ vertices is called a *simplex* of *rank r* if each undirected edge $\hat{ab}$ of S is contained in a different color class of G. For example, each undirected edge $\hat{ab}$ of E is itself a simplex of rank 1. A simplex is *maximal* if it is not properly contained in any larger simplex.

The *multiplex* generated by a simplex S of rank r is defined to be the following undirected (partial) subgraph: (V_M, M), where

$$M = \{ab \in E \,|\, ab \; \Gamma^* \; xy \text{ for some } xy \in S\},$$

or alternatively,

$$M = \bigcup \hat{A},$$

where the union is over all color classes $\hat{A} \in \hat{\mathscr{I}}(G)$ satisfying $\hat{A} \cap S \neq \varnothing$. Thus, M is the union of the $\frac{1}{2}r(r + 1)$ color classes represented by the edges of the simplex S. (This number is due to S being a complete graph on $r + 1$ vertices.) Anticipating Corollary 5.15 we say that the multiplex M also has rank r. A multiplex is *maximal* if it is not properly contained in any larger multiplex. We will soon see that M is a maximal multiplex if and only if S is a maximal simplex.

Remark.   If we actually assign a different color to each class of $\hat{\mathscr{I}}(G)$ and paint the edges of G accordingly, then a complete subgraph S whose edges are each painted a different color is a *simplex*. The collection of edges of E painted the same color as some edge of S is a *multiplex*. For example, if there is a red, white, and blue triangle in the graph, then the set of all red, white, and blue edges is a multiplex of rank 2. The graph in Figure 5.2 has two disjoint maximal multiplexes, one of rank 2 and one of rank 1. The expressions *tricolored triangle* and simplex of rank 2 are synonymous. Finally, notice that *the edges and implication classes of a tricolored triangle satisfy the hypotheses of the Triangle Lemma 5.3.*

An *isomorphism* between two simplices (V_1, S_1) and (V_2, S_2) of an undirected graph is a bijection $f: V_1 \rightarrow V_2$ such that $ab \; \Gamma^* \; f(a)f(b)$ for each distinct pair $a, b \in V_1$. It is thus possible to lay S_1 on top of S_2 so that the colors of their edges match.

Theorem 5.14 (Golumbic [1977a]).   Let (V_T, T) be a simplex generating the multiplex M, and let (V_S, S) be a simplex contained in M. Then (V_S, S) is isomorphic to a subsimplex of (V_T, T).

Proof.   Choose an edge $bc \in S$. Since T generates M, there exists an edge $b'c' \in T$ such that $bc \; \Gamma^* \; b'c'$. Define $f(b) = b'$ and $f(c) = c'$. If rank $S = 1$, then the theorem is proved.

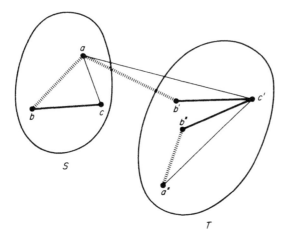

Figure 5.7.   From a tricolored triangle in S we find an isomorphic tricolored triangle in T. The vertices b' and b'' must be equal since T is a simplex.

Otherwise, consider any other vertex $a \in V_S$, and let A, B, and C denote the implication classes such that $bc \in A$, $ac \in B$, and $ab \in C$. Since T generates M, there exists an edge $a''b'' \in T \cap C$. Applying the Triangle Lemma 5.3(i) twice we obtain (1) $ab' \in C$ and $ac' \in B$ and (2) $a''c' \in B$ and $b''c \in A$ (Figure 5.7). But the simplex T cannot contain two different edges $b'c'$ and $b''c'$ which are the same color; hence $b' = b''$. Define $f(a) = a''$. In this manner f is defined for all vertices of V_S. Choose distinct vertices a and d of V_S, different from b and c. Since $ab \ \Gamma^* \ f(a)f(b)$ and $db \ \Gamma^* \ f(d)f(b)$, then $f(a) = f(d)$ would imply $a = d$, since S is a simplex. Thus f is injective. Moreover, the Triangle Lemma 5.3(ii) implies that $ad \ \Gamma^* \ f(a)f(d)$. Therefore, f is an isomorphism from S to a subsimplex of T.   ∎

The following is an immediate result of the preceding theorem.

Corollary 5.15.   Simplices generating the same multiplex are isomorphic.

The next lemma shows us how to construct simplices.

Lemma 5.16.   Let (V_S, S) be a simplex of an undirected graph $G = (V, E)$ generating a multiplex M. If G contains a tricolored triangle on vertices a, b, c such that $ab \notin M$ but $bc \in M$, then we may adjoin the vertex a to (V_S, S) to obtain the larger simplex (V_T, T) containing (V_S, S), where

$$V_T = V_S \cup \{a\},$$
$$T = S \cup \{\widehat{ad} \mid d \in V_S\}.$$

Proof.   Let us assume that G contains a tricolored triangle on a, b, c satisfying $ab \notin M$ and $bc \in M$. Since S generates M, there is some edge $b'c' \in S$ for which $b'c'$ Γ^* bc. The Triangle Lemma 5.3(i) implies that ab' and ac' are in the same two distinct color classes of G, respectively, as are ab and ac. Thus, $ab' \notin M$. Next we shall show that $ac' \notin M$ as well.

Suppose that $ac' \in M$; then ac' Γ^* xy for some $xy \in S$ (because ac' must be the same color as some edge in S). Again by the Triangle lemma, $b'a$ Γ^* $b'x$, however, $b'a \notin M$ while $b'x \in S$, a contradiction. Thus, $ac' \notin M$. This argument actually proves the stronger claim:

Fact 1.   If a tricolored triangle has one side in M and another side not in M, then the third side is also not in M.

Next let $d \in V_S$, $d \neq b', c'$. Certainly $ad \in E$ since ab' and $b'd$ are in different color classes. Whereas the edges $b'c$ and $c'd$ are in different color classes, the edge ad is in a different class than at least one of them. Therefore, at least one of the triangles $G_{\{a, b', d\}}$ or $G_{\{a, c', d\}}$ is tricolored and satisfies the hypothesis of Fact 1, implying that $ad \notin M$. Thus, the set $\{\hat{ad} \mid d \in V_S\}$ shares no color classes with S.

Since ab' and ac' are in different color classes, to conclude the proof that (V_T, T) is a simplex it suffices to show the following claim:

Fact 2.   Either the undirected edges $\hat{ad}$ (for $d \in V_S$) are all in different color classes, or all of the edges ad (for $d \in V_S$) are Γ^*-related.

Suppose that $ad, ad' \in \hat{A} \in \mathscr{I}(G)$. If $\hat{A}$ has no transitive orientation, then Theorem 5.4(i) implies that ad $\Gamma^* ad'$. If A has a transitive orientation, then Theorem 5.4(ii) implies that ad Γ^* ad' since $dd' \notin \hat{A}$. Now let d'' be any vertex of V_S other than d or d'. If $ad'' \notin \hat{A}$, then $G_{\{a, d, d''\}}$ and $G_{\{a, d', d''\}}$ are both tricolored triangles sharing two common colors. So by the Triangle Lemma 5.3(i), dd'' Γ^* $d'd''$, which contradicts the definition of a simplex. Thus, $ad'' \in \hat{A}$ and, as before, ad Γ^* ad''. This proves Fact 2 and concludes the proof of the theorem. Obviously, rank $T = 1 + \text{rank } S$.   ∎

Lemma 5.3(ii) tells us that if an undirected graph contains a red, white, and blue triangle, then anywhere in the graph where we find a red edge ab and a white edge bc, the edge ac will be blue. Suppose there is a multiplex M containing a red, white, and blue triangle. The next theorem shows, in particular, that every red, white, and blue triangle is part of a simplex generating M.

Theorem 5.17 (Golumbic [1977a]).   Let S be a simplex contained in a multiplex M. There exists a simplex S_M generating M such that $S \subseteq S_M$.

Proof. If rank S = rank M, then S itself generates M. We proceed by reverse induction, assuming the theorem to be true for any simplex of rank greater than rank S.

Let U be any simplex generating M. Since rank U = rank M, only some of the edges of U have "cousins" in S of the same color. These are the ones contained in M_1, defined here as the multiplex generated by S. Thus $M_1 \subset M$. Since U is connected it has a tricolored triangle on a, b, c with $bc \in M_1$, $ab \notin M_1$. By Lemma 5.16, we can adjoin the vertex a to S creating a simplex T containing S with rank $T = 1 +$ rank S. Thus, by induction, there is a simplex S_M generating M such that $S \subset T \subseteq S_M$.   ∎

Theorems 5.14 and 5.17 can be summarized as follows:

Corollary 5.18. Let M_1, M_2 be multiplexes with $M_1 \subseteq M_2$.

(i) Every simplex generating M_1 is contained in a simplex generating M_2.

(ii) Every simplex generating M_2 contains a subsimplex which generates M_1.

Theorem 5.19. Let M be the multiplex generated by a simplex S. Then, M is a maximal multiplex if and only if S is a maximal simplex.

Proof. ($\Rightarrow$) This implication follows directly from the definition of multiplex.

($\Leftarrow$) Suppose S is maximal and $M \subseteq M'$, where M' is another multiplex. Since $S \subseteq M \subseteq M'$, Theorem 5.17 implies the existence of a simplex S' containing S with S' generating M'. But the maximality of S yields $S = S'$, so $M = M'$.   ∎

By virtue of the preceding theorem and corollary we can now locate a maximal multiplex by a *local search* of the edges. We pick an edge at random and build up successively large simplices each containing its predecessor until the simplex we have is maximal. It then generates a maximal multiplex.

The next theorem implies that *the maximal multiplexes partition the edges of G*.

Theorem 5.20. If M_1 and M_2 are maximal multiplexes of an undirected graph G, then either $M_1 \cap M_2 = \varnothing$ or $M_1 = M_2$.

Proof. Let S_1 and S_2 be simplices generating M_1 and M_2, respectively. By Theorem 5.19, S_1 and S_2 are maximal. Suppose $M_1 \cap M_2 \neq \varnothing$ and $M_1 \neq M_2$, then some edges of S_2 are in M_1 and some are not. Because S_2 is connected, it must contain a tricolored triangle $G_{\{a,b,c\}}$ with $bc \in M_1$ and

$ab \notin M_1$. By Lemma 5.16, we can construct a large simplex T containing S_1, contradicting the maximality of S_1. Thus, one of the alternatives of the theorem must hold.   ∎

Theorem 5.21.   If A is an implication class of an undirected graph $G = (V, E)$ such that $A = \hat{A}$, then A itself is a maximal multiplex of rank 1.

The proof of Theorem 5.21 follows directly from the Triangle lemma and the definition of multiplex. It is left as an exercise for the reader.   ∎

A simplex of rank r has $(r + 1)!$ transitive orientations, as we have seen in the example at the beginning of this section. Moreover, in the proof of the next theorem we will show that a transitive orientation of the simplex extends uniquely to a transitive orientation of the multiplex generated by it, *except* when the multiplex is itself an implication class and hence not transitively orientable (by Theorem 5.4). Conversely, a transitive orientation of a multiplex restricts uniquely to a transitive orientation of any simplex contained in it.

Theorem 5.22.   Let M be a multiplex of rank r. If M is transitively orientable, then $t(M) = (r + 1)!$.

Remark.   Theorem 5.21 shows that the *only* case in which M might fail to be transitively orientable is when $r = 1$.

Proof.   Let S be a simplex of rank r generating M, and let F_S be a transitive orientation of S. Finally, let $A_1, \ldots, A_k$ [$k = \frac{1}{2}r(r + 1)$] be the *implication classes* containing the edges of F_S. The corresponding color classes $\hat{A}_i$ are distinct, and $\hat{A}_1 + \cdots + \hat{A}_k = M$. If $r = 1$, then A_1 is a transitive orientation of $M = \hat{A}_1$ if and only if $A_1 \neq \hat{A}_1$ if and only if $t(M) = 2$. If $r > 1$, then $F = A_1 + \cdots + A_k$ is certainly an orientation of M by Theorems 5.4 and 5.21. We must show that F is transitive. Let $ab \in A_i$, $bc \in A_j$. If $i = j$, then $ac \in A_i$ by the transitivity of A_i [Theorem 5.4(ii)]. If $i \neq j$, then $ac \in E$ since $\hat{A}_i \cap \hat{A}_j = \varnothing$. Suppose $ca \in F$, then the individual transitivity of A_i and A_j implies that $ca \in A_t$ for some $i \neq t \neq j$. Theorem 5.14, however, implies that there exist edges $a'b', b'c', c'a' \in S$ such that $a'b' \in A_i$, $b'c' \in A_j$, and $c'a' \in A_t$, contradicting the transitivity of F_S. Therefore, $ac \in F$ and F is transitive. Thus, for each transitive orientation of S we obtain a unique transitive orientation of M, so $t(M) \geq t(S) = (r + 1)!$.

Conversely, given a transitive orientation F_2 of M, consider its restriction $F_2 \cap S$ to S. The three facts, ab, $bc \in F_2 \cap S$, F_2 being transitive and S being complete, collectively imply that $ac \in F_2 \cap S$. So $F_2 \cap S$ is a transitive orientation of S. Therefore, $t(S) \geq t(M)$ and Theorem 5.22 is proved.   ∎

The partition of an undirected graph $G = (V, E)$ into its maximal multiplexes $E = M_1 + \cdots + M_k$ will be referred to as its *M-decomposition*. It is unique up to the order of the M_i. Having just examined the transitive orientability of a multiplex, let us now investigate the transitive orientability of all of E. The next major theorem shows a one-to-one correspondence between the transitive orientations of the M_i and those of E.

Theorem 5.23 (Golumbic [1977a]).   Let $G = (V, E)$ be an undirected graph, and let $E = M_1 + \cdots + M_k$, where each M_i is a maximal multiplex of E.
(i)   If F is a transitive orientation of G, then $F \cap M_i$ is a transitive orientation of M_i.
(ii)   If $F_1, \ldots, F_k$ are transitive orientations of $M_1, \ldots, M_k$, respectively, then $F_1 + \cdots + F_k$ is a transitive orientation of G.
(iii) $t(G) = t(M_1)t(M_2) \cdots t(M_k)$.
(iv)   If G is a comparability graph and $r_i = \text{rank } M_i$, then $t(G) = \prod_{i=1}^{k} (r_i + 1)!$.

Proof.   Statement (iii) follows from (i) and (ii), while (iv) is implied by (iii) and Theorem 5.22.

(i)   Assume F is a transitive orientation of G and let $ab, bc \in F \cap M_i$. Suppose that $ac \notin M_i$; then $G_{\{a, b, c\}}$ must not be a tricolored triangle. Therefore, $ab, bc \in \hat{A}$ for some $\hat{A} \in \mathscr{F}(G)$. Thus $ab, bc \in F \cap \hat{A}$, and $F \cap \hat{A}$ equals either A or A^{-1}, both of which are transitive by Theorems 5.1 and 5.4. Hence $ac \in \hat{A}$, which is a contradiction.

(ii)   Assume that $F_1, \ldots, F_k$ are transitive orientations of $M_1, \ldots, M_k$, respectively. We shall show that $F_1 + \cdots + F_k$ is transitive. Let $ab \in F_i$, $bc \in F_j$. If $i = j$, then $ac \in F_i$ by transitivity of F_i. If $i \neq j$, then ab and bc are in different color classes, so $ac \in E$. Since $G_{\{a, b, c\}}$ cannot be a tricolored triangle and hence cannot be contained in a single multiplex, it follows that $ac \in M_i + M_j$. But if $ca \in F_i + F_j$, then transitivity gives a contradiction. Thus, $ac \in F_i + F_j$.   ∎

Summarizing the results of this section, we have shown that the maximal multiplexes partition the edges and act independently with respect to transitive orientation. They are generated by maximal simplices which can be built up from a single edge by a local search. Simplices generating the same multiplex are isomorphic. Finally, the number of transitive orientations of an undirected graph is a product of factorials depending on the ranks of its maximal multiplexes. Thus, every comparability graph behaves as if it were a disjoint collection of complete graphs.

4.   Schemes and G-Decompositions—An Algorithm for Assigning Transitive Orientations

In this section we describe an algorithm for calculating transitive orientations and for determining whether or not a graph is a comparability graph. This technique is a modification of one first presented by Pnueli, Lempel, and Even [1971]. Our version uses the notions introduced in Section 5.1; the proof of its correctness relies on some of the results of Section 5.3. A discussion of its computational complexity will follow in Section 5.6.

Let $G = (V, E)$ be an undirected graph. A partition of the edge set $E = \hat{B}_1 + \hat{B}_2 + \cdots + \hat{B}_k$ is called a G-*decomposition* of E if B_i is an implication class of $\hat{B}_i + \cdots + \hat{B}_k$ for all $i = 1, 2, \ldots, k$. A sequence of edges $[x_1 y_1, x_2 y_2, \ldots, x_k y_k]$ is called a *decomposition scheme* for G if there exists a G-decomposition $E = \hat{B}_1 + \hat{B}_2 + \cdots + \hat{B}_k$ satisfying $x_i y_i \in B_i$ for all $i = 1, 2, \ldots, k$. In this chapter the term *scheme* will always mean a decomposition scheme.

For a given G-decomposition there will be many corresponding schemes (any set of representatives from the B_i). However, for a given scheme there exists exactly one corresponding G-decomposition. A scheme and G-decomposition can be constructed by the following procedure:

Algorithm 5.1 (Decomposition Algorithm).
Let $G = (V, E)$ be an undirected graph.
Initially, let $i = 1$ and $E_1 = E$.

Step (1): Arbitrarily pick an edge $e_i = x_i y_i \in E_i$.
Step (2): Enumerate the implication class B_i of E_i containing $x_i y_i$.
Step (3): Define $E_{i+1} = E_i - \hat{B}_i$.
Step (4): If $E_{i+1} = \varnothing$, then let $k = i$ and Stop; otherwise, increase i by 1 and go back to Step (1).

Clearly, the decomposition algorithm yields a scheme $[x_1 y_1, \ldots, x_k y_k]$ and corresponding G-decomposition $\hat{B}_1 + \cdots + \hat{B}_k$ for any undirected graph G. Moreover, if $y_i x_i$ had been chosen instead of $x_i y_i$ for some i, then B_i^{-1} would replace B_i in the G-decomposition. Applying the algorithm to the graph in Figure 5.2, the scheme $[ac, bc, dc]$ gives the G-decomposition for which $B_1 = A_3$, $B_2 = A_4 + A_1^{-1}$ and $B_3 = A_2^{-1}$ (see p. 106 and Figure 5.8*). In this example notice that although ba and bc were not Γ-related in the original graph, once $\hat{B}_1$ is removed they become Γ-related in the remaining subgraph and their implication classes merge. In general, each implication

---

* Another example is given in Exercise 8.

i	(V, E_i)	$x_i y_i$	(V, B_i)
1		ac	
2		bc	
3		dc	

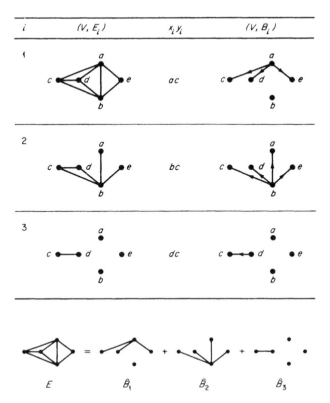

Figure 5.8.   An illustration of the decomposition algorithm.

class of E_{i+1} will be the union of *some* number of implication classes of E_i. We now examine exactly how the old classes merge.

Theorem 5.24 (Golumbic [1977a]).   Let A be an implication class of an undirected graph $G = (V, E)$, and let D be an implication class of $E - \hat{A}$. Either

   (i) D is an implication class of E, and A is an implication class of $E - \hat{D}$,

or

   (ii) $D = B + C$ where B and C are implication classes of E, and $\hat{A} + \hat{B} + \hat{C}$ is a multiplex of E of rank 2.

Proof.   Removing $\hat{A}$ from E may cause some implication classes of E to merge. Let D be the union of k implication classes of E.

Assume $k \geq 2$; then there exists a triangle on vertices a, b, c with $bc \in \hat{A}$ and either $ac \in B$ and $ab \in C$ or $ca \in B$ and $ba \in C$, where B and C are distinct implication classes of E contained in D. Without loss of generality we may assume $ac \in B$ and $ab \in C$ since the other case is identical for D^{-1}. Suppose $B = C^{-1}$, then ba, $ac \in B$. But $bc \notin B$, so by Theorem 5.4 $B = \hat{B} = B^{-1}$, implying $B = C$, a contradiction. Therefore $\hat{B} \cap \hat{C} = \emptyset$ and $G_{\{a,b,c\}}$ is a tricolored triangle, making $\hat{A} + \hat{B} + \hat{C}$ a multiplex of rank 2.

Furthermore, any Γ-chain in $E - \hat{A}$ containing edges from $\hat{B}$ and $\hat{C}$ could not contain edges from other implication classes since all triangles in E with one edge in $\hat{A}$ and a second edge in $\hat{B}$ (resp. $\hat{C}$) must have its third side in $\hat{C}$ (resp. $\hat{B}$) and would be isomorphic as a simplex to $G_{\{a,b,c\}}$. Thus $k = 2$ and $D = B + C$.

Finally, we shall show that if $k = 1$, then A is an implication class of $E - \hat{D}$. By what we have already proved, if A is not an implication class of $E - \hat{D}$, then $\hat{D} + \hat{A} + \hat{A}_1$ is a multiplex of rank 2 in E for some third implication class A_1 of E. However, this implies that D alone is not an implication class of $E - \hat{A}$, contradicting $k = 1$. So indeed A is an implication class of $E - \hat{D}$. ∎

Corollary 5.25. Let A be an implication class of an undirected $G = (V, E)$. If $A = \hat{A}$, then all other implication classes of E are again implication classes of $E - \hat{A}$.

Corollary 5.26. Let A be an implication class of an undirected graph $G = (V, E)$. Then $|\hat{\mathscr{I}}(E)| = r + |\hat{\mathscr{I}}(E - \hat{A})|$, where r is the rank of the maximal multiplex of E containing A.

The proof of the first corollary follows directly from Theorem 5.21, while the second corollary is a result of $\hat{A}$ being a part of exactly $r - 1$ different multiplexes of rank 2. ∎

The next theorem is of major importance since it legitimizes the use of G-decompositions as a constructive tool for deciding whether an undirected graph is a comparability graph, and if so, producing a transitive orientation. Condition (iv) is the traditional characterization due to Gilmore and Hoffman [1964] and Ghouila-Houri [1962].

Theorem 5.27 (TRO Theorem). Let $G = (V, E)$ be an undirected graph with G-decomposition $E = \hat{B}_1 + \cdots + \hat{B}_k$. The following statements are equivalent:

   (i) $G = (V, E)$ is a comparability graph;
   (ii) $A \cap A^{-1} = \emptyset$ for all implication classes A of E;
   (iii) $B_i \cap B_i^{-1} = \emptyset$ for $i = 1, \ldots, k$;
   (iv) every "circuit" of edges $v_1 v_2$, $v_2 v_3$, $\ldots$, $v_q v_1 \in E$ such that $v_{q-1} v_1$, $v_q v_2$, $v_{i-1} v_{i+1} \notin E$ (for $i = 2, \ldots, q - 1$) has even length.

Furthermore, when these conditions hold, $B_1 + \cdots + B_k$ is a transitive orientation of E.

Proof. (i) $\Rightarrow$ (ii) This is precisely Theorem 5.1.

(ii) $\Rightarrow$ (iii) We shall proceed by induction. Since B_1 is an implication class of E, we have $B_1 \cap B_1^{-1} = \varnothing$. If $k = 1$, then we are done. Assume the implication is true for all G-decompositions of graphs of length less than k. Then, in particular, it is true for $E - \hat{B}_1$.

Let D be an implication class of $E - \hat{B}_1$. By Theorem 5.24, either D is an implication class of E, in which case $D \cap D^{-1} = \varnothing$, or $D = B + C$, where B and C are implication classes of E such that $\hat{B} \cap \hat{C} = \varnothing$, implying that

$$D \cap D^{-1} = (B + C) \cap (B^{-1} + C^{-1})$$
$$= (B \cap B^{-1}) + (C \cap C^{-1})$$
$$= \varnothing.$$

Therefore, by induction, $B_i \cap B_i^{-1} = \varnothing$, for $i = 2, \ldots, k$.

(iii) $\Rightarrow$ (i) Let $E = \hat{B}_1 + \cdots + \hat{B}_k$ be a G-decomposition of E with $B_i \cap B_i^{-1} = \varnothing$. By Theorem 5.4, B_1 is transitive. If $k = 1$, then the implication holds. Assume the implication is true for all G-decompositions of graphs of length less than k. By this assumption, $F = B_2 + \cdots + B_k$ is a transitive orientation of $E - \hat{B}_1$. We must show that $B_1 + F$ is transitive.

Let $ab, bc \in B_1 + F$. If both these edges are in B_1 or both in F, then by the individual transitivity of B_1 and F, $ac \in B_1 + F$. Assume, therefore, that $ab \in B_1$ and $bc \in F$, which implies that $ab \; \Gamma^* \; cb$, so $ac \in E$. What would happen if $ac \notin B_1 + F$? Then $ca \in B_1 + F$. However,

$$ca \in B_1, ab \in B_1 \Rightarrow cb \in B_1, \qquad \text{a contradiction,}$$

and

$$ca \in F, bc \in F \Rightarrow ba \in F, \qquad \text{a contradiction.}$$

Thus $ac \in B_1 + F$. Similarly, $ab \in F$ and $bc \in B_1$ imply $ac \in B_1 + F$. So indeed $B_1 + \cdots + B_k$ is a transitive orientation of E.

(iv) $\Leftrightarrow$ (i) Suppose $v_1 v_2 \in A \cap A^{-1} \neq \varnothing$. By Lemma 5.2, there exists a Γ-chain

$$v_1 v_2 \; \Gamma \; v_3 v_2 \; \Gamma \; v_3 v_4 \; \Gamma \cdots \Gamma \; v_q v_{q-1} \; \Gamma \; v_q v_{q+1} = v_2 v_1.$$

By construction, q is odd, since all first coordinates have odd index. Furthermore, $v_1 v_2, v_2 v_3, \ldots, v_q v_1$ is such a circuit, a contradiction.

Conversely, if E has such a circuit of odd length q, then

$$v_1 v_2 \; \Gamma \; v_3 v_2 \; \Gamma \; v_3 v_4 \; \Gamma \cdots \Gamma \; v_q v_{q-1} \; \Gamma \; v_q v_1 \; \Gamma \; v_2 v_1$$

is a Γ-chain in E, implying that $A \cap A^{-1} \neq \emptyset$ for the implication class A containing $v_1 v_2$, a contradiction.   ∎

By combining the TRO theorem with the decomposition algorithm, we obtain an algorithm for recognizing comparability graphs and assigning a transitive orientation.

Algorithm 5.2 (TRO Algorithm).

Input: An undirected graph $G = (V, E)$.
Output: A transitive orientation F of edges of G, or a message that G is not a comparability graph.
Method: The entire algorithm is as follows:

```
    begin
        Initialize: i ← 1; Eᵢ ← E; F ← ∅ ;
1.      Arbitrarily pick an edge xᵢyᵢ ∈ Eᵢ;
2.      Enumerate the implication class Bᵢ of Eᵢ containing xᵢyᵢ;
            if Bᵢ ∩ Bᵢ⁻¹ = ∅ then
                add Bᵢ to F;
            else
                print "G is not a comparability graph";
                STOP;
3.      Define: Eᵢ₊₁ ← Eᵢ − B̂ᵢ;
4.      if Eᵢ₊₁ = ∅ then
                k ← i; output F;
                STOP;
            else
                i ← i + 1;
                go to 1;
    end
```

The sequence of free choices made in line 1 of the algorithm determines which of the many transitive orientations of G is produced by the algorithm. A different scheme may give a different transitive orientation. But when you try out a few different schemes you will notice a remarkable phenomenon: No matter how the free choices for G are made, the number of iterations k will always be the same. A proof that this is actually true for any graph G and, more importantly, a characterization of the underlying mathematical structure which causes it are the subject of the next section.

5. The Γ^*-Matroid of a Graph

The Decomposition Algorithm 5.1 emphasizes that the order in which the edges appear in a scheme is extremely important. The free choices made in earlier iterations affect which edges remain to be chosen in latter iterations. If the algorithm once gave us a scheme $[e_1, e_2, e_3, \ldots, e_k]$, what will happen if we rerun the algorithm by choosing e_2 first and e_1 second? Is there any reason for believing that e_3 will not have been removed and will therefore be available as the third free choice? The answer to the latter question is yes.

All the results in this section are due to Golumbic [1977a].

Theorem 5.28. Let $[e_1, e_2, \ldots, e_k]$ be a scheme for an undirected graph G, and let π be a permutation of the numbers $\{1, \ldots, k\}$. Then $[e_{\pi(1)}, e_{\pi(2)}, \ldots, e_{\pi(k)}]$ is also a scheme for G.

Proof. If $k = 1$, then there is nothing to prove. Assume therefore that $k \geq 2$. Let $\hat{B}_1 + \hat{B}_2 + \cdots + \hat{B}_k$ be the G-decomposition corresponding to the given scheme. Theorem 5.24 allows us to commute edges occurring next to each other in a scheme in the following manner. Fix $i < k$. Let

$$E_i = \hat{B}_i + \cdots + \hat{B}_k,$$

$$C_i = \text{implication class of } E_i \text{ containing } e_{i+1},$$

$$C_{i+1} = \text{implication class of } E_i - \hat{C}_i \text{ containing } e_i.$$

By Theorem 5.24, either (i) $B_{i+1} = C_i$ and $B_i = C_{i+1}$, so that $\hat{B}_i + \hat{B}_{i+1} = \hat{C}_i + \hat{C}_{i+1}$, or (ii) there exists an implication class A of E_i such that $\hat{B}_{i+1} = \hat{A} + \hat{C}_i$ and $\hat{C}_{i+1} = \hat{A} + \hat{B}_i$, also implying that $\hat{B}_i + \hat{B}_{i+1} = \hat{C}_i + \hat{C}_{i+1}$. Consequently, in either case, $\hat{B}_i + \cdots + \hat{C}_i + \hat{C}_{i+1} + \cdots + \hat{B}_k$ is a G-decomposition of E with scheme $[e_1, \ldots, e_{i+1}, e_i, \ldots, e_k]$.

However, every permutation can be expressed as a composition of such local commutations (often called transpositions), from which the theorem follows. ∎

Theorem 5.29 (Golumbic [1977a]). Let $G = (V, E)$ be an undirected graph.

(i) Each scheme for G has the same length.
(ii) Each G-decomposition of G has the same length.
(iii) If $[e_1, e_2, \ldots, e_k]$ and $[f_1, f_2, \ldots, f_k]$ are schemes for G, then for any e_i there exists f_j such that $[e_1, \ldots, e_{i-1}, f_j, e_{i+1}, \ldots, e_k]$ is also a scheme for G.

Proof. If G has an implication class A such that $E = \hat{A}$, then any scheme has length 1 and any edge can be chosen as a scheme. Therefore, assume that the theorem is true for all graphs having fewer implication classes than G, and let $[e_1, e_2, \ldots, e_k]$ and $[f_1, f_2, \ldots, f_m]$ be schemes for G with $k, m \geq 2$. Choose e_i and (using Theorem 5.28 if necessary) make sure that it is *not* in the first position. If $E = \hat{C}_1 + \hat{C}_2 + \cdots + \hat{C}_m$ is the G-decomposition corresponding to $[f_1, f_2, \ldots, f_m]$, then $e_1 \in \hat{C}_p$ for some p. Thus $[f_1, \ldots, f_{p-1}, e_1, f_{p+1}, \ldots, f_m]$ is also a scheme. Theorem 5.28 then implies that $[e_1, f_1, \ldots, f_{p-1}, f_{p+1}, \ldots, f_m]$ is a scheme for G.

Finally, both $[e_2, \ldots, e_i, \ldots, e_k]$ and $[f_1, \ldots, f_{p-1}, f_{p+1}, \ldots, f_m]$ are schemes for $E - \hat{B}$, where B is the implication class of E containing e_1. Since $E - \hat{B}$ has fewer implication classes than E, by induction the lengths $k - 1$ and $m - 1$ are equal and there exists some f_j which can replace e_i in its scheme. In conclusion, since corresponding G-decompositions and schemes have the same length, all G-decompositions must have the same length. ∎

Thus we have found a number associated with an undirected graph G which is invariant over all schemes and G-decompositions of the graph, namely the length of any scheme or G-decomposition of G. We shall denote this number by $r(G)$.

Theorem 5.30. Let $G = (V, E)$ be an undirected graph, and let $E = M_1 + \cdots + M_k$, where M_i is a maximal multiplex of E of rank r_i. Then $r(G) = r_1 + \cdots + r_k$.

Proof. Let $\hat{A} \in \mathcal{I}(G)$ satisfy $\hat{A} \subseteq M_1$. Now $M_1 - \hat{A}$ is a multiplex of rank $r_1 - 1$, and $E - \hat{A} = (M_1 - \hat{A}_1) + M_2 + \cdots + M_k$ is an M-decomposition of $G' = (V, E - \hat{A})$. Since $|\mathcal{I}(G)| > |\mathcal{I}(G')|$, we may assume by induction that $r(G') = (r_1 - 1) + r_2 + \cdots + r_k$. Therefore, $r(G) = r_1 + r_2 + \cdots + r_k$. ∎

Let $G = (V, E)$ be a comparability graph with G-decomposition $E = \hat{B}_1 + \cdots + \hat{B}_k$ and corresponding scheme $[e_1, \ldots, e_k]$. By Theorem 5.27, $B_1 + \cdots + B_k$ will be a transitive orientation of G. Replacing e_i by e_i^{-1} in the scheme will have the effect of replacing B_i by B_i^{-1}, thus giving a new transitive orientation of G. In this manner we obtain $2^{r(G)}$ transitive orientations of G, since $k = r(G)$. There may, however, be others; the scheme $[e_{\pi(1)}, \ldots, e_{\pi(k)}]$ may even give a transitive orientation of G different from the $2^{r(G)}$ above. In fact, the only time when these $2^{r(G)}$ represent all the transitive orientations of G is when each maximal multiplex is of rank one. (This follows from Theorems 5.22 and 5.30 and the inequality $2^r < (r + 1)!$ for $r > 1$.) For example, $r(K_{r+1}) = r$ and $t(K_{r+1}) = (r + 1)!$ for the complete graph on $r + 1$ vertices. On the other hand, the graph G in Figure 5.9 has $t(G) = 2^{r(G)}$.

Story

The owner of a large railroad decided to introduce his sons into the business. He asked his eldest to choose any two cities between which they provide train service, and the father would give him control of that run. The

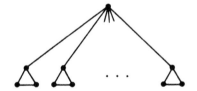

Figure 5.9.   The number of triangles is $r(G) - 1$.

lad chose New York and Philadelphia. But the boy was clever and reasoned with his father saying, "Since you operate service between Harrisburg and Philadelphia and I operate the New York–Philadelphia trains, and since we don't offer any direct service between Harrisburg and New York, why not give me also the Harrisburg–Philadelphia run for the convenience of our passengers who would otherwise be burdened with their heavy luggage in changing trains!"

The father was convinced by the son's argument and gave him the extra rail link. The son, encouraged by his success, continued this type of reasoning for triples of cities that fit the above pattern and accumulated more rail lines until finally no more triples of that form were left. His father handed him the corresponding deeds; they embraced and the son left to go out on his own.

The father continued the same process with his other sons, giving one rail line and then also giving any other link A–B when the son already controlled B–C provided they did not operate A–C between the two of them. Finally, the father had given away his entire rail system.

Theorem 5.29 shows that no matter how each son chooses his initial free choice, *exactly* $r(G)$ sons get portions of the railroad, where G is the graph whose vertices are the cities and edges the rail links.   ∎

We will now describe the underlying mathematical structure that causes the invariant $r(G)$ to arise.

A matroid $\langle E, \mathscr{B} \rangle$ consists of a nonempty (finite) set E of *elements* together with a nonempty collection $\mathscr{B}$ of subsets of E, called *bases*, satisfying the following axioms.

(i)   No base properly contains another base.

(ii)   If $\beta_1, \beta_2 \in \mathscr{B}$ and $x \in \beta_1$, then there exists an element $y \in \beta_2$ such that $(\beta_1 - \{x\} + \{y\}) \in \mathscr{B}$.

Theorem 5.31.   Let $G = (V, E)$ be an undirected graph.

(i) $\langle E, \mathscr{B} \rangle$ is a matroid, where $\{e_1, \ldots, e_k\} \in \mathscr{B}$ if and only if $[e_1, \ldots, e_k]$ is a scheme for G.

(ii) $\langle \hat{\mathscr{I}}(G), \mathscr{B}(G) \rangle$ is a matroid, where $\hat{\mathscr{I}}(G)$ is the set of color classes of G and $\{\hat{A}_1, \ldots, \hat{A}_k\} \in \mathscr{B}(G)$ if and only if $\{e_1, \ldots, e_k\} \in \mathscr{B}$ for $e_i \in \hat{A}_i$.

Proof.   The order in which the edges appear in a scheme is important for the G-decomposition it will produce. Theorem 5.28, however, allows us to treat schemes as sets of chosen representative edges in which order is not relevant. By Theorem 5.29, these subsets satisfy the axioms of a matroid. This proves (i). Condition (ii) follows easily from (i).   ∎

The matroid $\langle \hat{\mathscr{I}}(G), \mathscr{B}(G) \rangle$ may be regarded as the quotient of the matroid $\langle E, \mathscr{B} \rangle$. For those readers familiar with matroids, the invariant $r(G)$ equals the rank (in the usual matroid sense) of $\langle E, \mathscr{B} \rangle$ and of $\langle \hat{\mathscr{I}}(G), \mathscr{B}(G) \rangle$. These matroids are of a very special type. Let us see exactly what class of matroids is produced in this manner.

By Theorem 5.24, the free choices taken from one maximal multiplex in no way influence choices taken from any other maximal multiplex. Therefore, it suffices to restrict our attention to applying the decomposition algorithm to a maximal simplex (V_S, S). Let $r =$ rank S. Its free choices (r of them) constitute the edges of a spanning tree of (V_S, S). Why is that? It is certainly true if $r = 1$ or $r = 2$. If it were false, then there would be a scheme β containing a simple cycle of edges $v_1 v_2, v_2 v_3, \ldots, v_l v_1$ of minimal length l over all schemes. By Theorem 5.24, $l \neq 3$. Again by Theorem 5.24, $v_2 v_3$ could be replaced by $v_1 v_3$ in β, forming another scheme with a cycle of length less than l, contradicting minimality. Therefore, the r edges contain no simple cycles and must be a spanning tree of (V_S, S), since there are r edges and $r + 1$ vertices. Furthermore, any spanning tree of (V_S, S) is a scheme since it contains r edges, and for every other edge ab the tree provides a path $e_1, e_2, \ldots, e_q$ from a to b which, when used successively in the construction of a G-decomposition, will also eliminate the edge ab.

Two matroids $\langle E_1, \mathscr{B}_1 \rangle$ and $\langle E_2, \mathscr{B}_2 \rangle$ are *isomorphic* if there exists a bijection $f: E_1 \to E_2$ such that

$$f(\beta_1) \in \mathscr{B}_2 \qquad \text{for all} \quad \beta_1 \in \mathscr{B}_1$$

and

$$f^{-1}(\beta_2) \in \mathscr{B}_1 \qquad \text{for all} \quad \beta_2 \in \mathscr{B}_2.$$

Let $\mathscr{M}$ denote the family of matroids

$$\mathscr{M} = \{\langle \hat{\mathscr{I}}(G), \mathscr{B}(G) \rangle \,|\, G \text{ is an undirected graph}\}.$$

From the above discussion we may state the following characterization of the matroids in $\mathscr{M}$.

Theorem 5.32. A matroid is in the family $\mathcal{M}$ if and only if it is isomorphic to the matroid of spanning trees of a set of disjoint complete graphs.

6. The Complexity of Comparability Graph Recognition

A version of the decomposition algorithms of Section 5.4 is presented here in a pseudo-computer-language. It will suggest to us how we may actually enumerate the implication classes of a graph. We shall show that one can find a G-decomposition and test for transitive orientability of an undirected graph $G = (V, E)$ in $O(\delta \cdot |E|)$ time and $O(|V| + |E|)$ space, where δ is the maximum degree of a vertex.

Let $G = (V, E)$ be an undirected graph with vertices $v_1, v_2, \ldots, v_n$. In the algorithm below we use the function

$$\mathrm{CLASS}(i, j) = \begin{cases} 0 & \text{if } v_i v_j \notin E, \\ k & \text{if } v_i v_j \text{ has been assigned to } B_k, \\ -k & \text{if } v_i v_j \text{ has been assigned to } B_k^{-1}, \\ \text{undefined} & \text{if } v_i v_j \in E \text{ has not yet been assigned,} \end{cases}$$

and $|\mathrm{CLASS}(i, j)|$ denotes the absolute value of $\mathrm{CLASS}(i, j)$. As usual, the set E is always assumed to be a collection of ordered pairs and the *degree* d_i of vertex v_i is taken here to mean the number of edges with v_i as first coordinate (i.e., the out-degree). We freely use the identity

$$|E| = \sum_{i=1}^{n} d_i$$

in our analysis.

Algorithm 5.3 (Decomposition Algorithm — Alternate Version).

Input: An undirected graph $G = (V, E)$ with vertices $v_1, v_2, \ldots, v_n$ whose adjacency sets obey $j \in \mathrm{Adj}(i)$ if and only if $v_i v_j \in E$.

Output: A G decomposition of the graph given by the final values of CLASS and a variable FLAG which is 0 if the graph is a comparability graph and 1 otherwise. If the algorithm terminates with FLAG equal to zero, then a transitive orientation of G is obtained by combining all edges having positive CLASS.

Method: The algorithm proceeds until all edges have been explored. In the kth iteration an unexplored edge is placed in B_k. (Its CLASS is changed to k.) Whenever an edge is placed into B_k it is explored using the recursive procedure of Figure 5.10 by adding to B_k those edges Γ-related to it in the

```
procedure EXPLORE(i, j):
for each m ∈ Adj(i) such that [m ∉ Adj(j) or |CLASS(j, m)| < k] do
   begin
      if CLASS (i, m) is undefined then
         begin
            CLASS(i, m) ← k; CLASS(m, i) ← −k;
            EXPLORE(i, m);
         end
      else
         if CLASS(i, m) = −k then
            begin
               CLASS(i, m) ← k; FLAG ← 1;
               EXPLORE(i, m);
            end
   end
for each m ∈ Adj(j) such that [m ∉ Adj(i) or |CLASS(i, m)| < k] do
   begin
      if CLASS(m, j) is undefined then
         begin
            CLASS(m, j) ← k; CLASS(j, m) ← −k;
            EXPLORE(m, j);
         end
      else
         if CLASS(m, j) = −k then
            begin
               CLASS(m, j) ← k; FLAG ← 1;
               EXPLORE(m, j);
            end
   end
return
```

Figure 5.10.

graph E_k. (Notice that $v_i v_j \in E_k$ if and only if either $|CLASS(i, j)|$ equals k or is undefined throughout the kth iteration.)

The variable FLAG is changed from 0 to 1 the first time a B_k is found such that $B_k \cap B_k^{-1} \neq \emptyset$. At that point it is known that G is not a comparability graph (by Theorem 5.27).

The algorithm is as follows:

```
begin
   initialize: k ← 0; FLAG ← 0;
   for each edge vᵢvⱼ in E do
      if CLASS (i, j) is undefined then
         begin
            k ← k + 1;
            CLASS (i, j) ← k; CLASS (j, i) ← −k;
            EXPLORE (i, j);
         end
end
```

Complexity Analysis

We begin by specifying an appropriate data structure. The adjacency sets are stored as linked lists sorted into increasing order. The element of the list Adj(i) which represents edge $v_i v_j$ will have four fields containing j, CLASS(i, j), pointer to CLASS(j, i), and pointer to next element on Adj(i) (see Figure 5.11). The storage requirement for this data structure is $O(|V| + |E|)$, and if sorting the lists is done using Algorithm 2.1, then the entire initialization of the data structure can be accomplished in linear time.

The crucial factor in the analysis of our algorithm is the time required to access or assign the CLASS function. Ordinarily finding CLASS(i, m) could take $O(d_i)$ steps by scanning Adj(i), but if a temporary pointer happened to be in the neighborhood, then a reference to CLASS(i, m) or CLASS(m, i) would take a fixed number of steps. Consider the first loop of EXPLORE(i, j). Two temporary pointers simultaneously scan Adj(i) and Adj(j) looking for values of m which satisfy the condition in the **for** statement. Since the lists are sorted and thanks to these neighborly pointers, this loop can be executed in $O(d_i + d_j)$ steps. The second loop is done similarly; hence the time complexity of EXPLORE(i, j) is $O(d_i + d_j)$.

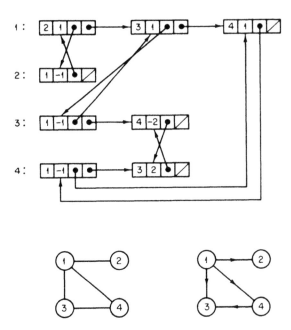

Figure 5.11. An undirected graph, the transitive orientation generated by the scheme [(1, 2), (4, 3)] and its data structure after running the algorithm.

In the main program, a temporary pointer scans each adjacency list successively in the **for** loop, implying a time complexity of $O(|E|)$. Finally, the algorithm calls EXPLORE once for each edge or its reversal (both if their implication classes are not disjoint). Therefore, since

$$\sum_{v_i v_j \in E} (d_i + d_j) = 2 \sum_{i=1}^{n} d_i^2 \le 2\delta \sum_{i=1}^{n} d_i = 2\delta|E|,$$

it follows that the time complexity for the entire algorithm (including pre-processing the input) is at most $O(\delta \cdot |E|)$. Thus we have proved the following:

Theorem 5.33.   Comparability graph recognition and finding a transitive orientation can be done in $O(\delta \cdot |E|)$ time and $O(|V| + |E|)$ space, where δ is the maximum degree of a vertex.

The algorithm as presented in this section explores the edges in a depth-first search. Replacing each recursive call EXPLORE(x, y) by placing xy in a queue of edges to be explored would change the algorithm to breadth-first search. Some future application may lead us to prefer one over the other.

7.   Coloring and Other Problems on Comparability Graphs

To any *acyclic* orientation F (not necessarily transitive) of an undirected graph $G = (V, E)$ we may associate a strict partial ordering of the vertices, namely, $x > y$ iff there exists a nontrivial path in F from x to y. A *height function* h can then be placed on V as follows: $h(v) = 0$ if v is a sink; otherwise, $h(v) = 1 + \max\{h(w)|vw \in F\}$. We have already seen, in Chapter 2, Exercise 8, that the height function can be assigned in linear time using a recursive depth-first search. The function h is always a proper vertex coloring of G, but it is not necessarily a minimum coloring. The number of colors used will be equal to the number of vertices in the longest path of F. This is also equal to $1 + \max\{h(v)|v \in V\}$ since we started at height (color) zero. A poor choice of F may result in an overly colorful coloring. However, the situation is guaranteed to be better if F happens also to be transitive.

Suppose that G is a comparability graph, and let F be a transitive orientation of G. In such a case, every path in F corresponds to a clique of G because of transitivity. Thus, the height function will yield a coloring which uses exactly $\omega(G)$ colors, which is the best possible. Moreover, since being a comparability graph is a hereditary property, we find that $\omega(G_A) = \chi(G_A)$ for all induced subgraphs G_A of G. This proves the following result.

Theorem 5.34. Every comparability graph is a perfect graph.

Theorem 5.34 coupled with the Perfect Graph Theorem 3.3 implies that the stability number of a comparability graph is equal to the clique cover number of the graph. This proves the following classical result.

Theorem 5.35 (Dilworth [1950]). Let $(X, \leq)$ be a partially ordered set. The minimum number of linearly ordered subsets (usually called *chains*) needed to partition X is equal to the maximum cardinality of a subset of X having no two members comparable (usually called an *antichain*).

Many proofs of Dilworth's theorem can be found in the literature. Among them, those of Fulkerson [1956] and Perles [1963] seem most elegant. The reader is referred also to Dilworth [1950], Pretzel [1979], Trotter [1975], and Tverberg [1967]. Greene and Kleitman [1976] have recently extended Dilworth's theorem to more general partitions of a poset into chains. Some related references include Greene [1974, 1976], Griggs [1979], and Hoffman and Schwartz [1977].

We direct our attention next to some algorithmic aspects of problems on comparability graphs. In Section 5.6 we showed that a transitive orientation F could be constructed for a comparability graph G in $O(\delta e + n)$ steps, where δ is the maximum degree of a vertex, e is the number of edges, and n is the number of vertices. From the transitive orientation F we can assign a minimum coloring of G using the height function in $O(n + e)$ additional steps. At the same time a maximum clique could also be calculated. We shall illustrate this by solving a slightly more general problem.

MAXIMUM WEIGHTED CLIQUE.
Instance: An undirected graph G and an assignment of a weight $w(v)$ to each vertex v.
Question: Find a clique of G for which the sum of the weights of its vertices is largest possible.

If all vertices have the same weight, then the problem is reduced to the usual problem of finding a clique of maximum cardinality. In general the MAXIMUM WEIGHTED CLIQUE problem is NP-complete, but when restricted to comparability graphs it becomes tractable.

Algorithm 5.4. Maximum weighted clique of a comparability graph.
Input: A transitive orientation F of a comparability graph $G = (V, E)$ and a weight function w defined on V.
Output: A clique K of G whose weight is maximum.

```
procedure EXPLORE(v):
    if Adj(v) = ∅ then
        W(v) = w(v);
        POINTER(v) ← Λ;
        return;;
    for all x ∈ Adj(v) do
        if x is unexplored then
            EXPLORE(x);
    end for all;
    select y ∈ Adj(v) such that W(y) = max{W(x) | x ∈ Adj(v)};
    W(v) ← w(v) + W(y);
    POINTER(v) ← y;
    return
end
```

Figure 5.12.

Method: We use a modification of the height calculation technique employing the recursive depth-first search procedure EXPLORE in Figure 5.12. To each vertex v we associate its cumulative weight $W(v)$, which equals the weight of the heaviest path from v to some sink. A pointer is assigned to v designating its successor on that heaviest path. Lines 4–10 calculate K once the cumulative weights are assigned. The algorithm is given as a procedure.

```
    procedure MAXWEIGHT CLIQUE(V, F):
1.  for all v ∈ V do
2.      if v is unexplored then
3.          EXPLORE (v);
    end for all;
4.  select y ∈ V such that W(y) = max{W(v) | v ∈ V};
5.  K ← {y};
6.  y ← POINTER (y);
7.  while y ≠ Λ do
8.      K ← K ∪ {y};
9.      y ← POINTER (y);;
10. return K;
    end
```

Proving the correctness of Algorithm 5.4 and displaying an implementation whose complexity is linear in the size of the graph (assuming that F is provided to the algorithm in the proper data structure) are left as exercises for the reader.

We conclude with an interesting polynomial-time method for finding $\alpha(G)$, the size of the largest stable set of a comparability graph G. We transform a transitive orientation (V, F) of G into a transportation network by adding two new vertices s and t and edges sx and yt for each source x and sink y of F. Assigning a lower capacity of 1 to each vertex, we initialize a compatible integer-valued flow and then call a minimum-flow algorithm. The value of the minimum flow will equal the size of the smallest covering of the vertices by cliques, which in turn will equal the size of the largest independent set since every comparability graph is perfect. Such a minimum-flow algorithm can run in polynomial time. (See Figure 2.1 for the complexities of various maximum-flow algorithms.)

8.   The Dimension of Partial Orders

Szpilrajn [1930] first noted that any partial order (X, P) could always be extended to a linear ordering L of X. In Section 2.4 we called such a linear extension a topological sorting. Let $\mathscr{L}(P)$ denote the collection of all linear extensions of P. Any subset $\mathscr{L} \subseteq \mathscr{L}(P)$ satisfying $\bigcap_{L \in \mathscr{L}} L = P$ is called a *realizer* of P, and its *size* is $|\mathscr{L}|$. The intersection is that of sets of ordered pairs, that is,

$$ab \in \bigcap_{L \in \mathscr{L}} L \Leftrightarrow ab \in L \qquad \text{for every} \quad L \in \mathscr{L}.$$

Clearly, $\mathscr{L}(P)$ itself is a realizer of P. We define the *dimension* of P, dim P, to be the size of the smallest possible realizer for P. Such a realizer is called a *minimum realizer* for P. The notion of dimension of a partial order first appeared in Dushnik and Miller [1941].

Examples.   The partial order P whose Hasse diagram is illustrated in Figure 5.13 has dimension 2. A minimum realizer for P is also shown. Notice

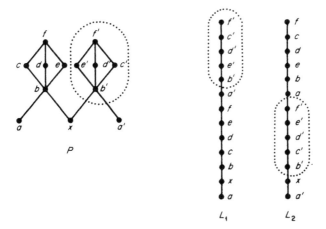

Figure 5.13.   A partial order P of dimension 2. We have $P = L_1 \cap L_2$.

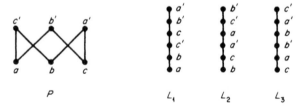

Figure 5.14. A partial order P of dimension 3. We have $P = L_1 \cap L_2 \cap L_3$. Why would two linear orders be insufficient to realize P?

that the subposet P' which is circled also has dimension 2 and that it must appear above element a in one of the linear orders and below element a in the other. Figure 5.14 shows the Hasse diagram of a partial order whose dimension is 3 (see Exercise 16).

Lemma 5.36. Let (X, P) be a poset. For each $Y \subseteq X$, we have

$$\dim P_Y \leq \dim P.$$

Proof. Clearly, restricting the linear extensions in a realizer $\mathscr{L}$ of P to the elements of Y yields a realizer (not necessarily minimum) of P_Y. Choosing $\mathscr{L}$ to be minimum for P we obtain the result. ∎

Theorem 5.37 (Hiraguchi [1951]). Let $P = P_0[P_1, P_2, \ldots, P_k]$ be the composition of disjoint partial orders (X_i, P_i) $(0 \leq i \leq k)$. Then

$$\dim P = \max\{\dim P_i | 0 \leq i \leq k\}.$$

Proof. For each i, let $L_{i, 1}, L_{i, 2}, \ldots, L_{i, m}$ be a realizer for P_i, where $m = \max\{\dim P_i | i = 0, 1, \ldots, k\}$. Define

$$\Lambda_j = L_{0, j}[L_{1, j}, L_{2, j}, \ldots, L_{k, j}].$$

Then $\{\Lambda_j | j = 1, 2, \ldots, m\}$ is a realizer of P, so $\dim P \leq m$.

Next, observe that P contains each of the P_i as a subposet. (To obtain P_0 take a set of representatives from $X_1, \ldots, X_k$.) Hence, by Lemma 5.36, $m \leq \dim P$. ∎

As noted earlier, the dimension of a partial order was introduced by Dushnik and Miller [1941]. They showed that there exist partial orders of dimension d for all positive integers d, and they gave the first characterization of the posets of dimension 2. We shall briefly mention some other known results on dimension theory. A special bibliography on the subject appears at the end of this chapter. In addition, W. T. Trotter is currently completing a book on the subject.

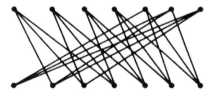

Figure 5.15. The Hasse diagram of the crown B_7^3.

Let S be a nonempty set and let $\mathscr{P}(S)$ denote its power set ordered by inclusion. Komm [1948] proved that dim $\mathscr{P}(S) = |S|$. Hiraguchi [1951] showed that dim $P \leq \frac{1}{2}|X|$ for any partial order (X, P) and gave examples of posets for which equality holds. Another proof of this result can be found in Bogart [1973].

Sedmak [1952–1954] investigated the poset $P(\pi)$ consisting of the empty set and the points, lines, faces, etc., of a polyhedron π in $\mathbb{R}^k$. He proved the following implications.

(1) If π is a polygon in $\mathbb{R}^2$, then dim $P(\pi) = 3$.
(2) If π is a polyhedron in $\mathbb{R}^3$, then dim $P(\pi) \geq 4$, with equality holding for regular polyhedra, pyramids, prisms, and their duals in $\mathbb{R}^3$.
(3) There exist polyhedra in $\mathbb{R}^3$ with arbitrarily high dimension.

This problem was originally posed by Kurepa [1951].

Ducamp [1967] showed that finding a minimum realizer for a partial order is equivalent to a certain bipartite covering problem. However, for all but small posets the method is intractable.

Let G be a connected undirected graph, and let $P(G)$ denote the collection of connected induced subgraphs of G ordered by inclusion. Trotter and Moore [1976a] proved that dim $P(G)$ equals the number of nonarticulation vertices of G. (A nonarticulation vertex is one whose removal from G leaves it connected.) This result generalizes a result of Leclerc [1976], namely, the dimension of the collection of subtrees of a tree T ordered by inclusion equals the number of leaves of T. The special case of dim $P(G) = 2$ was done by Dushnik and Miller [1941].

Trotter [1974a] studied the class of partial orders called *crowns*, obtaining an exact formula for their dimension. Briefly, let B_m^l be a poset on $2m$ elements split into an incomparable set $\{x_0, x_1, \ldots, x_{m-1}\}$ and another incomparable set $\{y_0, y_1, \ldots, y_{m-1}\}$ with $x_i y_j \in B_m^l$ for $j = i + 1, i + 2, \ldots, i + l$ (addition modulo m) (see Figure 5.15). Trotter proved that for $0 < l < m$ and $m \geq 3$

$$\dim B_m^l = \lceil 2m/(m - l + 1) \rceil.$$

Baker, Fishburn, and Roberts [1972] used the family $\{B_m^2\}_{m \geq 3}$ to show that, for any $n \geq 1$, the collection of all posets of dimension $\leq n$ is not axiomatizable by a sentence in first-order logic and cannot be characterized by a finite collection of forbidden subconfigurations.

Ore [1962] observed that the dimension of a partial order could be viewed in another, equivalent manner. The points in the Euclidean space $\mathbb{R}^k$ of dimension k can be partially ordered in a natural way: $(x_1, x_2, \ldots, x_k) \leq (y_1, y_2, \ldots, y_k)$ iff $x_i \leq y_i$ for each i. Then the dimension of a poset P is the smallest nonnegative integer k for which P can be embedded in $\mathbb{R}^k$. In some sense this justifies the choice of the term *dimension* for partial orders.

Rabinovitch [1973, 1978a] has shown that the dimension of a semiorder is at most three. Semiorders arise naturally in psychology* and are discussed in Chapter 8. Kelly [1977] and Trotter and Moore [1976b] have characterized all posets of dimension 3.

Application.   Let (X, P) be a partially ordered set, perhaps obtained as the transitive closure of an acyclic graph, and let $|X| = n$. The dim P may be regarded as the minimum number k of attributes needed to distinguish between the comparability and incomparability of pairs from X. The technique is the following: To each item $x \in X$ we associate a k-tuple $(x_1, x_2, \ldots, x_k) \in \mathbb{R}^k$, where x_i is the relative position of x in L_i and $\mathcal{L} = \{L_i\}$ is a minimum realizer of P. In such a setup, (X, P) would be stored using $O(kn)$ storage locations, and a query of the form "Is $xy \in P$?" will require at most k comparisons. This technique is advantageous when n is large and k is very small provided that the preprocessing needed to obtain a minimum realizer is not too expensive. This is always the case when dim $P \leq 2$.†

Theorem 5.38 (Dushnik and Miller [1941]).   Let G be the comparability graph of a poset P. Then dim $P \leq 2$ if and only if the complementary graph $\bar{G}$ is transitively orientable.

Proof.   Let F be a transitive orientation of $\bar{G}$. It is easy to show that $\mathcal{L} = \{P + F, P + F^{-1}\}$ is a realizer of P. Conversely, if $\mathcal{L} = \{L_1, L_2\}$ is any realizer of P, then $F = L_1 - P = (L_2 - P)^{-1}$ is a transitive orientation of $\bar{G}$. For, suppose $ab, bc \in F$ but $ac \notin F$. The transitivity of L_1 implies that $ac \in P$; similarly, the transitivity of L_2 implies that $ca \in P$, a contradiction.   ∎

---

* Some psychologists believe that preference is based on a single criterion with some degree of fuzziness; this viewpoint is modeled in Section 8.5. Other psychologists believe that the brain is actually comparing multiple criteria; this viewpoint is modeled by the realizers described in this section.

† To date the complexity of computing dim P for an arbitrary poset P is unknown. It may or may not be NP-complete.

From the preceding theorem it follows that two partial orders which have the same comparability graph either both have dimension ≤ 2 or both have dimension > 2. A stronger result holds, which we shall now present.

Theorem 5.39 (Trotter, Moore, and Sumner [1976]). If two partial orders P and Q have the same comparability graph G, then dim $P = $ dim Q.

Proof. The theorem is certainly true for posets of one element. We proceed by induction. Let (X, P) and (X, Q) be partial orders having the same comparability graph G, and let us assume that for all proper subsets Y of X, dim $Q_Y = $ dim P_Y. There are two cases to consider.

Case 1: G is indecomposable. In this case, Corollary 5.13 implies that G is UPO. Therefore, either $P = Q$ or $P = Q^{-1}$, both implying that dim $P = $ dim Q.

Case 2: G is decomposable. Let $G = G_R[G_{V_1}, \ldots, G_{V_r}]$ be a proper decomposition of G. By Corollary 5.7, $P = P_R[P_{V_1}, \ldots, P_{V_r}]$ and $Q = Q_R[Q_{V_1}, \ldots, Q_{V_r}]$. Applying Theorem 5.37 and the induction hypothesis, we can obtain

$$\begin{aligned} \dim P &= \max\{\dim P_R, \dim P_{V_1}, \ldots, \dim P_{V_r}\} \\ &= \max\{\dim Q_R, \dim Q_{V_1}, \ldots, \dim Q_{V_r}\} \\ &= \dim Q \end{aligned} \qquad \blacksquare$$

Theorem 5.39 also appears in Gysin [1977].

In a personal communication, Richard Stanley has reported that two partial orders P and Q having the same comparability graph also have the same number of linear extensions, i.e., $|\mathcal{L}(P)| = |\mathcal{L}(Q)|$. His proof is based on the results of Section 5.3.

EXERCISES

1. (i) Prove that the forcing relation Γ^* is an equivalence relation.
    (ii) Prove that the following properties hold:

$$ab \; \Gamma \; a'b' \Leftrightarrow ba \; \Gamma \; b'a'$$

$$ab \; \Gamma^* \; a'b' \Leftrightarrow ba \; \Gamma^* \; b'a'$$

2. The complete graph K_2 has two implication classes. Give a formula for $|\mathscr{I}(K_n)|$ for $n \geq 2$.

3. Which of the graphs in Figure 5.16 are comparability graphs? How many implication classes and color classes do they have?

Figure 5.16.

4. Let G be a connected comparability graph whose complement $\bar{G}$ is connected and contains no induced subgraph isomorphic to $K_{1,3}$. Prove that G is UPO. [Hint: Use Theorem 5.12 (Aigner and Prins [1971]).]

5. Prove the following result for an undirected graph G. If F_1 and F_2 are transitive orientations of G and $\bar{G}$, respectively, then $F_1 + F_2$ is a transitive tournament.

6. Draw the graph $G = H_0[H_1, H_2, H_3, H_4]$ for the graphs in Figure 5.17. Verify that it has 16 color classes: 9 within the internal factors, 6 among the external edges connecting H_1 with H_2, and 1 consisting of the remaining external edges. Prove that G has 1440 transitive orientations.

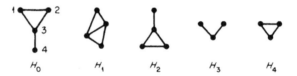

Figure 5.17.

7. Show that if an undirected graph G has no induced subgraph isomorphic to the path P_4, then both G and $\bar{G}$ are comparability graphs.

8. Verify that the graph in Figure 5.18 has four color classes partitioned into two maximal multiplexes of rank 1 and 2, respectively. Use the decomposition algorithm of Section 5.4 to obtain a G-decomposition of this graph. (One solution is given in Appendix D.) Is this graph a comparability graph?

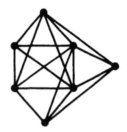

Figure 5.18.

9. Calculate $r(G)$ for the graphs in Exercises 3, 6, and 8.

10. Let $\alpha(G)$ be the stability number of an undirected graph $G = (V, E)$. Prove that $r(G) \leq |V| - \alpha(G)$ (Golumbic [1977a]).

11. A binary relation R is vacuously transitive if $R^2 = \varnothing$. (Vacuously transitive relations have been studied by Sharp [1973].) Prove that an undirected graph has a vacuously transitive orientation if and only if it is bipartite.

12. Prove that every transitive orientation of a comparability graph G is obtainable from some G-decomposition of G.

13. Let $G = (V, E)$ be an undirected graph, and consider the equivalence relation $\sim$, defined on V as follows:

$$a \sim a' \qquad \text{iff} \qquad \text{Adj}(a) = \text{Adj}(a').$$

By irreflexivity, equivalent vertices are not adjacent. We form the quotient graph $\tilde{G} = (\tilde{V}, \tilde{E})$ by merging equivalent vertices. Formally, let $\tilde{V}$ be the set of all equivalence classes under $\sim$, and let $\tilde{a}$ denote the $\sim$-class containing the vertex a. For any subset of edges $A \subseteq E$ we define

$$\tilde{A} = \{\tilde{a}\tilde{b} \mid ab \in A\}.$$

(i) Prove that $\tilde{a}\tilde{b} \in \tilde{E} \Leftrightarrow ab \in E$. Give an example of a graph G and a subset of edges A such that $cd \in A$ but $\tilde{c}\tilde{d} \notin \tilde{A}$ for some edge cd.

(ii) Prove that the following conditions are equivalent:

  (1) $\tilde{a}\tilde{b} = \tilde{c}\tilde{d}$,
  (2) $a \sim c$ and $b \sim d$,
  (3) $\tilde{a} = \tilde{c}$ and $\tilde{b} = \tilde{d}$.

(iii) Prove that (1)–(3) above imply that $ab \, \Gamma^* \, cd$ but not conversely (Golumbic [1977b]).

14. Let $G = (V, E)$ be an undirected graph and let $\tilde{G} = (\tilde{V}, \tilde{E})$ be its quotient graph as defined in Exercise 13. Prove the following.

(i) If $A \in \mathscr{I}(G)$, then $\tilde{A} \in \mathscr{I}(\tilde{G})$.

(ii) If $[e_1, e_2, \ldots, e_k]$ is a scheme for G with corresponding G decomposition $\hat{B}_1 + \hat{B}_2 + \cdots + \hat{B}_k$, then $[\tilde{e}_1, \tilde{e}_2, \ldots, \tilde{e}_k]$ is a scheme for $\tilde{E}$ with corresponding G-decomposition $\hat{\tilde{B}}_1 + \hat{\tilde{B}}_2 + \cdots + \hat{\tilde{B}}_k$.

(iii) If (V, F) is a transitive orientation of G, then $(\tilde{V}, \tilde{F})$ is a transitive orientation of $\tilde{G}$.

(iv) Every implication class, scheme, G-decomposition, and transitive orientation of G is of the form indicated in (i)–(iii) (Golumbic [1977b]).

15. Prove that Algorithm 5.4 correctly computes a maximum weighted clique of a comparability graph. Show that the algorithm can be implemented to run in linear time in the size of the graph.

16. Prove that the partial order in Figure 5.14 has dimension 3.

17. Let Q be a subset of $\mathbb{R}^k$, and for $1 \le i \le k$ let Q_i consist of those numbers which appear as the ith coordinate in some k-tuple in Q. Consider the natural partial order on Q as defined in Section 8.

(i)   Show that if Q is the Cartesian product $Q = Q_1 \times Q_2 \times \cdots \times Q_k$ and $|Q_i| \geq 2$ for each i, then dim $Q = k$.

(ii) Prove Komm's theorem, namely, that dim $\mathscr{P}(S) = k$ for a set S.

18. Complete the proof of Theorem 5.38.

19. A partial order (X, P) is an *interval inclusion order* if X can be put into one-to-one correspondence with a family $\{I_x\}_{x \in X}$ of intervals on a linearly ordered set such that

$$x < y \quad \text{iff } I_x \subset I_y \quad (\forall x, y \in X).$$

Prove the following: dim $P \leq 2$ if and only if P is an interval inclusion order (Dushnik and Miller [1941]).

20. Let $G = (V, E)$ be an undirected graph. Show the following statements are equivalent:

(i) G has a transitive orientation whose Hasse diagram is a rooted tree;

(ii) G is a comparability graph and the Hasse diagram of every transitive orientation of G is a rooted tree;

(iii) if $a, b, c, d \in V$ are distinct vertices satisfying $ab, bc, cd \in E$, then either $ac \in E$ or $bd \in E$;

(iv) G contains no induced subgraph isomorphic to C_4 or P_4.

Give an example of a comparability graph which is triangulated and whose complement is a comparability graph but which fails to satisfy the conditions above (Wolk [1962, 1965]). Arditti [1975b] has investigated comparability graphs whose Hasse diagram is a tree.

21. Show that the leaves of a rooted tree can be linearly ordered so that the set of decendent leaves of any vertex occur consecutively. Use this result to show that any graph G which is the comparability graph of a rooted tree is also an interval graph.

22. If $\bar{B}_1 + \bar{B}_2 + \cdots + \bar{B}_k$ is a G-decomposition of an undirected graph $G = (V, E)$, then $\bar{B}_1 + \cdots + \bar{B}_j$ is called a *partial G-decomposition* for each $j = 0, 1, \ldots, k$. Show that the subgraphs of G obtained as partial G-decompositions (including $\emptyset$ and E) form a lattice. Show that this lattice is modular but not necessarily distributive.

Bibliography

General References

Aigner, Martin

[1969] Graphs and partial orderings, *Monatsh. Math.* **73**, 385–396. MR41 #1561.
        Discusses minimal noncomparability graphs and when the line graph $L(G)$ is a comparability graph.

Aigner, Martin, and Prins, Geert
  [1971] Uniquely partially orderable graphs, *J. London Math. Soc. 2* **3**, 260–266.
          MR43 #1866.
          A connected comparability graph whose complement is connected and does not
          contain an induced $K_{1,3}$ is UPO.

Arditti, Jean-Claude
  [1973a] Hamiltonisme et pancyclisme dans les graphes de comparabilité d'arbres orientés,
          Colloq. sur la Théorie des Graphes, Bruxelles, 1973, *Cahiers Centre Études Rech. Opér.*
          15, 265–284. MR50 #9644.

  [1973b] Dénombrement des arborescences dont le graphe de comparabilité est Hamiltonien,
          Discrete Math. **5**, 189–200. MR47 #4848.
          Using the results of Arditti and Cori [1970] the author gives a method for calculating
          the number of arborescences with n points. Using Polya's method he obtains a
          generating function.

  [1975a] Cheminements dans le graphe de comparabilité d'un arbre partition des sommets en
          cycles, *Cahiers Centre Études Rech. Opér.* **17**, 111–116. MR53 #5344.
          Extends results of Arditti and Cori [1970].

  [1975b] Graphes de comparabilité d'arbres et d'arborescences, Thèse d'Etat, *Publ. Math.
          Orsay* No. 127-7531.

  [1976a] Graphes de comparabilité et dimension des ordres, Note de recherches CRM 607,
          Centre Rech. Math. Univ. Montréal.

  [1976b] Partially ordered sets and their comparability graphs, their dimension and their
          adjacency, *Proc. Colloq. Int. CNRS., Problèmes Combinatoires et Théorie des Graphes*,
          Orsay, France.

Arditti, Jean-Claude, and Cori, Robert
  [1970] Hamiltonian circuits in the comparability graph of a tree, *in* "Combinatorial Theory
          and its Applications I," *Proc. Colloq. Balatonfüred, 1969*, pp. 41–53. North-Holland,
          Amsterdam. MR46 #3361.

Arditti, Jean-Claude, and de Werra, D.
  [1976] A note on a paper by D. Seinsche, *J. Combin. Theory B* **21**, 90. MR54 #2510.

Bryant, V. W., and Harris, K. G.
  [1975] Transitive graphs, *J. London Math. Soc. Ser. 2* **11**, 123–128. MR55 #5476.
          The authors rediscover many of the results of Gilmore and Hoffman [1964] and
          Ghouilà-Houri [1962].

Dilworth, R. P.
  [1950] A decomposition theorem for partially ordered sets, *Ann. Math. Ser. 2* **51**, 161–166.
          MR11, p. 309.

Even, Shimon
  [1973] "Algorithmic Combinatorics," Macmillan, New York. MR49 #48.

Even, Shimon, Pnueli, Amir, and Lempel, Abraham
  [1972] Permutation graphs and transitive graphs. *J. Assoc. Comput. Mach.* **19**, 400–410.
          MR47 #1675.

Filippov, N. D.
  [1968] σ-isomorphisms of partially ordered sets (Russian), *Ural. Gos. Univ. Mat. Zap.* **6**,
          71–85. MR42 #4452.

Fulkerson, D. R.
  [1956] Note on Dilworth's decomposition theorem for partially ordered sets, *Proc. Amer.
          Math. Soc.* **7**, 701–702. MR17 #1176.

Gallai, Tibor
  [1967]   Transitiv orientierbare graphen, *Acta Math. Acad. Sci. Hungar.* **18**, 25–66. MR36
           #5026.
           Contains many results on the structure of comparability graphs.
Ghouilà-Houri, Alain
  [1962]   Caractérisation des graphes non orientés dont on peut orienter les arrêtes de manière
           à obtenir le graphe d'une relation d'ordre, *C.R. Acad. Sci. Paris* **254**, 1370–1371.
           MR30 #2495.
Gilmore, Paul C., and Hoffman, Alan J.
  [1964]   A characterization of comparability graphs and of interval graphs, *Canad. J. Math.*
           16, 539–548; abstract in *Int. Congr. Math.* (Stockholm), 29 (A) (1962). MR31 #87.
Golumbic, Martin Charles
  [1975]   Comparability graphs and a new matroid, extended abstract, *Proc. Conf. Algebraic
           Aspects of Combinatorics*, Univ. Toronto, January 1975, "Congressus Numeran-
           tium," XIII, Utilitas Math., Winnipeg. pp. 213–217. MR53 #10653.
  [1976]   Recognizing comparability graphs in SETL, *SETL Newsletter No. 163*, Courant
           Institute, New York Univ.
  [1977a] Comparability graphs and a new matroid, *J. Combin. Theory B* **22**, 68–90.
           MR55 #12575.
  [1977b] The complexity of comparability graph recognition and coloring. *Computing* **18**,
           199–208.
Green, C. D.
  [1975]   The detection of mistakes in the comparability graph of a tree, *Proc. British Combin.
           Conf.*, Univ. Aberdeen, 1975, "Congress Numerantium," No. XV, pp. 255–260.
           Utilitas Math., Winnipeg. MR54 #5023.
Greene, Curtis
  [1974]   Sperner families and partitions of a partially ordered set, *in* "Combinatorics, Part 2,"
           Proc. Adv. Study Inst. on Combinatorics, Nijenrode Castle, Breukelen, The Nether-
           lands, July 1974 (M. Hall and J. H. VanLint, eds.), pp. 91–106. Mathematisch
           Centrum, Amsterdam. MR50 #9606.
  [1976]   Some partitions associated with a partially ordered set. *J. Combin. Theory A* **20**,
           69–79. MR53 #2763.
Greene, Curtis, and Kleitman, Daniel J.
  [1976]   The structure of Sperner *k*-families, *J. Combin. Theory A* **20**, 41–68. MR53 #2695.
Griggs, J. R.
  [1979]   On chains and Sperner *k*-families in ranked posets, *J. Combin. Theory* (to be pub-
           lished).
Hoffman, Alan J., and Schwartz, D. E.
  [1977]   On partitions of a partially ordered set, *J. Combin. Theory B* **23**, 3–13.
Johnson, C. S., Jr., and McMorris, F. R.
  [1979]   A note on two comparability graphs, Bowling Green State Univ. Res. Report.
Jung, H. A.
  [1968]   Zu einem Satz von E. S. Wolk über die Vergleichbarkeitsgraphen von ordnungs-
           theoretischen Bäumen, *Fund. Math.* **53**, 217–219. MR38 #3167.
  [1978]   On a class of posets and the corresponding comparability graphs, *J. Combin. Theory
           B* **24**, 125–133.
           Generalizes some notions of Wolk. See also Johnson and McMorris [1979].
Perles, M. A.
  [1963]   On Dilworth's theorem in the infinite case, *Israel J. Math.* **1** 108–109. MR29 #5758.

Pnueli, Amir, Lempel, Abraham, and Even, Shimon
  [1971] Transitive orientation of graphs and identification of permutation graphs, *Canad. J. Math.* **23**, 160–175. MR45 #1800.
Pretzel, Oliver
  [1979] Another proof of Dilworth's decomposition theorem, *Discrete Math.* **25**, 91–92.
Sankoff, David, and Sellers, Peter H.
  [1973] Shortcuts, diversions and maximal chains in partially ordered sets, *Discrete Math.* **4**, 287–293. MR47 #1690.
          Contains some interesting applications of posets to molecular genetics, critical path scheduling, bipartite graph theory, and traffic routing.
Seinsche, D.
  [1974] On a property of the class of *n*-colorable graphs, *J. Combin. Theory B* **16**, 191–193. MR49 #2448.
          A graph which does not contain a chain of length 3 (i.e., P_4) without chords is perfect. As pointed out by Arditti and deWerra [1976], this is immediate from Wolk [1962, 1965].
Sharp, Henry, Jr.
  [1973] Enumeration of vacuously transitive relations, *Discrete Math.* **4**, 185–196. MR47 #47.
Shevrin, L. N., and Filippov, N. D.
  [1970] Partially ordered sets and their comparability graphs, *Siberian Math. J.* **11**, 497–509. MR42 #4451.
Stanley, Richard P.
  [1973] A Brylawski decomposition for finite ordered sets, *Discrete Math.* **4**, 77–82. MR46 #8918.
Trotter, William T., Jr.
  [1975] A note on Dilworth's embedding theorem, *Proc. Amer. Math. Soc.* **52**, 33–39. MR51 #10188.
Trotter, William T., Jr., Moore, John I., Jr., and Sumner, David P.
  [1976] The dimension of a comparability graph, *Proc. Amer. Math. Soc.* **60**, 35–38. MR54 #5062.
Tverberg, Helge
  [1967] On Dilworth's decomposition theorem for partially ordered sets, *J. Combin. Theory* **3**, 305–306. MR35 #5366.
Wolk, E. S.
  [1962] The comparability graph of a tree, *Proc. Amer. Math. Soc.* **13**, 789–795. MR30 #2493.
  [1965] A note on the comparability graph of a tree, *Proc. Amer. Math. Soc.* **16**, 17–20. MR30 #2494.

The Dimension of Partial Orders

Adnadević, Dušan
  [1961] Dimenzije neikih razvrstanih skupova sa primenama, *Bull. Soc. Math. Phys. Serbie* **13**, 49–106, 225–262. R. Z. Mat 1963 #9A224, 1964 #2A392.
  [1964] On the dimension of the product of partially ordered sets (Serbo-Croatian, English summary), *Mat. Vesnik* **1** (16), 9–12. MR34 #7413.
  [1966] On the representations of finite partially ordered sets (Serbo-Croatian, English summary) *Mat. Vesnik* **3** (18), 17–21. MR35 #1510.

Arditti, Jean-Claude
[1976a] Graphes de comparabilité et dimension des ordres, Note de recherches CRM 607, Centre de Recherche Mathématique de l'Université de Montréal.
[1976b] Partially ordered sets and their comparability graphs, their dimension and their adjacency, *Proc. Colloq. Int. CNRS, Problemes Combinatoires et Theorie des Graphes*, Orsay, France.
Baker, K. A., Fishburn, P. C., and Roberts, F. S.
[1970] A new characterization of partial orders of dimension two, *Ann. N.Y. Acad. Sci.* **175**, 23–24. MR42 #140.
[1972] Partial orders of dimension 2, *Networks* **2**, 11–28. MR46 #104.
Bogart, Kenneth P.
[1973] Maximal dimensional partially ordered sets I. Hiraguchi's theorem, *Discrete Math.* **5**, 21–31. MR47 #6562.
Bogart, Kenneth P., and Trotter, William T., Jr.
[1973] Maximal dimensional partially ordered sets II. Characterization of $2n$-element posets with dimension n, *Discrete Math.* **5**, 33–43. MR47 #6563.
Bogart, Kenneth P., Rabinovitch, I., and Trotter, William T., Jr.
[1976] A bound on the dimension of interval orders, *J. Combin. Theory A* **21**, 319–328. MR54 #5059.
Ducamp, A.
[1967] Sur la dimension d'un ordre partiel, *in* "Theory of Graphs," Proc. Symp. Rome (P. Rosenstiehl, ed.), pp. 103–112. Gordon & Breach, New York. MR36 #3684.
Dushnik, Ben
[1950] Concerning a certain set of arrangements, *Proc. Amer. Math. Soc.* **1**, 788–796. MR12, p. 470.
Dushnik, Ben, and Miller, E. W.
[1941] Partially ordered sets. *Amer. J. Math.* **63**, 600–610. MR3, p. 73.
Ginsburg, S.
[1954] On the λ-dimension and the A-dimension of partially ordered sets, *Amer. J. Math.* **76**, 590–598. MR15, p. 943.
Gysin, R.
[1977] Dimension transitiv orientierbaren graphen, *Acta Math. Acad. Sci. Hungar.* **29**, 313–316.
Harzheim, E.
[1970] Ein Endlichkeitssatz über die Dimension teil weise geordneter Mengen, *Math. Nachr.* **46**, 183–188. MR43 #113.
Hiraguchi, Toshio
[1951] On the dimension of partially ordered sets, *Sci. Rep. Kanazawa Univ.* **1**, 77–94. MR17, p. 19.
[1953] A note on Mr. Komm's theorems, *Sci. Rep. Kanazawa Univ.* **2**, 1–3. MR17, p. 937.
[1955] On the dimension of orders, *Sci. Rep. Kanazawa Univ.* **4**, 1–20. MR17, p. 1045.
[1956] On the λ-dimension of the product of orders, *Sci. Rep. Kanazawa Univ.* **5**, 1–5. MR20 #1638.
Kelly, David
[1977] The 3-irreducible partially ordered sets, *Canad. J. Math.* **29**, 367–383. MR55 #205.
Kelly, David, and Rival, Ivan
[1975] Certain partially ordered sets of dimension three, *J. Combin. Theory A* **18**, 239–242. MR50 #12828.
Kimble, R.
[1973] Extremal problems in dimension theory for partially ordered sets, Ph.D. thesis, MIT, Cambridge, Massachusetts.

Komm, H.
 [1948]   On the dimension of partially ordered sets, *Amer. J. Math.* **70**, 507–520. MR10, p. 22.
Kurepa, Georges
 [1950]   Ensembles partiellement ordonnés et ensembles partiellement bien ordonnés, *Acad. Serbe Sci. Publ. Inst. Math.* **3**, 119–125. MR12, p. 683.
 [1951]   "Teorija skupova," p. 205, Problem 16.8.1. Školska Knjiga, Zagreb. MR12, p. 683. A textbook on set theory.
Leclerc, B.
 [1976]   Arbres et dimension des ordres, *Discrete Math.* **14**, 69–76. MR52 #7979.
Moore, J. I., Jr.
 [1977]   Interval hypergraphs and *D*-interval hypergraphs, *Discrete Math.* **17**, 173–179. MR55 #10333.
Novák, V.
 [1962]   A note on a problem of T. Hiraguchi, *Spisy Přírod. Fak. Univ. Brno*, 147–149. MR29 #4710.
 [1963]   On the pseudodimension of ordered sets, *Czech. Math. J.* **13**, 587–597. MR31 #4742.
Novák, V. and Novotný, M.
 [1974]   Abstrakte Dimension von Strukturen, *Z. Math. Logik Grundlagen Math.* **20**, 207–220. MR53 #2765.
Ore. O.
 [1962]   "Theory of Graphs," Section 10.4. *Amer. Math. Soc. Colloq. Publ.* **38**, Providence, Rhode Island. MR27 #740.
Perfect, Hazel
 [1974]   Addendum to a theorem of O. Pretzel, *J. Math. Anal. Appl.* **46**, 90–92. MR49 #156.
Pretzel, Oliver
 [1967]   A representation theorem for partial orders, *J. London Math. Soc.* **42**, 507–508. MR35 #6588.
 [1977]   On the dimension of partially ordered sets. *J. Combin. Theory A* **22**, 146–152. MR55 #206.
Rabinovitch, Issie B.
 [1973]   The dimension theory of semiorders and interval orders, Ph.D. thesis, Dartmouth.
 [1978a] The dimension of semiorders, *J. Combin. Theory A* **25**, 50–61.
 [1978b] An upper bound on the dimension of interval orders, *J. Combin. Theory A* **25**, 68–71.
Sedmak, Victor
 [1952]   Dimension des ensembles partiellement ordonnés associés aux polygones et polyèdres (Serbo-Croatian. French summary), *Hrvatsko Prirod. Drustvo. Glasnik Mat.-Fiz. Astronom. Ser. II* **7**, 169–182. MR14, p. 783.
 [1953]   Quelques applications des ensembles partiellement ordonnés, *C.R. Acad. Sci. Paris* **236**, 2139–2140. MR15, p. 50.
 [1954]   Quelques applications des ensembles ordonnés, *Bull. Soc. Math. Phys. Serbie* **6**, 12–39, 131–153. MR18, p. 186.
 [1959]   Sur les réseaux de polyèdres *n*-dimensionnels, *C.R. Acad. Sci. Paris* **248**, 350–352. MR23A #1559.
Szpilrajn, E.
 [1930]   Sur l'extension de l'ordre partiel, *Fund. Math.* **16**, 386–389.
Trotter, William T., Jr.
 [1974a] Dimension of the crown S_n^k, *Discrete Math.* **8**, 85–103. MR49 #158.
 [1974b] Irreducible posets with large height exist, *J. Combin. Theory A* **17**, 337–344. MR50 #6935.
 [1974c] Some families of irreducible partially ordered sets, Univ. of South Carolina Math. Tech. Rep. 06A10-2.

[1975a] Inequalities in dimension theory for posets, *Proc. Amer. Math. Soc.* **47**, 311–316. MR51 #5427.

[1975b] A note on Dilworth's embedding theorem, *Proc. Amer. Math. Soc.* **52**, 33–39. MR51 #10188.

[1975c] Embedding finite posets in cubes, *Discrete Math.* **12**, 165–172. MR51 #5426.

[1976a] A forbidden subposet characterization of an order-dimension inequality, *Math. Syst. Theory* **10**, 91–96. MR55 #7856.

[1976b] A generalization of Hiraguchi's: Inequality for posets, *J. Combin. Theory A* **20**, 114–123. MR52 #10515.

[1977] Some combinatorial problems for permutations, *Proc. 8th Southeastern Conf., on Combinatorics, Graph Theory and Computing.*

Trotter, William T., Jr., and Bogart, Kenneth P.

[1976a] On the complexity of posets, *Discrete Math.* **16**, 71–82. MR54 #2553.

[1976b] Maximal dimensional partially ordered sets III: A characterization of Hiraguchi's inequality for interval dimension. *Discrete Math.* **15**, 389–400. MR54 #5061.

Trotter, William T., Jr., and Moore, J. I., Jr.

[1976a] Some theorems on graphs and posets, *Discrete Math.* **15**, 79–84. MR54 #5060.

[1976b] Characterization problems for graphs, partially ordered sets, lattices and families of sets, *Discrete Math.* **16**, 361–381. MR56 #8437.

[1977] The dimension of planar posets, *J. Combin. Theory B* **22**, 54–67. MR55 #7857.

Trotter, William T., Jr., Moore, John I., Jr. and Sumner, David P.

[1976] The dimension of a comparability graph, *Proc. Amer. Math. Soc.* **60**, 35–38. MR54 #5062.

Wille, Rudolf

[1974] On modular lattices of order dimension two, *Proc. Amer. Math. Soc.* **43**, 287–292. MR48 #8327.

[1975] A note on the order dimension of partially ordered sets, *Algebra Universalis* **5**, 443–444. MR52 #13536.

CHAPTER 6

Split Graphs

1. An Introduction to Chapters 6–8: Interval, Permutation, and Split Graphs

An undirected graph G may possess one or more of these familiar properties:

Property C: G is a comparability graph.
Property $\bar{C}$: $\bar{G}$ is a comparability graph (i.e., G is a *cocomparability* graph).
Property T: G is a triangulated graph.
Property $\bar{T}$: $\bar{G}$ is a triangulated graph (i.e., G is a *cotriangulated* graph).

These four properties are independent of one another. Examples of all 16 possible combinations are given in Appendix F.

Chapters 6–8 deal with the classes of graphs which have been characterized in terms of these four properties. In particular, we shall show the following:

$$\text{interval graphs} \equiv T + \bar{C};$$
$$\text{permutation graphs} \equiv C + \bar{C};$$
$$\text{split graphs} \equiv T + \bar{T}.$$

We begin our study with split graphs, which are defined in the next section. Chapters 6–8 are independent of one another; they may be read in any order without loss of continuity.

2. Characterizing Split Graphs

An undirected graph $G = (V, E)$ is defined to be *split* if there is a partition $V = S + K$ of its vertex set into a stable set S and a complete set K. There is

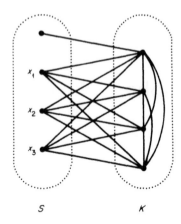

Figure 6.1.   A split graph with one of its four partitions indicated. The other partitions are $(S - \{x_i\}) + (K \cup \{x_i\})$ for $i = 1,2,3$.

no restriction on edges between vertices of S and vertices of K. In general, the partition $V = S + K$ of a split graph will not be unique; neither will S (resp. K) necessarily be a maximal stable set (resp. clique). For example, the graph G in Figure 6.1 has four partitions, one of which is indicated. Notice also that $|S| = \alpha(G) = 4$ whereas $4 = |K| < \omega(G) = 5$; S is the only maximum stable set of G, and $K \cup \{x_i\}$ (for $i = 1, 2, 3$) are the only maximum cliques.

Since a stable set of G is a complete set of the complement $\bar{G}$ and vice versa, we have an immediate result.

Theorem 6.1.   An undirected graph G is a split graph if and only if its complement $\bar{G}$ is a split graph.

The next theorem follows from the work of Hammer and Simeone [1977].

Theorem 6.2.   Let G be a split graph whose vertices have been partitioned into a stable set S and a complete set K. Exactly one of the following conditions holds:

   (i) $|S| = \alpha(G)$ and $|K| = \omega(G)$
(in this case the partition $S + K$ is unique),
   (ii) $|S| = \alpha(G)$ and $|K| = \omega(G) - 1$
(in this case there exists an $x \in S$ such that $K + \{x\}$ is complete),
   (iii) $|S| = \alpha(G) - 1$ and $|K| = \omega(G)$
(in this case there exists a $y \in K$ such that $S + \{y\}$ is stable).

Proof.   Since a stable set and a complete set can have at most one common vertex, it follows that a split graph has the sum $\alpha(G) + \omega(G)$ equal to either $|V|$ or $|V| + 1$.

If $\alpha(G) + \omega(G) = |V|$, then we are in case (i). Suppose, in this case, there is another partition $V = S' + K'$. Let $\{x\} = S \cap K'$ and $\{y\} = S' \cap K$. If x and y are adjacent in G, then $\{x\} + K$ is a clique of size $\omega(G) + 1$, which is impossible. If x and y are not adjacent in G, then $\{y\} + S$ is a stable set of size $\alpha(G) + 1$, which is impossible. Hence, the partition $V = S + K$ must be unique.

If $\alpha(G) + \omega(G) = |V| + 1$, then we are in either case (ii) or case (iii). We will prove the claim in case (ii) only, case (iii) being analogous. Let $|S| = \alpha(G)$, $|K| = \omega(G) - 1$ and let K' be a clique of size $\omega(G)$. Since $S + K$ is a partition and K' is larger than K, $S \cap K'$ must be nonempty and therefore of cardinality 1. Let $\{x\} = S \cap K'$, it follows that $K' = K + \{x\}$, which is complete.   ∎

Theorem 6.3 (Földes and Hammer [1977b]).   Let G be an undirected graph. The following conditions are equivalent:

   (i) G is a split graph,
  (ii) G and $\bar{G}$ are triangulated graphs,
 (iii) G contains no induced subgraph isomorphic to $2K_2$, C_4, or C_5.

Proof.   (i) ⇒ (ii)   Let $G = (V, E)$ have vertex partition $V = S + K$ with S stable and K complete. Suppose G contained a chordless cycle C of length ≥ 4. At least one and at most two (adjacent) vertices of C would be in K. Both cases would imply that S contains a pair of adjacent vertices, a contradiction. Therefore, G must be triangulated. By Theorem 6.1, $\bar{G}$ is split, so $\bar{G}$ is triangulated.

(ii) ⇒ (iii)   Immediate.

(iii) ⇒ (i)   Let K be a maximum clique of G chosen (among all maximum cliques) so that G_{V-K} has fewest possible edges. We must show that $S = V - K$ is stable.

Suppose, on the contrary, that G_S has an edge xy. By the maximality of K, no vertex of S could be adjacent to every member of K. Moreover, if both x and y were adjacent to every vertex of K with the exception of the same single vertex z, then $K - \{z\} + \{x\} + \{y\}$ would be a complete set larger than K. Thus, there must exist distinct vertices u, $v \in K$ such that $xu \notin E$ and $yv \notin E$.

Since G contains neither an induced copy of $2K_2$ nor C_4, it follows that exactly one of the edges xv or yu is in G. Assume, without loss of generality, that $xv \notin E$ and $yu \in E$. For any $w \in K - \{u, v\}$, if $yw \notin E$ and $xw \notin E$, then $G_{\{x, y, v, w\}} \cong 2K_2$, whereas if $yw \notin E$ and $xw \in E$, then $G_{\{x, y, u, w\}} \cong C_4$. Thus,

y is adjacent to every vertex of $K - \{v\}$, and $K' = K - \{v\} + \{y\}$ is a maximal clique.

Since $G_{V-K'}$ can have no fewer edges than G_{V-K} has, it follows from the fact that x is adjacent to y but not to v that there exists a vertex $t \neq y$ in $V - K$ which is adjacent to v but not to y. Now tx must be an edge of G, for otherwise $\{t, x, y, v\}$ would induce a copy of $2K_2$. Similarly, $tu \notin E$, for otherwise $\{t, x, y, u\}$ would induce a copy of C_4. However, this implies that $\{t, x, y, u, v\}$ induces a copy of C_5, a contradiction. Therefore, $S = V - K$ is stable, and G is a split graph.   ∎

A characterization of when a split graph is also a comparability graph appears in Chapter 9 (Theorem 9.7).

3. Degree Sequences and Split Graphs

A sequence $\Delta = [d_1, d_2, \ldots, d_n]$ of integers, $n - 1 \geq d_1 \geq d_2 \geq \cdots \geq d_n \geq 0$, is called *graphic* if there exists an undirected graph having Δ as its degree sequence. For example, the sequence $[2, 2, 2, 2]$ corresponds to the chordless 4-cycle C_4, while the sequence $[2, 2, 2, 2, 2, 2]$ corresponds to both $2K_3$ and C_6. It is easy to construct sequences which are not graphic, such as $[1, 1, 1]$ and $[4, 4, 2, 1, 1]$.

A simple necessary condition for a sequence to be graphic comes from Euler's theorem: The sum $\sum d_i$ must be even. However, as the preceding example shows, an even sum is not sufficient to insure graphicness. Two classical theorems characterizing graphic sequences will now be stated.

Theorem 6.4 (Havel [1955], Hakimi [1962]).   A sequence Δ of integers $n - 1 \geq d_1 \geq d_2 \geq \cdots \geq d_n \geq 0$ is graphic if and only if the modified sequence

$$\Delta' = [d_2 - 1, d_3 - 1, \ldots, d_{d_1+1} - 1, d_{d_1+2}, \ldots, d_n]$$

(sorted into decreasing order) is graphic.

Theorem 6.5 (Erdös and Gallai [1960]).   A sequence of integers $n - 1 \geq d_1 \geq d_2 \geq \cdots \geq d_n \geq 0$ is graphic if and only if

(i) $\sum\limits_{i=1}^{n} d_i$ is even,   and

(ii) $\sum\limits_{i=1}^{r} d_i \leq r(r - 1) + \sum\limits_{i=r+1}^{n} \min\{r, d_i\}$,

for $r = 1, 2, \ldots, n - 1$.

The inequality (ii) will be called the *rth Erdös–Gallai inequality* (EGI). Shortly, we shall give a characterization of split graphs in terms of these inequalities. We shall not prove Theorem 6.4 or Theorem 6.5 here since a very readable treatment can be found in Harary [1969, Chapter 6]. Both of these theorems suggest algorithms for testing whether or not a given sequence is graphic (Exercise 6).

A third classical theorem on graphic sequences depends partly on the following observation. Let x, y, z, and w be distinct vertices of G with xy and zw edges of G and xz and yw nonedges of G. If we replace the two edges by the two nonedges, the resulting graph G' will have the same degree sequence as G (see Figure 6.2). Such a replacement will be called an *interchange*. A stronger result holds, which we now state.

Remark 6.6.   Provided that we allow graphs to have multiple edges, if two graphs have the same degree sequence, then each can be obtained from the other by a finite sequence of interchanges.

A proof of Remark 6.6 can be found in Ryser [1963, Chapter 6, Theorem 3.1] by applying his technique to the edges-versus-vertices incidence matrix of G.

A general question arises: What graph theoretic properties can be determined solely from the degree sequence? In Section 2.5 we remarked that transitive tournaments could be recognized by the in-degrees of the vertices. Also, a characterization of trees in terms of degree sequences is known. We will now discuss this problem as applied to split graphs.

Let $\Delta = [d_1, d_2, \ldots, d_n]$ be an integer sequence with $n - 1 \geq d_1 \geq d_2 \geq \cdots \geq d_n \geq 0$, and let $\zeta = [0, 1, 2, \ldots, n - 1]$. Comparing the decreasing sequence Δ with the increasing sequence ζ, let us draw attention to the position just prior to ζ overtaking Δ. Let m be the largest index i such that $d_i \geq i - 1$. Thus, either $m = n$ and Δ is the degree sequence of K_n, or $d_m \geq m - 1$ and $d_{m+1} < m$.

The next result characterizes split graphs as those for which equality holds in the mth Erdös–Gallai inequality, where m is defined as above.

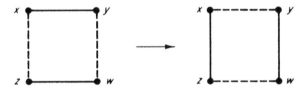

Figure 6.2.   A solid line denotes an edge of G; a broken line denotes a nonedge of G. An *interchange* replaces the two edges with the two nonedges.

Theorem 6.7 (Hammer and Simeone [1977]). Let $G = (V, E)$ be an undirected graph with degree sequence $d_1 \geq d_2 \geq \cdots \geq d_n$, and let $m = \max\{i \mid d_i \geq i - 1\}$. Then, G is a split graph if and only if

$$\sum_{i=1}^{m} d_i = m(m - 1) + \sum_{i=m+1}^{n} d_i.$$

Furthermore, if this is the case, then $\omega(G) = m$.

Proof.   The theorem is true if G is a complete graph, so we may assume that $d_m \geq m - 1$ and $d_{m+1} < m$. Since Δ is nonincreasing, $\min\{m, d_i\} = d_i$ for $i \geq m + 1$. Therefore, the mth EGI simplifies to

$$s = \sum_{i=1}^{m} d_i \leq m(m - 1) + \sum_{i=m+1}^{n} d_i. \tag{1}$$

Let K denote the first m vertices of largest degree. The left summand of (1) splits into two contributions $s = s_1 + s_2$, where

$$s_1 = \sum_{x \in K} |\{z \in K \mid xz \in E\}| \leq m(m - 1), \tag{2}$$

$$s_2 = \sum_{x \in K} |y \notin K \mid xy \in E\}|$$

$$= \sum_{y \notin K} |\{x \in K \mid xy \in E| \leq \sum_{i=m+1}^{n} d_i. \tag{3}$$

Equality holds in (2) if and only if K is complete. Equality holds in (3) if and only if $V - K$ is stable. Therefore, if equality holds in (1), then G is a split graph.

Conversely, assume that $G = (V, E)$ is a split graph. By Theorem 6.2 we can partition V into a stable set S and a complete set K such that $|K| = \omega(G)$. Every vertex in K has degree at least $|K| - 1$, and, since K is maximum, every vertex in S has degree at most $|K| - 1$. Therefore, we may assume that the vertices are ordered so that $K = \{v_1, \ldots, v_{|K|}\}$ and $S = \{v_{|K|+1}, \ldots, v_n\}$, where $\deg v_i = d_i$. Moreover, $d_{|K|} \geq |K| - 1$ and $d_{|K|+1} \leq |K| - 1 < |K|$, so $\omega(G) = |K| = m$. Finally, since K is complete and $S = V - K$ is stable, we conclude that equality holds in (2) and (3) and therefore also in (1).   ∎

Corollary 6.8.   If G is a split graph, then every graph with the same degree sequence as G is also a split graph.

Remark.   Hammer and Simeone [1977] investigated a more general problem on graphs. They define the *splittance* of an arbitrary undirected graph to be the minimum number of edges to be added or erased in order to produce a split graph. Of course, split graphs are just those graphs whose splittance is zero. Their main result shows that the splittance depends only on the degree sequence of the graph, and is given by the expression

$$\frac{1}{2}\left[m(m-1) - \sum_{i \leq m} d_i + \sum_{i \geq m+1} d_i \right],$$

where m and the d_i are as in Theorem 6.7.

Those who have further interest in the topic of graphs and their degree sequences are encouraged to read the survey paper by Hakimi and Schmeichel [1978].

EXERCISES

1. Give necessary and sufficient conditions for a tree to be a split graph. Prove that your answer is correct.
2. Prove that the Hamiltonian circuit problem is NP-complete for split graphs. (Hint. Use the fact that the Hamiltonian circuit problem is NP-complete for bipartite graphs.)
3. How many nonisomorphic graphs are there with the following degree sequences: (i) $[3, 3, 2, 2, 1, 1]$, (ii) $[5, 5, 5, 4, 3, 2]$, (iii) $[5, 5, 4, 3, 3, 3, 2, 1]$?
4. Give an example of two nonisomorphic split graphs having the same degree sequence.
5. What is the splittance of graphs C_n, $K_{m,n}$, mK_n, and P_n?
6. Give an $O(n)$ time algorithm for determining whether or not a nonincreasing integer sequence $n - 1 \geq d_1 \geq d_2 \geq \cdots \geq d_n \geq 0$ is graphic. Prove that your algorithm is correct and that its complexity is linear.
7. Let $\Delta = [d_1, d_2, \ldots, d_n]$ be an integer sequence, and define $\bar{\Delta} = [\bar{d}_1, \bar{d}_2, \ldots, \bar{d}_n]$ by the formula

$$\bar{d}_i = n - 1 - d_{n-i+1} \qquad (i = 1, \ldots, n).$$

Show that Δ is graphic if and only if $\bar{\Delta}$ is graphic. What can you say about the graphs corresponding to Δ and $\bar{\Delta}$?

8. Let $m = \max\{i \mid d_i \geq i - 1\}$ where $n - 1 \geq d_1 \geq \cdots \geq d_n \geq 0$. Show that if the mth EGI holds, then the rth EGI automatically holds for $r = m - 1, \ldots, n$ (Hammer, Ibaraki, and Simeone [1978]).
9. Prove Corollary 6.8 directly from Theorem 6.6.

Research problem. Characterize those graphs which are uniquely determined up to isomorphism by their degree sequence. R. H. Johnson has solved this problem for trees; the solution is the class obtained in Exercise 1.

Bibliography

Burkard, R., and Hammer, Peter L.
  [1977] On the Hamiltonicity of split graphs, Univ. of Waterloo, Dept. of Combinatorics
     and Optimization, Res. Report CORR 77–40.
Erdös, Paul, and Gallai, Tibor
  [1960] Graphen mit Punkten vorgeschriebenen Grades, *Mat. Lapok* **11**, 264–272.
Földes, Stephane, and Hammer, Peter L.
  [1977a] Split graphs having Dilworth number two, *Canad. J. Math.* **29**, 666–672.
     MR57 # 3005.
  [1977b] Split graphs, *Proc. 8th Southeastern Conf. on Combinatorics, Graph Theory and
     Computing* (F. Hoffman *et al.*, eds.), Louisiana State Univ., Baton Rouge, Louisiana,
     311–315.
  [1978] The Dilworth number of a graph, *Ann. Discrete Math.* **2**, 211–219.
Hakimi, S. L.
  [1962] On the realizability of a set of integers as degrees of the vertices of a graph, *SIAM J.
     Appl. Math.* **10**, 496–506. MR26 # 5558.
Hakimi, S. L., and Schmeichel, E. F.
  [1978] Graphs and their degree sequences: A survey, *in* "Theory and Applications of
     Graphs," Lecture Notes in Math. 642, pp. 225–235. Springer-Verlag, Berlin.
Hammer, Peter L., Ibaraki, T., and Simeone, B.
  [1978] Degree sequence of threshold graphs, Univ. of Waterloo, Dept. of Combinatorics and
     Optimization Res. Report CORR 78–10.
Hammer, Peter L., and Simeone, Bruno
  [1977] The splittance of a graph, Univ. of Waterloo, Dept. of Combinatorics and Optimiza-
     tion, Res. Report CORR 77–39.
Hanlon, Phil
  [1979] Enumeration of graphs by degree sequence, *J. Graph Theory* **3**, 295–299.
Harary, Frank
  [1969] "Graph Theory," Addison-Wesley, Reading, Massachusetts. Chapter 6.
Havel, Václav
  [1955] A remark on the existence of finite graphs (Czech), *Časopis Pěst Mat.* **80**, 477–480.
Ryser, Herbert J.
  [1963] "Combinatorial Mathematics," Carus Monograph No. 14, Chapter 6, Theorem 3.1.
     Math. Assoc. of America, Washington, D.C.

CHAPTER 7

Permutation Graphs

1. Introduction

In this chapter we consider a class of perfect graphs which has a large number of applications. Suppose π is a permutation of the numbers $1, 2, \ldots, n$. Let us think of π as the sequence $[\pi_1, \pi_2, \ldots, \pi_n]$, so, for example, the permutation $\pi = [4, 3, 6, 1, 5, 2]$ has $\pi_1 = 4$, $\pi_2 = 3$, etc. Notice that $(\pi^{-1})_i$, denoted here as π_i^{-1}, is the *position in the sequence* where the number i can be found; in our example $\pi_4^{-1} = 1$, $\pi_3^{-1} = 2$, etc.

We can construct an undirected graph $G[\pi]$ from π in the following manner: $G[\pi]$ has vertices numbered from 1 to n; two vertices are joined by an edge if the larger of their corresponding numbers is to the left of the smaller in π (that is, they occur out of their proper order reading left to right). In our example, both 4 and 3 are connected to 1 since they are each larger and to the left of 1, whereas neither 5 nor 2 is connected to 1 (see Figure 7.1). The graph $G[\pi]$ is sometimes called the *inversion graph* of π.

More formally, if π is a permutation of the numbers $1, 2, \ldots, n$, then the graph $G[\pi] = (V, E)$ is defined as follows:

$$V = \{1, 2, \ldots, n\}$$

and

$$ij \in E \Leftrightarrow (i - j)(\pi_i^{-1} - \pi_j^{-1}) < 0.$$

An undirected graph G is called a *permutation graph* if there exists a permutation π such that $G \cong G[\pi]$.

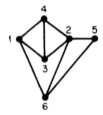

Figure 7.1.   The graph $G[4, 3, 6, 1, 5, 2]$.

2.   Characterizing Permutation Graphs

Permutation graphs have many interesting properties. Notice what happens when we *reverse* the sequence π. Each pair of numbers which occurred in the correct order in π is now in the wrong order, and vice versa. Thus, the permutation graph we obtain is the complement of $G[\pi]$. In other words, if π^{ν} is the permutation obtained by reversing the sequence π, then

$$G[\pi^{\nu}] = \overline{G[\pi]}.$$

This shows that *the complement of a permutation graph is also a permutation graph*.

Another property of the graph $G[\pi]$ (which you may have already guessed) is that it is transitively orientable. If we orient each edge toward its larger endpoint, then we will obtain a transitive orientation F. For, suppose $ij \in F$ and $jk \in F$, then $i < j < k$ and $\pi_i^{-1} > \pi_j^{-1} > \pi_k^{-1}$, which implies that $ik \in F$. This result is only half of the story; we actually have the following:

Theorem 7.1 (Pnueli, Lempel, and Even [1971]).   An undirected graph G is a permutation graph if and only if G and $\bar{G}$ are comparability graphs.

Proof.   Suppose $G \cong G[\pi]$; then G is a comparability graph since $G[\pi]$ has a transitive orientation. Likewise, $\bar{G}$ is a comparability graph since $\bar{G} \cong G[\pi^{\nu}]$.

Conversely, let (V, F_1) and (V, F_2) be transitive orientations of $G = (V, E)$ and $\bar{G} = (V, \bar{E})$, respectively. We claim that $(V, F_1 + F_2)$ is an *acyclic* orientation of the complete graph $(V, E + \bar{E})$. For suppose $F_1 + F_2$ had a cycle $[v_0, v_1, v_2, \ldots, v_l, v_0]$ of the smallest possible length l. If $l > 3$, then the cycle can be shortened either by $v_0 v_2$ or $v_2 v_0$, contradicting minimality. If $l = 3$, then at least two of the edges of the cycle are in the same F_i, implying that F_i is not transitive. Thus $(V, F_1 + F_2)$ is acyclic. Similarly $(V, F_1^{-1} + F_2)$ is acyclic.

We conclude the proof by constructing a permutation π such that $G \cong G[\pi]$. An acyclic orientation of a complete graph is transitive, and it determines a unique linear ordering of the vertices. (See Section 2.5 on transitive tournaments.) Consider the following procedure.

Step I. Label the vertices according to the order determined by $F_1 + F_2$; namely, the vertex x of in-degree $i - 1$ gets label $L(x) = i$.

Step II. Label the vertices according to the order determined by $F_1^{-1} + F_2$; namely, the vertex x of in-degree $i - 1$ gets label $L'(x) = i$.

Notice that

$$xy \in E \Leftrightarrow [L(x) - L(y)][L'(x) - L'(y)] < 0, \tag{1}$$

since it is the edges of E which have their orientations reversed between steps I and II. This is the key to our argument.

Step III. Define π as follows: For each vertex x, if $L(x) = i$, then $\pi_i^{-1} = L'(x)$. The relationship is depicted in the commuting diagram below.

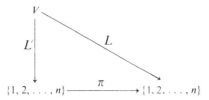

Therefore, by (1), π is the desired permutation and L is the desired isomorphism.   ∎

Remark.   In terms of the nomenclature of Section 5.8, G is a permutation graph if and only if the transitive orientations of G, when regarded as partial orders have dimension at most 2.

The construction technique presented above is illustrated in Figure 7.2.

Theorem 7.1 suggests an algorithm for recognizing permutation graphs, namely, applying the transitive orientation algorithm to the graph and to its complement. If we succeed in finding transitive orientations, then the graph is a permutation graph. To find a suitable permutation we can follow the construction procedure in the proof of the theorem. The entire method requires $O(n^3)$ time and $O(n^2)$ space for a graph with n vertices.

We conclude this section with a remark which follows from transitive orientability.

Remark.   The decreasing subsequences of π and the cliques of $G[\pi]$ are in one-to-one correspondence. The increasing subsequences of π and the stable sets of $G[\pi]$ are in one-to-one correspondence.

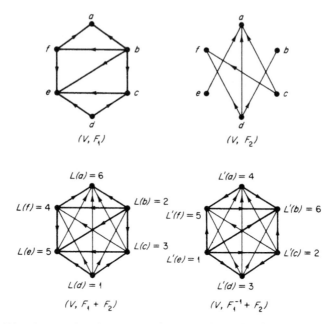

Figure 7.2.   Construction of the permutation $\pi = [5, 3, 1, 6, 4, 2]$ from the transitive orientations F_1 and F_2. Vertex a gives $\pi_6^{-1} = 4$, vertex b gives $\pi_2^{-1} = 6$, etc.

3.   Permutation Labelings

A related, but simpler, problem is that of testing whether a *given* labeling of the vertices of a graph is a permutation labeling. Let $G = (V, E)$ be an undirected graph, and let $L: V \to \{1, 2, \ldots, n\}$ be a bijection labeling the vertices. We call L a *permutation labeling* if there exists a permutation π of $\{1, 2, \ldots, n\}$ such that

$$xy \in E \Leftrightarrow [L(x) - L(y)][\pi^{-1}(L(x)) - \pi^{-1}(L(y))] < 0.$$

Clearly, G is a permutation graph if and only if it has at least one permutation labeling.

Figure 7.3 shows two labelings of the same graph. The first is the permutation labeling already constructed in Figure 7.2. The second is not a permutation labeling for the following reason. Since $\text{Adj}(1) = \{5, 6\}$, both 5 and 6 would be on the left of 1 while 2–4 would be on the right of 1 in any permutation π that might work. However, this implies that 3 and 4 would be to the right of 6—yet they are not connected to 6. Hence, no such permutation π exists for this labeling.

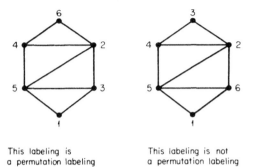

This labeling is
a permutation labeling

This labeling is not
a permutation labeling

Figure 7.3.

Theorem 7.2 (Gill and Acharya [1977]). Let $G = (V, E)$ be an undirected graph. A bijection $L: V \rightarrow \{1, 2, \ldots, n\}$ is a permutation labeling of G if and only if the mapping

$$F: x \rightarrow L(x) - d^-(x) + d^+(x) \qquad (x \in V)$$

is an injection, where

$$d^-(x) = |\{y \in \text{Adj}(x) | L(y) < L(x)\}|$$

and

$$d^+(x) = |\{y \in \text{Adj}(x) | L(y) > L(x)\}|.$$

Proof. ($\Rightarrow$) Let π be a permutation corresponding to the labeling L. Then $d^-(x)$ is the number of integers in π smaller than and to the right of $L(x)$, and $d^+(x)$ is the number of integers in π larger than and to the left of $L(x)$. By Exercise 4, $f(x) = \pi^{-1}(L(x))$, and since π^{-1} and L are injective (indeed bijective), so too is f.

($\Leftarrow$) Assuming that f is injective, we will construct the desired permutation. Since $d^-(x) \le L(x) - 1$ and $d^+(x) \le n - L(x)$, it follows that

$$1 \le f(x) \le n \qquad (x \in V). \tag{2}$$

But f is injective and integer valued, so (2) implies that f is a bijection from V to $\{1, 2, \ldots, n\}$. Define π as follows:

$$\pi(i) = L(f^{-1}(i))$$

(see Figure 7.4).

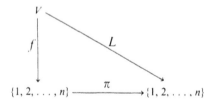

Figure 7.4.

Now, π is a permutation, since L and f^{-1} are bijective. Furthermore,

$$\pi^{-1}(L(x)) = f(x) \qquad (x \in V),$$

so we must verify that

$$xy \in E \Leftrightarrow [L(x) - L(y)][f(x) - f(y)] < 0.$$

This is left as an exercise for the reader.   ∎

4.   Applications

Permutation graphs can be regarded as a class of intersection graphs in the following manner. Write the numbers 1, 2, ..., n horizontally from left to right; underneath them write the numbers $\pi_1, \pi_2, \ldots, \pi_n$ in sequence, again horizontally left to right; finally, draw n straight line segments joining the two 1's, the two 2's, etc. We call this the *matching diagram* of π (see Figure 7.5). Notice that the ith segment intersects the jth segment if and only if i and j appear in reversed order in π; this is the same criterion for the vertices i and j of $G[\pi]$ to be adjacent. Therefore, the intersection graphs of the segments of matching diagrams are exactly the permutation graphs.

The reason for our introducing these matching diagrams is to assist us in studying some applications of permutation graphs.

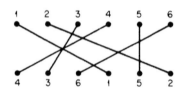

Figure 7.5.   The matching diagram of [4, 3, 6, 1, 5, 2].

Application 7.1. Suppose we have two collections of cities, the X cities and the Y cities, lying, respectively, on two parallel lines. Suppose also, that there are airline routes connecting various X cities with various Y cities, all scheduled to be utilized at the same time of the day. Our mission, should we decide to accept it, will be to assign altitudes to each flight path so that intersecting routes will be at different altitudes. We will thereby assure that no midair collisions will occur. Being clever graph theorists, we recognize this as a coloring problem.

The data, as given, provides us with a bipartite graph embedded in the plane, as pictured in Figure 7.6. We number the flight paths by traversing the northern cities from west to east. From this we can extract a matching diagram, or go straight to the corresponding permutation graph $G[\pi]$. Assigning altitudes to the flight paths so that intersecting paths receive different altitudes is equivalent to coloring the vertices of $G[\pi]$ so that adjacent vertices receive different colors. An efficient coloring algorithm for permutation graphs is given in the next section.

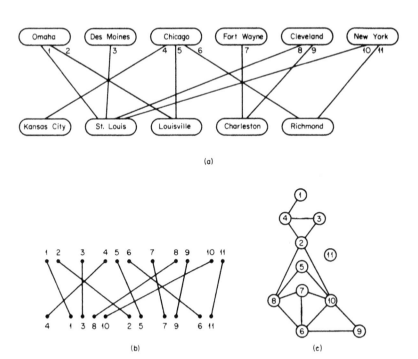

Figure 7.6. (a) A bipartite graph B representing flight paths between cities. (b) The matching diagram of a permutation π extracted from the bipartite graph B. (c) The graph $G[\pi]$. Color the vertices of $G[\pi]$ and solve the altitude assignment problem for B.

Application 7.2 (Shifted Intervals). Let $\mathscr{I} = \{I_i | i = 1, 2, \ldots, n\}$ be a collection of intervals on a line, where $I_i = (x_i, y_i)$ and $|I_i| = y_i - x_i$ denotes the length of I_i. Assume that the intervals, which may overlap, have been ordered such that $x_1 \leq x_2 \leq \cdots \leq x_n$. Let w_i represent the cost of shifting the interval I_i (assumed to be independent of the distance shifted). Find the cheapest shifting of intervals so that (1) the order is preserved and (2) no overlap remains. (In Even, Pnueli, and Lempel [1972], the intervals correspond to the memory requirements of n programs at a certain time in a multiprogramming computer.)

A solution to this problem is as follows. Consider the oriented graph $(\mathscr{I}, F)$ where

$$(I_i, I_j) \in F \Leftrightarrow \sum_{i \leq k \leq j} |I_k| \leq y_j - x_i \qquad (i < j).$$

Two intervals are thus related in F if and only if the intervals between them can be shifted in such a way that none of these $j - i + 1$ intervals (including the fixed I_i and I_j) will intersect. It is routine to show that F is a transitively oriented graph (see Exercise 5). The solution to our problem will be to find a chain of F having maximum weight (to remain, all others are shifted); in other words, find a maximum weighted clique of the graph $E = F + F^{-1}$ which is not only a comparability graph but is even a permutation graph.

5. Sorting a Permutation Using Queues in Parallel

A *queue* is a linear storage device in which items are loaded at one end and unloaded at the other end in a first-in–first-out fashion (FIFO). Let us consider the problem of sorting a permutation π of the numbers $1, 2, \ldots, n$

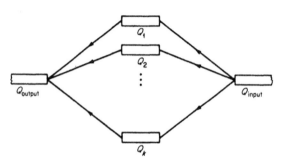

Figure 7.7. A network of k queues in parallel.

using a *network of k queues arranged in parallel* (see Figure 7.7). The permutation sits in the input queue initially. Each number, in turn, passes along to one of the k internal queues where it is stored temporarily until it is moved onto the output queue. We assume that each queue has unbounded capacity and that backing up along an edge, counter to its direction, is forbidden. One can easily imagine a station master directing railroad cars through such a switchyard in order to rearrange the cars of a freight train. Typically, the number of sidings (queues) will be limited, so we are led to the following problem. Given a network of k queues in parallel, characterize the permutations which can be sorted on it. Or similarly, given a permutation π, how many queues will we need? In addition, find an optimal sorting method.

Example. Suppose $\pi = [4, 3, 6, 1, 5, 2]$. The 4 is placed in Q_1. The 3 cannot go in Q_1 because it will be forever stuck behind the 4, so put it in Q_2. Next comes the 6, which can go either behind 4 on Q_1 or behind 3 on Q_2. Put 6 behind 4 on Q_1. How about 1? It must go on Q_3. The 5 cannot go on Q_1 because 6 is already there; put 5 on Q_2 behind 3. Finally 2 cannot go on Q_1 or Q_2, but it can go on Q_3. Now that everything is stored (Figure 7.8), we unload the numbers 1–6 from their respective storage places.

We call your attention to a few obvious facts. The contents of each Q_i must be in increasing order, for otherwise it would be impossible to successfully unload all the numbers in proper order. Furthermore, it is easy to show that it makes no difference whatsoever whether we (a) require loading *all* input numbers onto the queues before unloading any of them or (b) allow unloading anytime it is possible.

What is it that forces two numbers to go into different queues? Answer: The numbers occur in reversed order in π. Thus, if i and j are adjacent in $G[\pi]$, then they *must* go through different queues.

Proposition 7.3. Let $\pi = [\pi_1, \pi_2, \ldots, \pi_n]$ be a permutation of the integers $\{1, 2, \ldots, n\}$. There is a one-to-one correspondence between the

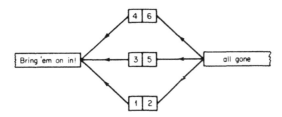

Figure 7.8. A network which is sorting $\pi = [4, 3, 6, 1, 5, 2]$.

proper k-colorings of $G[\pi]$ and the successful sorting strategies for π in a network of k parallel queues.

Proof. Assign painters to each Q_i, each with a different color paint. Now sort π in the k-network and have every number painted as it enters its corresponding queue. Since connected vertices i and j of $G[\pi]$ pass through different queues, they receive different colors.

Conversely, given a proper coloring of $G[\pi]$ using colors $1, 2, \ldots, k$, assign a traffic director to the input queue. If the color of x is c, then the traffic director sends x to Q_c. Suppose this strategy is unsuccessful. There must be a bottleneck in some queue, say Q_m; i.e., Q_m has a pair of numbers x and y stored in reversed order. However, x and y enter Q_m in the same order that they appear in π, namely, reversed. Thus, x and y are adjacent in $G[\pi]$, and yet they are both colored the same, a contradiction. Clearly, this correspondence is a bijection. ∎

Corollary 7.4. Let π be a permutation. The following numbers are equal:

   (i)   the chromatic number of $G[\pi]$,
  (ii) the minimum number of queues required to sort π,
 (iii) the length of a longest decreasing subsequence of π.

Proof. The equivalence of (i) and (ii) follows immediately from Proposition 7.3, and its proof suggests a method for transforming a solution of one problem into a solution of the other. Equality between (i) and (iii) holds since a longest subsequence of π corresponds to a maximum clique of $G[\pi]$, which will be of size $\chi(G[\pi])$ since permutation graphs are perfect. ∎

The *canonical sorting strategy* for π places each number in the first available queue. (Our example was done that way.) From this strategy, we obtain the *canonical coloring* of $G[\pi]$. The following algorithm simulates the process. It yields a minimum coloring.

Algorithm 7.1. Canonical coloring of a permutation.
Input: A permutation $\pi = [\pi_1, \pi_2, \ldots, \pi_n]$ of the numbers $\{1, 2, \ldots, n\}$.
Output: A coloring of the vertices of $G[\pi]$ and the chromatic number χ of $G[\pi]$.
Method: During the jth iteration, π_j is transferred onto the queue Q_i having smallest index i satisfying $\pi_j \geq$ last entry of Q_i (i.e., the first allowable Q_i). We do not *actually* save the entire contents of Q_i. Instead, an array LAST(i) holds the last number in Q_i. The counter k keeps track of the actual number of queues (colors) used. The entire algorithm is as follows:

```
    begin
1.    k ← 0;
2.    for j ← 1 to n do
        begin
3.          i ← FIRST allowable queue;
4.          COLOR(π_j) ← i;
5.          LAST(i) ← π_j;
6.          k ← max{k, i};
        end
7.    χ ← k;
    end
```

In order to execute statement 3 efficiently, a type of binary insertion can be used. One such subroutine is given in Figure 7.9. The next result shows the correctness of Algorithm 7.1.

Theorem 7.5. Let π be a permutation of the numbers $\{1, 2, \ldots, n\}$. The canonical coloring of $G[\pi]$, as produced by Algorithm 7.1, is a minimum coloring.

Proof. Clearly, Algorithm 7.1 produces a proper χ-coloring of $G[\pi]$. We must show that $\chi = \chi(G[\pi])$. It is sufficient to show that π has a decreasing subsequence of length χ. Consider the predecessor function p defined as follows: If $COLOR(\pi_j) = i \geq 2$, then $\pi_{p(j)}$ equals the value of $LAST(i - 1)$ during the jth iteration. Clearly, $\pi_{p(j)} > \pi_j$ and $p(j) < j$ since it was $\pi_{p(j)}$ sitting on Q_{i-1} which forced π_j to go down to Q_i. Then

$$\pi_{j_1}, \pi_{j_2}, \ldots, \pi_{j_\chi}$$

where

$$COLOR(\pi_{j_\chi}) = \chi$$

and

$$\pi_{j_{i-1}} = \pi_{p(j_i)} \qquad \text{for} \quad i = \chi, \chi - 1, \ldots, 2$$

is the desired decreasing subsequence. ∎

```
procedure FIRST allowable queue:
    begin
        i ← 1; t ← k + 1;
        until i = t do
            begin
                r ← ⌊(i + t)/2⌋;
                if π_j ≥ LAST(r)
                    then t ← r;
                    else i ← r + 1;
            end
        return i;
    end
```

Figure 7.9.

Remark. To find a minimum clique cover of $G[\pi]$, apply Algorithm 7.1 to the reversal π^ρ of π.

Algorithm 7.1 can be used to color any permutation graph G in time proportional to $n \log n$ provided we have the permutation π and the isomorphism $G \to G[\pi]$. Notice that this complexity is independent of the number of edges of G. If we do not have π, then we would revert to the coloring algorithm for comparability graphs.

Remark. If we apply the algorithm in Exercise 8 of Chapter 2 to the orientation F of $G[\pi]$ where each edge is oriented toward its larger endpoint, then the coloring we obtain will be exactly the same as the canonical coloring, namely,

$$\text{COLOR}(\pi_j) = i \Leftrightarrow \text{HEIGHT}(\pi_j) = i - 1.$$

EXERCISES

1. For what permutation π of the numbers $1, 2, 3, \ldots, n$ is $G[\pi]$ the following:
  (a) the complete graph on n vertices;
  (b) the graph of n isolated vertices;
  (c) two disjoint complete graphs on r and $n - r$ vertices, respectively?
2. Find a permutation π whose graph $G[\pi]$ is isomorphic to G_1. Do the same for G_2 (see Figure 7.10).

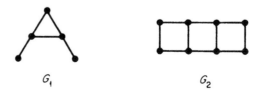

G_1 $\qquad\qquad\qquad\qquad\qquad$ G_2

Figure 7.10.

3. Let π^{-1} be the inverse of the permutation π. Prove that $G[\pi] \cong G[\pi^{-1}]$.
4. Let $\pi = [\pi_1, \pi_2, \ldots, \pi_n]$ be a permutation of $\{1, 2, \ldots, n\}$. Let p_i denote the number of integers less than and to the right of i in π, and let q_i denote the number of integers greater than and to the left of i in π. Prove that the following equality holds:

$$\pi_i^{-1} + p_i = i + q_i.$$

5. Let F be defined as in Application 7.2 (the shifted interval problem), and let $E = F \cup F^{-1}$.
  (i) Show that F is a transitive orientation of E.
  (ii) Prove that E is a permutation graph.

6. Give an application similar to Application 7.2 which uses the fact proved in Exercise 5(ii), namely, that E is not only a comparability graph, but is even a permutation graph.

7. A permutation graph G is *uniquely representable* if there is only one permutation π such that $G[\pi] \cong G$. Characterize the uniquely representable permutation graphs.

8. Let G be a permutation graph with n vertices. Given transitive orientations F_1 and F_2 of G and $\bar{G}$, respectively, write an algorithm which calculates a permutation π such that $G[\pi] \cong G$. Show that your algorithm can run in $O(n^2)$ time.

9. Using the canonical sorting strategy, give a minimum coloring of the graph $G[\pi]$ for $\pi = [9, 8, 2, 5, 6, 1, 7, 4, 3]$.

10. In sorting, using a network of parallel queues, prove that it makes no difference whether we (a) require loading *all* input numbers onto the queues before unloading any of them or (b) allow unloading anytime it is possible.

11. Prove the following: At any point during the execution of Algorithm 7.1,

$$\text{LAST}(1) > \text{LAST}(2) > \cdots > \text{LAST}(k).$$

Why is this fact needed to justify the correctness of the subroutine in Figure 7.9? Analyze the time complexity of Algorithm 7.1.

12. Let G be a permutation graph with n vertices. Show that either G or $\bar{G}$ contains a clique of size $\lceil n^{1/2} \rceil$ (Erdös and Szekeres [1935]).

Bibliography

Alter, Ronald, Curtz, Thaddeus B., and Wang, Chung C.
  [1974] Permutations with fixed index and number of inversions, *Proc. Fifth Southeastern Conf. on Combinatorics, Graph Theory and Computing,* Congressus Numerantium No. X, Utilitas Math, Winnipeg, 3–38. MR51 #2932.
Baker, Kirby A., Fishburn, Peter C., and Roberts, Fred S.
  [1970] A new characterization of partial orders of dimension two, *Ann. N.Y. Acad. Sci.* **175**, 23–24. MR42 #140.
  [1972] Partial orders of dimension 2, *Networks* **2**, 11–28. MR46 #104.
de Bruijn, N. G.
  [1974] Sorting by means of swappings, *Discrete Math.* **9**, 333–339. MR49 #8869.
Erdös, P:, and Szekeres, G.
  [1935] A combinatorial problem in geometry, *Compositio Math.* **2**, 463–470.
Even, Shimon
  [1973] "Algorithmic Combinatorics," Macmillan, New York. MR49 #48.
Even, Shimon, and Itai, Alon
  [1971] Queues, stacks, and graphs, *in* "Theory of Machines and Computations," pp. 71–86. Academic Press, New York.

Even, Shimon, Pnueli, Amir, and Lempel, Abraham
  [1972]   Permutation graphs and transitive graphs, *J. Assoc. Comput. Mach.* **19**, 400–410.
          MR47 #1675.
Gill, M. K., and Acharya, B. D.
  [1977]   A new characterization of permutation graphs (unpublished).
Pnueli, Amir, Lempel, Abraham, and Even, Shimon
  [1971]   Transitive orientation of graphs and identification of permutation graphs, *Canad. J. Math.* **23**, 160–175. MR45 #1800.

CHAPTER 8

Interval Graphs

1. How It All Started

In 1957 G. Hajös posed the following problem:

Given a finite number of intervals on a straight line, a graph associated with this set of intervals can be constructed in the following manner: each interval corresponds to a vertex of the graph, and two vertices are connected by an edge if and only if the corresponding intervals overlap at least partially. The question is whether a given graph is isomorphic to one of the graphs just characterized (Hajös [1957, p. 65, translated by M.C.G.]).

Independently, the well-known molecular biologist, Seymour Benzer, during his investigations of the fine structure of the gene, asked a related question.

From the classical researches of Morgan and his school, the chromosome is known as a linear arrangement of hereditary elements, the *genes*. These elements must have an internal structure of their own. At this finer level, within the *gene* the question arises again: . . . Are *they* [the subelements within the gene] linked together in a linear order analogous to the higher level of integration of the genes in the chromosome?

A crucial examination of the question should be made from the point of view of *topology*, since it is a matter of how parts of the structure are connected to each other, rather than of the distances between them. Experiments to explore the topology should ask *qualitative* questions (e.g., do two parts of the structure touch each other or not?) rather than *quantitative* ones (how far apart are they?). (Benzer [1959].)

The solution to this question would be found by studying those graphs which represent intersecting intervals on a line, and then verifying whether or not the data that was gathered was consistent with the linear genetic hypothesis.

Our treatment of interval graphs began in Chapter 1. Let us continue looking into the properties of this interesting and useful class of graphs. The reader may wish to review Section 1.3 at this point.

2. Some Characterizations of Interval Graphs

The following theorem and its corollary will establish where the class of interval graphs belongs in the world of perfect graphs.

Theorem 8.1 (Gilmore and Hoffman [1964]). Let G be an undirected graph. The following statements are equivalent.

(i) G is an interval graph.

(ii) G contains no chordless 4-cycle and its complement $\bar{G}$ is a comparability graph.

(iii) The maximal cliques of G can be linearly ordered such that, for every vertex x of G, the maximal cliques containing x occur consecutively.

Proof. (i) $\Rightarrow$ (ii) This was proved in Chapter 1, Propositions 1.2 and 1.3.

(ii) $\Rightarrow$ (iii) Let us assume that $G = (V, E)$ contains no chordless 4-cycle, and let F be a transitive orientation of the complement $\bar{G}$.

Lemma A. Let A_1 and A_2 be maximal cliques of G.

(a) These exists an edge in F with one endpoint in A_1 and the other endpoint in A_2.

(b) All such edges of $\bar{E}$ connecting A_1 with A_2 have the same orientation in F.

Proof of Lemma A. (a) If no such edge exists in F, then $A_1 \cup A_2$ is a clique of G, contradicting maximality.

(b) Suppose $ab \in F$ and $dc \in F$ with $a,c \in A_1$ and $b,d \in A_2$. We must show a contradiction. If either $a = c$ or $b = d$, then transitivity of F immediately gives a contradiction; otherwise, these four vertices are distinct (Figure 8.1), and ad or bc is in $\bar{E}$, since E may not have a chordless 4-cycle. Assume, without loss of generality, that $ad \in \bar{E}$; which way is it oriented? Using the

Figure 8.1.   Solid edges are in E; broken edges are in $\bar{E}$. Arrows denote the orientation F.

transitivity of F, $ad \in F$ (resp. $da \in F$) would imply $ac \in F$ (resp. $db \in F$), which is impossible, and Lemma A is proved.

Consider the following relation on the collection $\mathscr{C}$ of maximal cliques: $A_1 < A_2$ iff there is an edge of F connecting A_1 with A_2 which is oriented toward A_2. By Lemma A, this defines a tournament (complete orientation) on $\mathscr{C}$. We claim that $(\mathscr{C}, <)$ is a transitive tournament, and hence linearly orders $\mathscr{C}$. For suppose $A_1 < A_2$ and $A_2 < A_3$; then there would be edges $wx \in F$ and $yz \in F$ with $w \in A_1, x,y \in A_2$, and $z \in A_3$. If either $xz \notin E$ or $wy \notin E$, then $wz \in F$ and $A_1 < A_3$. Therefore, assume that the edges wy, yx, and xz are all in E (see Figure 8.2). Since G contains no chordless 4-cycle, $wz \notin E$, and the transitivity of F implies $wz \in F$. Thus $A_1 < A_3$, which proves the transitive tournament claim.

Next, assume that $\mathscr{C}$ has been linearly ordered $A_1, A_2, \ldots, A_m$ according to the relation above (i.e., $i < j$ iff $A_i < A_j$). Suppose there exist cliques $A_i < A_j < A_k$ with $x \in A_i$, $x \notin A_j$, and $x \in A_k$. Since $x \notin A_j$, there is a vertex $y \in A_j$ such that $xy \notin E$. But $A_i < A_j$ implies $xy \in F$, whereas $A_j < A_k$ implies $yx \in F$, contradiction. This proves (iii).

(iii) $\Rightarrow$ (i)   For each vertex $x \in V$, let $I(x)$ denote the set of all maximal cliques of G which contain x. The sets $I(x)$, for $x \in V$, are intervals of the linearly ordered set $(\mathscr{C}, <)$. It remains to be shown that

$$xy \in E \Leftrightarrow I(x) \cap I(y) \neq \varnothing \qquad (x, y \in V).$$

This obviously holds, since two vertices are connected if and only if they are both contained in some maximal clique. ∎

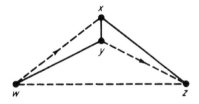

Figure 8.2.

Corollary 8.2.   An undirected graph G is an interval graph if and only if G is a triangulated graph and its complement $\bar{G}$ is a comparability graph.

Remark.   The coloring, clique, stable set, and clique cover problems can be solved in polynomial time for interval graphs by using the algorithms of Chapters 4 and 5. A recognition algorithm can be obtained by combining Algorithms 4.1 and 5.2, although the recognition algorithm to be presented in Section 8.3 will be asymptotically more efficient.

Statement (iii) of the Gilmore–Hoffman theorem has an interesting matrix formulation. A matrix whose entries are zeros and ones, is said to have the *consecutive 1's property for columns* if its rows can be permuted in such a way that the 1's in each column occur consecutively. In Figure 8.3 the matrix M_1 has the consecutive 1's property for columns since its rows can be permuted to obtain M_2. Matrix M_3 does not possess the property. Consider the *clique matrix* M (maximal cliques-versus-vertices incidence matrix) of a graph G. The following corollary to Theorem 8.1 is immediate.

Theorem 8.3 (Fulkerson and Gross [1965]).   An undirected graph G is an interval graph if and only if its clique matrix M has the consecutive 1's property for columns.

Proof.   An ordering of the maximal cliques of G corresponds to a permutation of the rows of M. This theorem follows from Theorem 8.1.   ∎

The earliest characterization of interval graphs was obtained by Lekerkerker and Boland. Their result embodies the notion that an interval graph cannot branch into more than two directions, nor can it circle back onto itself.

Theorem 8.4 (Lekkerkerker and Boland [1962]).   An undirected graph G is an interval graph if and only if the following two conditions are satisfied:

   (i) G is a triangulated graph, and
   (ii)   any three vertices of G can be ordered in such a way that every path from the first vertex to the third vertex passes through a neighbor of the second vertex.

Three vertices which fail to satisfy (ii) are called an *astroidal triple*. They would have to be pairwise nonadjacent, but any two of them would have to be connected by a path which avoids the neighborhood of the remaining vertex. Thus, G is an interval graph if and only if G is triangulated and contains no astroidal triple. Condition (ii) illustrates a well-known law of the business world: Every shipment from a supplier to the consumer must pass by the middle man.

$$
\begin{array}{c}
\textcircled{1} \\
\textcircled{2} \\
\textcircled{3} \\
\textcircled{4} \\
\textcircled{5}
\end{array}
\begin{pmatrix}
1 & 0 & 0 & 1 \\
1 & 1 & 1 & 0 \\
0 & 1 & 0 & 0 \\
1 & 0 & 1 & 1 \\
1 & 1 & 0 & 0
\end{pmatrix}
\qquad
\begin{array}{c}
\textcircled{1} \\
\textcircled{4} \\
\textcircled{2} \\
\textcircled{5} \\
\textcircled{3}
\end{array}
\begin{pmatrix}
1 & 0 & 0 & 1 \\
1 & 0 & 1 & 1 \\
1 & 1 & 1 & 0 \\
1 & 1 & 0 & 0 \\
0 & 1 & 0 & 0
\end{pmatrix}
\qquad
\begin{pmatrix}
1 & 1 & 1 & 1 \\
1 & 0 & 0 & 0 \\
0 & 1 & 0 & 0 \\
0 & 0 & 1 & 0 \\
0 & 0 & 0 & 1
\end{pmatrix}
$$

$$ \mathbf{M}_1 \qquad\qquad \mathbf{M}_2 \qquad\qquad \mathbf{M}_3 $$

Figure 8.3. Matrix $\mathbf{M}_1$ has the consecutive 1's property for columns since it can be transformed into $\mathbf{M}_2$. Matrix $\mathbf{M}_3$ does not have the consecutive 1's property for columns since it cannot be suitably transformed.

3. The Complexity of Consecutive 1's Testing

Interval graphs were characterized as those graphs whose clique matrices satisfy the consecutive 1's property for columns (Theorem 8.3). We may apply this characterization to a recognition algorithm for interval graphs $G = (V, E)$ in a two-step process. First, verify that G is triangulated and, if so, enumerate its maximal cliques. This can be executed in time proportional to $|V| + |E|$ (Corollary 4.6, Theorem 4.17) and will produce at most $n = |V|$ maximal cliques (Proposition 4.16). Second, test whether or not the cliques can be ordered so that those which contain vertex v occur consecutively for every $v \in V$. Booth and Leuker [1976] have shown that this step can also be executed in linear time. We shall look at the main ideas behind their algorithm and its implementation. The interested reader should consult their very readable paper for additional details. Subject to Corollary 8.8 and Exercise 3 we have the following.

Theorem 8.5 (Booth and Leuker [1976]).   Interval graphs can be recognized in linear time.

The general consecutive arrangement problem is the following: *Given a finite set X and a collection $\mathscr{I}$ of subsets of X, does there exist a permutation π of X in which the members of each subset $I \in \mathscr{I}$ appear as a consecutive subsequence of π?* In the interval graph problem, X is the set of maximal cliques and $\mathscr{I} = \{I(v)\}_{v \in V}$, where $I(v)$ is the set of all maximal cliques containing v. The consecutive arrangement and consecutive 1's problems are equivalent: The rows of the matrix constitute X, and each column determines, or is determined by, a subset of X consisting of those rows containing a 1 in the specified column. Tucker [1972] has characterized the consecutive 1's problem in terms of forbidden configurations. Another characterization, due to Nakano [1973a], is stated as Exercise 12.

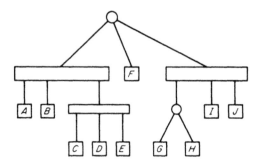

Figure 8.4.   A *PQ*-tree.

Besides its use in recognizing interval graphs, the consecutive 1's problem
has a number of other applications. These include a linear-time algorithm
for recognizing planar graphs (see Booth and Leuker [1976]), and a storage
allocation problem to be discussed in the next section (Application 8.4).

The data structure needed to solve the consecutive arrangement problem
most efficiently is the *PQ*-tree. *PQ*-trees were invented by Leuker [1975] and
Booth [1975] expressly for this purpose. They are used to represent *all* the
permutations of *X* which are consistent with the constraints of consecutivity
determined by $\mathscr{I}$. Most importantly, only a small amount of storage is
required for this representation.

A *PQ-tree T* is a rooted tree whose internal nodes are of two types: *P* and
Q. The children of a type *P*-node occur in no particular order, while those of a
Q-node appear in an order which must be locally preserved. This will be
explained in the next paragraph. We designate a *P*-node by a circle and a
Q-node by a wide rectangle. The leaves of *T* are labeled bijectively by the
elements of the set *X* (see Figure 8.4).

The *frontier* of a tree *T* is the permutation of *X* obtained by reading
the labels of the leaves from left to right. In our example, the frontier is
[*A B C D E F G H I J*]. Two *PQ*-trees *T* and *T'* are *equivalent*, denoted
$T \equiv T'$, if one can be obtained from the other by applying a sequence of the
following transformation rules.

1.   Arbitrarily permute the children of *P*-node.
2.   Reverse the children of a *Q*-node.

Figure 8.5 illustrates a *PQ*-tree which is equivalent to the tree in Figure 8.4.
Its frontier is [*F J I G H A B E D C*]. Parenthetically, we obtain an equivalent
tree by regarding *T* as a mobile and exposing it to a gentle summer breeze.

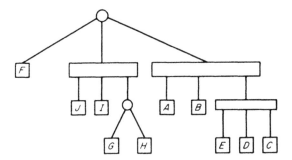

Figure 8.5. A *PQ*-tree equivalent to the tree in Figure 8.4.

Finally, any frontier obtainable from a tree equivalent with T is said to be *consistent* with T, and we define

$$\text{CONSISTENT}(T) = \{\text{FRONTIER}(T')| T' \equiv T\}.$$

It can be shown that the classes of consistent permutations of *PQ*-trees form a lattice. The null tree T_0 has no nodes and $\text{CONSISTENT}(T_0) = \emptyset$. The *universal tree* has one internal *P*-node, the *root*, and a leaf for every member of X (Figure 8.6).

Let us now relate *PQ*-trees to the consecutive arrangement problem. Let $\mathscr{I}$ be a collection of subsets of a set X, and let $\Pi(\mathscr{I})$ denote the collection of all permutations π of X such that the members of each subset $I \in \mathscr{I}$ occur consecutively in π. For example if $\mathscr{I} = \{\{A, B, C\}, \{A, D\}\}$, then $\Pi(\mathscr{I}) = \{[D \, A \, B \, C], [D \, A \, C \, B], [C \, B \, A \, D], [B \, C \, A \, D]\}$. We have the following important correspondence.

Theorem 8.6 (Booth and Leuker [1976]). (i) For every collection of subsets $\mathscr{I}$ of X there exists a *PQ*-tree T such that $\Pi(\mathscr{I}) = \text{CONSISTENT}(T)$.

(ii) For every *PQ*-tree T there exists a collection of subsets $\mathscr{I}$ such that $\Pi(\mathscr{I}) = \text{CONSISTENT}(T)$.

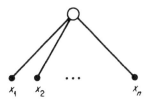

Figure 8.6. The universal tree T_u. CONSISTENT (T_u) includes all permutations of $X = \{x_1, x_2, \ldots, x_n\}$.

Note that the effect of Q-nodes is to restrict the number of permutations by making some of the brother relationships rigid. We leave it to the reader to verify that the tree in Figure 8.4 corresponds to the collection

$$\mathcal{I} = \{\{A, B\}, \{C, D\}, \{D, E\}, \{B, C, D, E\}, \{I, J\}, \{G, H\}, \{G, H, I\}\}.$$

The following procedure clearly calculates $\Pi(\mathcal{I})$.

```
procedure CONSECUTIVE(X, 𝒥):
begin
1.    Π ← {π|π is a permutation of X};
2.    for each I ∈ 𝒥 do
3.        Π ← Π ∩ {π|the members of I occur consecutively in π};;
4.    return Π;
end
```

Any naive implementation of this algorithm would be impractical because of the initially exponential size of Π. However, using PQ-trees we can represent Π with only $O(|X|)$ space. The equivalent program using PQ-trees is as follows.

```
procedure CONSECUTIVE(X, 𝒥):
begin
1.    T ← universal tree;
2.    for each I ∈ 𝒥 do
3.        T ← REDUCE(T, I);;
4.    return T;
end
```

This version makes use of a pattern matching routine **REDUCE** which attempts to apply a set of 11 *templates*. Each template consists of a *pattern* to be matched against the current PQ-tree and a *replacement* to be substituted for the pattern. The templates are applied from the bottom to the top of the tree. The null tree may be returned when no template applies. Two examples are illustrated in Figure 8.7. For details of the algorithm, the reader is directed to Booth and Leuker [1976]. There you will find the templates, a detailed version of the algorithm, a proof of correctness, and a proof of the following complexity theorem.

Theorem 8.7 (Booth and Leuker [1976]).   The class of permutations $\Pi(\mathcal{I})$ can be computed in $O(|\mathcal{I}| + |X| + \text{SIZE}|\mathcal{I}|)$ steps where $\text{SIZE}(\mathcal{I}) = \sum_{I \in \mathcal{I}} |I|$.

In the theorem the word *computed* means computed in its PQ-tree representation T. In the consecutive arrangements problem it is not necessary to calculate all of $\Pi(\mathcal{I})$. Rather, it is enough to produce one member of $\Pi(\mathcal{I})$ or to determine that $\Pi(\mathcal{I})$ is empty. This can be done very simply by calculating

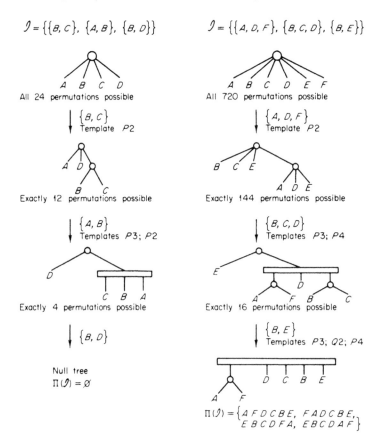

Figure 8.7. Two illustrations of the procedure CONSECUTIVE (*PQ*-tree version). The reductions make use of the templates in Booth and Leuker [1976].

FRONTIER(*T*). In the next section we suggest an application in which the permutations of $\Pi(\mathcal{I})$ may have to be compared according to secondary criteria.

Corollary 8.8. Let **M** be a (0, 1)-valued matrix with *m* rows, *n* columns, and *f* nonzero entries. Then, **M** can be tested for the consecutive 1's property in $O(m + n + f)$ steps.

Remark. If **M** is sparse ($f \ll mn$), then **M** would not be stored as an array. Rather, either a list of the nonzero entries or row lists of **M** would be used (Chapter 2).

$$
\begin{pmatrix}
1 & 0 & 0 & 1 & 1 & 1 \\
0 & 1 & 1 & 1 & 1 & 1 \\
1 & 1 & 1 & 0 & 0 & 1 \\
1 & 0 & 1 & 0 & 0 & 1 \\
0 & 1 & 0 & 1 & 1 & 1 \\
0 & 1 & 1 & 0 & 1 & 0
\end{pmatrix}
\rightarrow
\begin{pmatrix}
1 & 0 & 1 & 0 & 0 & 1 \\
1 & 0 & 0 & 1 & 1 & 1 \\
0 & 1 & 0 & 1 & 1 & 1 \\
0 & 1 & 1 & 1 & 1 & 1 \\
0 & 1 & 1 & 0 & 1 & 0 \\
1 & 1 & 1 & 0 & 0 & 1
\end{pmatrix}
\quad
\begin{pmatrix}
0 & 1 & 0 & 1 & 0 & 1 \\
1 & 1 & 0 & 0 & 1 & 0 \\
1 & 1 & 0 & 1 & 1 & 0 \\
1 & 1 & 1 & 1 & 1 & 0 \\
0 & 0 & 1 & 1 & 1 & 0 \\
0 & 0 & 0 & 1 & 1 & 0
\end{pmatrix}
$$
$$\mathbf{M}_1 \qquad\qquad\qquad \mathbf{M}_2 \qquad\qquad\qquad \mathbf{M}_3$$

Figure 8.8. Matrix $\mathbf{M}_1$ has the circular 1's property for columns since its rows can be permuted to yield $\mathbf{M}_2$. Matrix $\mathbf{M}_3$ does not have this property.

A (0, 1)-valued matrix has the *circular 1's property* for columns if its rows can be permuted in such a way that the 1's in each column occur in a circular consecutive order; regard the matrix as wrapped around a cylinder. In Figure 8.8 the matrix $\mathbf{M}_1$ has the circular 1's property since its rows can be permuted to obtain $\mathbf{M}_2$. However, $\mathbf{M}_3$ does not have the circular 1's property.

Remark 1.   **M** has circular 1's if and only if it has circular 0's.

Remark 2.   The circular 1's property is preserved under *complementation* of any column, i.e., interchanging ones and zeros.

The circular 1's property was introduced by Ryser [1969].

Clearly, consecutive 1's implies circular 1's, but not conversely. Nonetheless, one can verify the latter property using a test for the former, as follows.

Let **M** be a (0, 1)-valued matrix, and let **M′** be obtained from **M** by complementing those columns with a 1 in the kth row (k chosen arbitrarily).

Theorem 8.9 (Tucker [1970, 1971]).   Matrix **M** has the circular 1's property if and only if **M′** has the consecutive 1's property.

Proof.   By Remark 2, if **M** has the circular 1's property, then so does **M′**. By cyclically permuting the rows of **M′** so that the kth row (containing only zeros) is moved to the top, we shall obtain a matrix with consecutive 1's in each column. Conversely, if **M′** has the consecutive 1's property, then **M′** also has the circular 1's property. Hence, Remark 2 implies that **M** has the circular 1's property.   ∎

The efficiency of testing for circular 1's and consecutive 1's depends partly upon the sparseness of **M**. Thus, if **M** is sparse we shall want to choose k so that **M′** is also sparse. This can always be done provided **M** is stored as a list of its nonzero entries or by row lists.

Theorem 8.10 (Booth [1975]).   An $m \times n$ (0, 1)-valued matrix **M** with f nonzero entries can be tested for the circular 1's property in $O(m + n + f)$ steps.

Proof. Let **M** be given as a list L of its nonzero entries. Testing for circular 1's can be accomplished as follows.

Step I. Scan L once, setting up row lists for **M** and counting the number c_i of ones in each row i: $O(m + f)$.

Step II. Choose a row k having minimum number of 1's: $O(m)$.

Step III. Form **M**' by complementing the appropriate columns. This may be carried out by scanning each row in parallel with row k, or by using an auxiliary Boolean n-vector, as illustrated in Appendix B: $O(\sum_i(c_i + c_k)) = O(m + f)$.

Remark. **M**' has at most $2f$ nonzeros since each row is at most doubled in its number of ones.

Step IV. Test **M**' for consecutive 1's: $O(m + n + 2f) = O(m + n + f)$. ∎

We have seen that testing a given matrix **M** for the consecutive or circular 1's properties can be executed efficiently. It is natural to ask, if **M** does not satisfy one or both of these properties, whether certain columns of **M** can be deleted in order that the remaining matrix satisfies the property. In general this problem is very difficult to answer.

Theorem 8.11 (Booth [1975]). Let **M** be an $r \times c$ $(0, 1)$-valued matrix, and let k be an integer ($k < c$). Deciding whether or not there exists an $r \times k$ submatrix of **M** satisfying the consecutive 1's property (or the circular 1's property) is NP-complete.

A proof follows from Exercise 15.

Kou [1977] presents two other extensions of the consecutive 1's property which are also NP-complete:

(1) minimizing the number of consecutive blocks of 1's appearing in the columns;

(2) minimizing the number of times a row must be split into two pieces to obtain consecutive 1's.

4. Applications of Interval Graphs

Interval graphs are among the most useful mathematical structures for modeling real world problems. The line on which the intervals rest may represent anything that is normally regarded as one dimensional. The linearity may be due to *physical restriction*, such as blemishes on a micro-organism, speed traps on a highway, or files in sequential storage in a computer. It may arise from *time dependencies* as in the case of the life span of persons or cars, or jobs on a fixed time schedule. A *cost function* may be the reason as with the approximate worth of some fine wines or the potential for growth of a portfolio of securities. And so the list goes on.

The task to be performed on an interval graph will vary from problem to problem. If what is required is to find a coloring or a maximum weighted stable set or a big clique, then fast algorithms are available. If a Hamiltonian circuit must be found, then there are no known efficient algorithms (unless the graph has more structure than just being an interval graph). Also, the speed with which such a problem can be solved will depend partially on whether we are given simply the interval graph G or, in addition, an interval representation of G.

Let us direct our attention to a few interesting applications of interval graphs.

Application 8.1. Suppose $c_1, c_2, \ldots, c_n$ are chemical compounds which must be refrigerated under closely monitored conditions. If compound c_i must be kept at a *constant* temperature between t_i and t_i' degrees, how many refrigerators will be needed to store all the compounds?

Let G be the interval graph with vertices $c_1, c_2, \ldots, c_n$ and connect two vertices whenever the temperature intervals of their corresponding compounds intersect. By the Helly property (Section 4.5), if $\{c_{i_1}, c_{i_2}, \ldots, c_{i_k}\}$ is a clique of G, then the intervals $\{[t_{i_j}, t_{i_j}'] \mid j = 1, 2, \ldots, k\}$ will have a common point of intersection, say t. A refrigerator set at a temperature of t will be suitable for storing all of them. Thus, a solution to the minimization problem will be obtained by finding a minimum clique cover of G.

Application 8.2. Benzer's problem, as stated in the introduction to this chapter, asks if the subelements inside the gene are linked together in a linear arrangement. To answer this question data were gathered on mutations of the gene. For certain microorganisms a mutant form may be assumed to arise from the standard form by alteration of some *connected* part of the internal structure. By experiment it can be determined whether or not the blemished part of two mutant genes intersect. (We would hope to show that the blemished parts are linear.)

From a large collection of mutants we obtain the pairwise intersection data of their blemishes and consider its intersection graph G. Are the intersection data compatible with the hypothesis of linearity of subelements in the gene? Equivalently, is G an interval graph? A positive answer does not confirm linearity! However, if the data are correct, a negative answer definitely refutes the hypothesis. Benzer experimented on the virus Phage T4; his findings were consistent wih linearity (see Benzer [1959, 1962] and Roberts [1976]).

Cohen, Komlós, and Mueller [1979] have shown that the asymptotic probability $P_{n,e}$ that a random graph with n vertices and e edges is an interval graph satisfies

$$P_{n,e} \sim \exp(-\lambda)$$

for large n and e and not too large e^6/n^5 where $\lambda \approx 32e^6/3n^5$. From this result and from some Monte Carlo estimates, they suggest, "it appears that the chance that Benzer observed an interval graph by chance alone is nearly zero." For related results see Cohen [1968, 1978] and Hanlon [1979a, 1979b].

The phenomenon of overlap in biology has been brought to light again recently. Kolata [1977] surveys some of these developments. She writes,

> Since the early days of molecular biology, genes have been pictured as nonoverlapping sequences of DNA [within the chromosome]. Detailed studies of a few bacterial and viral genes confirmed this view, and most investigators did not question it. [Furthermore,] the hypothesis of non-overlapping genes is a keystone for many genetic theories. [However, recent evidence seems to suggest that] viral genes and possibly bacterial genes may overlap. None of the studies with bacteria provide incontrovertible evidence that genes overlap, but all suggest that this phenomenon occurs. [If overlapping genes do exist,] current views of gene organization and the control of gene expression, as well as views of the information content of DNA molecules and the effects of mutations in DNA, may have to be substantially revised. [*Science* **176**, 1187–1188 (1977), copyright 1977 by the American Association for the Advancement of Science.]

Application 8.3.   In archaeology *seriation* is the attempt to place a set of items in their proper chronological order. At the turn of the century, Flinders Petrie, a well-known archaeologist, formulated this problem, calling it "sequence dating," while studying 800 types of pottery found in 900 Egyptian graves. This problem has much in common with interval graphs and the consecutive 1's property. Let A be a set of artifacts (or aspects of artifacts) which have been discovered in various graves. To each artifact there ought to correspond a time *interval* (unknown to us) during which it was in use. To each grave there corresponds a *point* in time (also unknown) when its contents were interred. Our problem is to figure out these time relationships.

(a)   Consider the incidence matrix **M** whose rows represent the graves and whose columns represent the artifacts which either are or are not present in a given grave. Under the assumption that a grave contains *every*

member of A in use at the time of burial, the matrix **M** will have the consecutive 1's property for columns. Each permutation of the rows which gives
consecutive 1's corresponds to an acceptable seriation of the graves and
defines a possible interval assignment for A. Since there may be many of
these, other methods will also have to be used to further limit the possibilities.

   (b)   Consider the graph G whose vertices represent the artifacts with two
vertices being connected by an edge if their corresponding artifacts are found
in *some* common grave. Under the assumption that every pair of artifacts
whose usage intervals intersect are to be found together in some grave, we
have the G is an interval graph and any interval assignment for G would be a
candidate for the usage intervals of A. As before, additional techniques are
required to· choose the correct assignment. (See Kendall [1969a, 1969b],
Hodson, Kendall, and Tăutu [1971], and Roberts [1976].) One further
drawback to practical application is that there may be incomplete data so
that the assumptions are not satisfied.

   Application 8.4.   Let X represent a set of distinct data items (*records*) and
let $\mathscr{I}$ be a collection of subsets of X called *inquiries*. Can X be placed in
linear sequential storage in such a way that the members of each $I \in \mathscr{I}$ are
stored in consecutive locations? When this storage layout is possible, the
records pertinent to any inquiry can be accessed with two parameters, a
starting pointer and a length. Ghosh [1972, 1973] calls this the *consecutive
retrieval* property; it is clearly a restatement of the consecutive arrangement
property. Thus, the question can be answered efficiently using PQ-trees
(Section 8.3). For related results see Nakano [1973a, 1973b], Ghosh [1974,
1975], Waksman and Green [1974], Patrinos and Hakimi [1976], L. T.
Kou [1977], and Gupta [1979]. For an application of the circular 1's
property to cyclic staffing problems, see Bartholdi *et al.* [1977].

Commentary

   Application 8.5.   At the Typical Institute of Mathematical Sciences
(TIMS) each new faculty member visits the coffee lounge once during the
first day of the semester and meets everyone who is there at the time. How
can we assign the new faculty members to alcoves of the coffee lounge in
such a way that no one ever meets a new person during the entire remainder
of the semester? This is clearly a coloring problem on an interval graph. No
specific algorithm is needed, however, since it usually happens naturally.

   Additional applications of consecutive and circular 1's to such areas as
file organization and cyclic staffing appear in the bibliography at the end of
the chapter.

5. Preference and Indifference

Let V be a set. Let us assume that, for every pair of distinct members of V, a certain decision maker either clearly prefers one over the other or he feels indifferent about them. What is the nature of his preferences, and can they be quantified in an orderly manner? What does this imply about his decision processes?

We construct two graphs $H = (V, P)$ and $G = (V, E)$ as follows. For distinct $x, y \in V$,

$$xy \in P \Leftrightarrow x \text{ is preferred over } y,$$
$$xy \in E \Leftrightarrow \text{indifference is felt between } x \text{ and } y.$$

By definition, $H = (V, P)$ is an oriented graph, $G = (V, E)$ is an undirected graph, and $(V, P + P^{-1} + E)$ is complete.

What should we expect from the structure of H? If H has a cycle, then our decision maker is likely to be confused and is probably wasting time running around in circles.

Therefore, it is reasonable to require H to be acyclic. In fact, we would want H to be transitive. After all, if x is preferred over y and y is preferred over z, it is unlikely that a discriminating person would feel indifferent about x and z. Thus, we require that P be a partial order.*

Our example is not as whimsical as it may at first seem. The discussion above, and what will follow below, are important issues in decision theory and mathematical psychology. Analyzing how such preferences are made can enable us to understand and predict individual as well as group behavior. For example, how do we evaluate the decision making ability of a middle level corporate manager in order to determine if he is top management material?

The discipline of utility theory provides the mechanism for quantifying preference. One reasonable measure, due to Luce [1956], is the notion of a semiorder. We assign a real number $u(x)$ to each $x \in V$ so that for all x and y in V, x is preferred over y if and only if $u(x)$ is *sufficiently larger* than $u(y)$. Formally, letting $\delta > 0$, a real-valued function $u : V \to \mathbb{R}$ is called a *semiorder utility function* for a binary relation (V, P) if the following condition is satisfied:

$$xy \in P \Leftrightarrow u(x) \geq u(y) + \delta \qquad (x, y \in V). \tag{1}$$

---

* Krantz, Luce, Suppes, and Tversky [1971, p. 17] present an argument against transitivity of preference.

Clearly, a relation P satisfying (1) is a partial ordering of V. The quantity δ represents the amount of *fuzziness* that must be filtered out. This enables us to be indifferent about events that differ by a minuscule amount.

It is natural for us to ask the question, under what conditions does a preference relation (V, P) admit a semiorder utility function?

Theorem 8.12 (Scott and Suppes [1958]). There exists a semiorder utility function for a binary relation (V, P) if and only if the following conditions hold: For all $x, y, z, w \in V$,

    (S1) P is irreflexive;

    (S2) $xy \in P$ and $zw \in P$ imply $xw \in P$ or $zy \in P$.

    (S3) $xy \in P$ and $yz \in P$ imply $xw \in P$ or $wz \in P$.

Such a relation P is called a *semiorder*. The conditions (S1)–(S3) constitute a set of axioms for a semiorder.* Proof of the necessity of these three conditions is straightforward and is given as Exercise 9. For the sufficiency half of the theorem, the reader is directed to the constructive proof of Rabinovitch [1977] or to the existence proofs of Scott [1964] and Suppes and Zinnes [1963].

Dean and Keller [1968] prove that the number of nonisomorphic semiorders on an n-set is $\binom{2n}{n}/(n + 1)$. In particular, they show that each isomorphism class has a unique representative, called a *normal natural* partial order (NNPO), and they then demonstrate a one-to-one correspondence between (a) the NNPOs, (b) the normal subgroups of the upper triangular group of $n \times n$ matrices, and (c) the set of nondecreasing paths from $(0, 0)$ to (n, n) on a Cartesian grid which never rises above the line $x = y$. Rabinovitch [1978] shows that every semiorder may be expressed as the intersection of at most three linear orders. No similar result holds for orders satisfying only (S1) and (S2), i.e., *interval orders*. Jamison and Lau [1973] characterize the choice functions of semiorders. They also have a good table of references. For further investigation see the works of Fishburn [1970a–1970d, 1971, 1973, 1975] and the excellent book by Roberts [1979c].

Our attention has thus far been focused on semiorders from the standpoint of the preference relation (V, P). We now investigate the indifference relation $G = (V, E)$ of our semiorder (V, P). A number of characterizations are known for these undirected graphs. First, G is a special type of interval graph (Exercise 7). Second, a necessary condition easily follows from a semiorder utility function, namely, the existence of a real-valued function u on V satisfying

$$xy \in E \Leftrightarrow |u(x) - u(y)| < \delta \qquad (x \neq y).$$

---

* Unfortunately, the term semiorder was used in Ghouila-Houri [1962] and later in Berge [1973] in a different context.

We will see in the next theorem that this latter condition is also sufficient. The theorem provides a number of equivalent characterizations of *indifference graphs*, which are, simply stated, the class of cocomparability graphs of semiorders. Additional characterizations appear as Exercises 10 and 11.

Theorem 8.13 (Roberts [1969]).   Let $G = (V, E)$ be an undirected graph. The following conditions are equivalent.

(i)   There exists a real-valued function $u: V \to \mathbb{R}$ satisfying, for all distinct vertices $x, y \in V$,

$$xy \in E \Leftrightarrow |u(x) - u(y)| < 1.$$

(ii)   There exists a semiorder (V, P) such that $\bar{E} = P + P^{-1}$.

(iii) $\bar{G}$ is a comparability graph and every transitive orientation of $\bar{G} = (V, \bar{E})$ is a semiorder.

(iv) G is an interval graph containing no induced copy of $K_{1,3}$.

(v) G is a proper interval graph.

(vi) G is a unit interval graph.

Proof.   (i) $\Rightarrow$ (vi)   Let u be a real-valued function satisfying

$$xy \in E \Leftrightarrow |u(x) - u(y)| < 1 \qquad (x \neq y).$$

To each vertex $x \in V$ we associate the open interval $I_x = (u(x) - \frac{1}{2}, u(x) + \frac{1}{2})$. Clearly,

$$I_x \cap I_y \neq \varnothing \Leftrightarrow |u(x) - u(y)| < 1 \qquad (x \neq y).$$

Therefore, the collection $\{I_x\}_{x \in V}$ is a unit interval representation for the graph G.

(vi) $\Rightarrow$ (v)   Since no unit interval can properly contain another unit interval, a unit interval representation for G will be proper.

(v) $\Rightarrow$ (iv)   Let $\{I_x\}_{x \in V}$ be a proper interval representation of G. Suppose G contains an induced subgraph $G_{\{y, z_1, z_2, z_3\}}$ isomorphic to $K_{1,3}$ where $\{z_1, z_2, z_3\}$ is a stable set and y is adjacent to each z_i ($i = 1, 2, 3$). If I_{z_j} is that interval among the intervals $I_{z_1}, I_{z_2}, I_{z_3}$ which lies entirely between the other two, then I_y must properly contain I_{z_j}, a contradiction. Thus, G can have no induced copy of $K_{1,3}$.

(iv) $\Rightarrow$ (iii) (A. A. J. Marley [unpublished].)   Since G is an interval graph, its complement $\bar{G} = (V, \bar{E})$ is a comparability graph. Let F be a transitive orientation of $\bar{G}$. Using transitivity and Theorem 8.1 it is straightforward to show that F satisfies the axioms (S1) and (S2) of a semiorder (Exercise 7). We will show that (S3) also holds provided that G contains no induced copy of $K_{1,3}$. Suppose $xy \in F$ and $yz \in F$, while $xw \notin F$ and $wz \notin F$. By transitivity

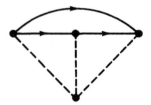

Figure 8.9. Solid edges are in the transitive orientation F of $\bar{G} = (V, \bar{E})$. Broken edges are in $G = (V, E)$.

of F, $wx \notin F$ and $zw \notin F$, and $wy \notin F$, $yw \notin F$, but $xz \in F$ (see Figure 8.9). Therefore, $G_{\{x, y, z, w\}}$ is isomorphic to $K_{1,3}$, a contradiction.

(iii) $\Rightarrow$ (ii)   Immediate.

(ii) $\Rightarrow$ (i)   If (V, P) is a semiorder, then there exists a real-valued function $u' : V \to \mathbb{R}$ and a number $\delta > 0$ such that

$$xy \in P \Leftrightarrow u'(x) - u'(y) \geq \delta.$$

Define $u(x) = u'(x)/\delta$. Since $P + P^{-1} = \bar{E}$, clearly

$$xy \in E \neq |u(x) - u(y)| < 1 \qquad (x \neq y). \qquad \blacksquare$$

6.   Circular-Arc Graphs

The intersection graphs obtained from collections of arcs on a circle are called *circular-arc graphs*. A circular-arc representation of an undirected graph G which fails to cover some point p on the circle will be topologically the same as an interval representation of G. Specifically, we can cut the circle at p and straighten it out to a line, the arcs becoming intervals. It is easy to see, therefore, that every interval graph is a circular-arc graph. The converse, however, is false. In fact, circular-arc graphs are, in general, not perfect graphs. For example, the chordless cycles C_5, C_7, C_9, ... are circular-arc graphs (see Figure 8.10).

As with interval graphs, it is immaterial whether we choose open arcs or closed arcs. The same class of intersection graphs will arise in either case (Exercise 13). We shall adopt the convention of open arcs. We call G a *proper circular-arc graph* if there exists a circular-arc representation for G in which no arc properly contains another.

In Section 1.2 we discussed an application of circular-arc graphs to the traffic light phasing problem due to Stoffers [1968]. The astute reader may well be able to adapt some of the applications of interval graphs given in Section 8.4 to the more general class of circular-arc graphs. Stahl [1967]

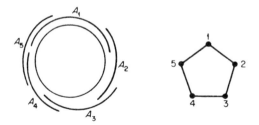

Figure 8.10. A circular-arc representation of the nonperfect graph C_5.

suggests such a problem in genetics. Other relevant papers on applications of circular-arc graphs include Luce [1971], Hubert [1974], Tucker [1978], and Trotter and Moore [1979].

A characterization of circular-arc graphs due to Tucker, originally formulated in terms of the augmented adjacency matrix of a graph, is equivalent to the following.

Theorem 8.14 (Tucker [1970b, 1971]).   An undirected graph $G = (V, E)$ is a circular-arc graph if and only if its vertices can be (circularly) indexed $v_1, v_2, \ldots, v_n$ so that for all i and j

$$v_i v_j \in E \Rightarrow \begin{cases} \text{either} & v_{i+1}, \ldots, v_j \in \mathrm{Adj}(v_i) \\ \text{or} & v_{j+1}, \ldots, v_i \in \mathrm{Adj}(v_j). \end{cases} \tag{2}$$

(If $i < j$, then $v_{j+1}, \ldots, v_i$ means $v_{j+1}, \ldots, v_n, v_1, \ldots, v_i$.)

Proof. Let G have a circular-arc representation (open arcs). We may assume, without loss of generality, that no pair of arcs share a common endpoint (Exercise 14). Moving clockwise once around the circle from an arbitrary starting point, index the vertices according to the order in which the counterclockwise endpoints of their corresponding arcs occur. Let A_i denote arc corresponding to v_i. Clearly, v_i is adjacent to v_j if and only if the counterclockwise endpoint of A_j lies within A_i or vice versa. In the former case, each of $A_{i+1}, \ldots, A_j$ intersects A_i, and in the latter case each of $A_{j+1}, \ldots, A_i$ intersects A_j. Thus (2) is satisfied.

Conversely, let the vertices be indexed as required in (2). We will construct a circular-arc representation for G. Let p_k be the kth hour marker on an n-hour clock. For each vertex v_i, let v_{m_i} be the first vertex in the cyclic sequence $v_{i+1}, v_{i+2}, \ldots, v_i$ which is *not* adjacent to v_i. Draw an open arc A_i clockwise from p_i to p_{m_i}. By construction, A_i intersects A_j ($i \neq j$) if and only if either $p_j \in A_i$ or $p_i \in A_j$. But also

$$p_j \in A_i \Leftrightarrow v_{i+1}, \ldots, v_j \in \mathrm{Adj}(v_i).$$

Therefore, by (2), $v_i v_j \in E$ if and only if $A_i \cap A_j \neq \varnothing$. ∎

Theorem 8.14 gives us a method for recognizing circular-arc graphs and constructing a circular-arc representation. However, since the characterization is quantified over all permutations of the vertices, this method will be impractical for all but very small graphs. Tucker [1978] approaches the problem of trying to find a more efficient recognition algorithm. Details of a polynomial time algorithm will appear in Tucker [1979].

In view of Theorem 8.3 it is tempting to guess that a circular-arc graph is characterized by the circular 1's property of its clique matrix or some other matrix. Unfortunately, this is not the case. Three related theorems, however, are stated here without proof.

We call G a *Helly circular-arc graph* if there exists a circular-arc representation for G which satisfies the Helly property.

Theorem 8.15 (Gavril [1974]).   An undirected graph G is a Helly circular-arc graph if and only if its clique matrix has the circular 1's property for columns.

The *augmented adjacency matrix* of G is obtained from the adjacency matrix by adding 1's along the main diagonal.

Theorem 8.16 (Tucker [1970b, 1971]).   An undirected graph G is a circular-arc graph if its augmented adjacency matrix has the circular 1's property for columns.

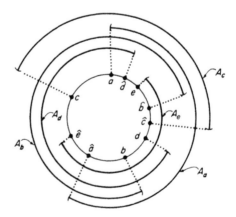

Figure 8.11. A collection of arcs of a circle with representing sequence of endpoints $\sigma = [a, \hat{d}, e, \hat{b}, \hat{c}, d, b, \hat{a}, \hat{e}, c]$.

Theorem 8.17 (Tucker [1970b, 1971]).   An undirected graph G is a proper circular-arc graph if and only if its augmented adjacency matrix has the circular 1's property for columns and, for every permutation of the rows and columns that is a cyclic shift or inversion of their circular 1's order, the last 1 in the first column does not occur after the last 1 of the second column, excluding columns which are either all zeros or all ones.

It is useful to regard a collection of arcs as a sequence σ of its endpoints listed clockwise. Without loss of generality, we shall assume that no two arcs share a common endpoint (Exercise 14). In σ the symbol x denotes the counterclockwise endpoint of arc A_x and $\hat{x}$ denotes its clockwise endpoint. For example, $\sigma = [a, \hat{a}, e, \hat{b}, \hat{c}, d, b, \hat{a}, \hat{e}, c]$ represents the collection of arcs in Figure 8.11. Any cyclic permutation of σ would be an equally valid representation. The manner in which two arcs A_x and A_y intersect is uniquely determined by the pattern of the subsequence of σ involving $\{x, \hat{x}, y, \hat{y}\}$. Some examples are shown in Table 8.1. We shall utilize this model in proving the next theorem.

Theorem 8.18.   If G is a proper circular-arc graph, then G has a proper circular-arc representation in which no two arcs share a common endpoint and no two arcs together cover the entire circle (i.e., they do not intersect at both ends).

Proof.   The proof will be induction on the number of "circle covering" pairs of arcs. Let $\mathscr{A} = \{A_x\}_{x \in V}$ be a proper circular-arc representation of $G = (V, E)$. We may assume that no two arcs share a common endpoint. Suppose A_a and A_b cover the entire circle, that is, they intersect in two

Table 8.1

Coding a family of arcs as a sequence of letters*

Pattern of subsequence	Interpretation
$[x, \hat{x}, y, \hat{y}]$	$A_x \cap A_y = \varnothing$
$[\hat{y}, y, x, \hat{x}]$	$A_x \subset A_y$
$[x, y, \hat{x}, \hat{y}]$	A_x and A_y overlap at one end
$[x, \hat{y}, y, \hat{x}]$	A_x and A_y overlap at both ends

* Some examples of how the pattern of the subsequence of σ involving $\{x, \hat{x}, y, \hat{y}\}$ determines the manner of intersection of arcs A_x and A_y. Any cyclic permutation of a pattern leaves the interpretation unchanged.

disjoint subarcs. Let σ be the sequence of endpoints of the arcs going clockwise from the counterclockwise endpoint of A_a. Thus, $[a, \hat{b}, b, \hat{a}]$ is the subsequence of σ involving these letters, and σ may be expressed as the concatenation $\sigma = \tau\rho$, where

$$\sigma = [\underbrace{a, \ldots, \hat{b}}_{\tau}, \underbrace{\ldots, b, \ldots, \hat{a}, \ldots}_{\rho}].$$

For any $x \in V$, it is impossible for x and $\hat{x}$ to appear in τ in the order $[x, \hat{x}]$ since such an appearance would imply $A_x \subseteq A_a$, contradicting the supposition that $\mathscr{A}$ is proper.

Consider the new sequence $\sigma' = \tau'\rho$, where τ' is obtained from τ by listing those entries of τ with hats followed by those without hats but preserving the relative order of each type. For any $x, y \in V$, this unshuffling operation will leave unchanged the subsequence of σ involving $\{x, \hat{x}, y, \hat{y}\}$ *unless* either $[x, \hat{y}]$ or $[y, \hat{x}]$ is a subsequence of τ. Since the cases are analogous, we assume that $[x, \hat{y}]$ is in τ. We allow the possibility that x equals a or that y equals b. Now $\hat{x}$ may either precede x or follow $\hat{a}$ in σ, and y must fall between $\hat{y}$ and b, for otherwise $\mathscr{A}$ would not be proper. This situation is depicted in Figure 8.12. Clearly, either $[\hat{x}, x, \hat{y}, y]$ or $[x, \hat{y}, y, \hat{x}]$ is a subsequence of σ, indicating that A_x and A_y overlap at both ends. After the transformation from σ or σ' occurs, these become, respectively, $[\hat{x}, \hat{y}, x, y]$ or $[\hat{y}, x, y, \hat{x}]$, which correspond to arcs which properly overlap at only one end.

Let $\mathscr{A}'$ be a set of arcs corresponding to σ'. We have just shown that (i) some doubly intersecting arcs in $\mathscr{A}$ are transformed into singly intersecting arcs in $\mathscr{A}'$, and (ii) all other pairs in $\mathscr{A}$, including nonintersecting, singly intersecting, the remaining doubly intersecting, and properly contained (of which there were none) arcs, were left unchanged. Thus, $\mathscr{A}'$ is a proper circular-arc representation of G with fewer circle covering pairs, and the theorem follows by induction. ∎

This theorem was used in Section 1.2 to show that every proper circular-arc graph is also the graph of intersecting chords of a circle. (See also Chapter 11.)

We conclude this section by presenting a polynomial-time algorithm which finds a maximum clique of a circular-arc graph. The algorithm appears in

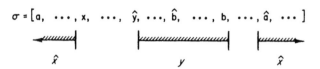

Figure 8.12.   A view of where in σ the letters $\hat{x}$ and y could be hiding.

Gavril [1974] along with efficient solutions of the stable set problem and the clique cover problem for circular-arc graphs. The complexity of the coloring problem is unknown for these graphs.

Let $\{A_x\}_{x \in V}$ be a circular-arc representation of $G = (V, E)$. Assume that no two arcs share an endpoint. Denote the counterclockwise and clockwise endpoints of A_x by $\tilde{x}$ and $\hat{x}$, respectively. For $x \in V$, we define

$$Y_x = \{v \in V \mid \tilde{x} \in A_v\} + \{x\},$$
$$Z_x = \{v \in V - Y_x \mid \hat{x} \in A_v\}.$$

Each of Y_x and Z_x are complete sets, so the induced subgraph $G_{Y_x + Z_x}$ is the complement of a bipartite graph. Thus, finding a maximum clique of $G_{Y_x + Z_x}$ can be done in polynomial time.

Let K be a maximum clique of G. Choose a vertex $x \in K$ such that A_x does not properly contain any arc A_w $(w \in K)$. Hence, for every $w \in K$, $x \neq w$, either $\tilde{x} \in A_w$ or $\hat{x} \in A_w$. Therefore, K is a clique of $G_{Y_x + Z_x}$.

A maximum clique of G can be obtained as follows: For each $x \in V$, construct and find a maximum clique $K(x)$ of $G_{Y_x + Z_x}$; then select the largest among the $K(x)$.

EXERCISES*

1. Discuss how interval graphs and the consecutive 1's property could be applied to the following problem. Several psychological traits are to be examined in children. Assign an age period to each trait representing the natural order in the development process during which the trait is present. What traits would be appropriate for such a study?

2. Let **M** be a symmetric $(0, 1)$-valued matrix. Prove that either **M** has the consecutive 1's property for columns and rows or **M** has neither property. Prove the same result for circular 1's.

3. Prove that the clique matrix of an interval graph $G = (V, E)$ has at most $O(|V| + |E|)$ nonzero entries. Is this equally true for triangulated graphs (Fulkerson and Gross [1965])?

4. Let $\mathscr{I}$ be a family of intervals on a line, and let k be the maximum possible number of pairwise disjoint intervals in $\mathscr{I}$. Prove that there exist k points on the line such that each interval contains at least one of these points (T. Gallai).

5. Let **A** and **B** be $(0, 1)$-valued matrices having the same shape. Prove that if $\mathbf{A}^T \mathbf{A} = \mathbf{B}^T \mathbf{B}$, then either both **A** and **B** have the consecutive 1's property for columns or neither has it (Fulkerson and Gross [1965, Theorem 2.1]). A

---

* Also review the exercises from Chapter 1.

stronger version of this is the following: If $A^TA = B^TB$ and A has no sub-configuration of either of the forms below, then $A = PB$ for some permutation matrix P

$$\begin{pmatrix} 1 & 1 & 1 \\ 1 & 0 & 0 \\ 0 & 1 & 0 \\ 0 & 0 & 1 \end{pmatrix}, \quad \begin{pmatrix} 0 & 0 & 0 \\ 0 & 1 & 1 \\ 1 & 0 & 1 \\ 1 & 1 & 0 \end{pmatrix}$$

(Ryser [1969, Theorem 4.1]).

6. Let P be a binary relation on a set V. A real-valued function $u: V \to \mathbb{R}$ is called an *ordinal utility function* for (V, P) if

$$xy \in P \Leftrightarrow u(x) > u(y).$$

(a) Show that (V, P) admits an ordinal utility function if and only if P is irreflexive, antisymmetric, and satisfies the negative transitivity condition (transitive indifference),

$$xy \notin P, \ yz \notin P \Rightarrow xz \notin P.$$

A (preference) relation satisfying the conditions in (a) is called a *weak order* in decision theory and a *preorder* in some mathematics literature. An ordinal utility function is like scores on an exam; this makes a weak ordering almost a total ordering (ties being allowed). Armstrong [1950] first observed that transitive indifference has important empirical shortcomings in a preference model.* To resolve these shortcomings, Luce [1956] introduced semiorders.

(b) What is the structure of the indifference graphs of weak orders?

7. Let $G = (V, E)$ be an undirected graph. Prove that the following conditions are equivalent:

(i) G is an interval graph.

(ii) $\bar{G}$ has a transitive orientation P satisfying axioms (S1) and (S2) of a semiorder.

(iii) Every transitive orientation P of $\bar{G}$ satisfies (S1) and (S2).

8. Consider the lexicographic ordering of the plane: A point (x, y) is strictly less than a point (x', y') if either $x < x'$ or both $x = x'$ and $y < y'$. Clearly for every pair of distinct points, one of them is strictly less than the other. Prove that there cannot exist a real-valued function f defined on the points of the plane which preserves the lexicographic ordering (i.e., $f(x, y) < f(x', y') \Leftrightarrow (x, y) < (x', y')$) (Debreu [1954]).

---

* He wrote,

  The nontransitiveness of indifference must be recognized and explained on [sic] any theory of choice, and the only explanation that seems to work is based on the imperfect powers of discrimination of the human mind whereby inequalities become recognizable only when of sufficient magnitude [1950, p. 122].

9. Prove the necessity half of the Scott–Suppes theorem.

10. Prove that G is a proper interval graph if and only if its augmented adjacency matrix satisfies the consecutive 1's property for columns (Roberts [1968]). (The *augmented adjacency matrix* of G is obtained from the adjacency matrix by adding 1's along the main diagonal.)

11. We say that vertices a and b are *equivalent*, denoted $a \approx b$, if their neighborhoods $N(a)$ and $N(b)$ are equal. A vertex x is called *extreme* if $N(x)$ is complete (i.e., x is a simplicial vertex) and $[a, b \in N(x), a \not\approx x, b \not\approx x]$ implies $[\exists z \in N(a) \cap N(b), z \notin N(x)]$ (see Figure 8.13). Finally, let G^* be the quotient graph obtained from G by coalescing the vertices of each $\approx$-equivalence class and preserving the adjacencies between classes.

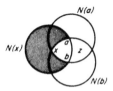

Figure 8.13. The shaded area is empty.

Prove that the following conditions are equivalent to those in Theorem 8.13 for an indifference graph $G = (V, E)$.

(vii) For every connected, induced subgraph H of G, either H^* has exactly one vertex (i.e., H is complete) or H^* has exactly two (nonadjacent) extreme vertices (Roberts [1969]).

(viii) G is triangulated and contains none of the graphs in Figure 8.14 as induced subgraph (Wegner [1967]).

Figure 8.14. Forbidden subgraphs.

12. Let $\mathbf{M} = [m_{ij}]$ be an incidence matrix and define the row sets A_i and column sets B_j as follows:

$$A_i = \{j \mid m_{ij} = 1\}, \qquad B_j = \{i \mid m_{ij} = 1\}.$$

(a) Show that the following are equivalent.

(i) *Row intersection property*: For every i, j, k,

$$A_i \cap A_j \subseteq A_k \quad \text{or} \quad A_j \cap A_k \subseteq A_i \quad \text{or} \quad A_k \cap A_i \subseteq A_j.$$

(ii)   *Column intersection property:* For every i, j, k,

$$B_i \cap B_j \subseteq B_k \quad \text{or} \quad B_j \cap B_k \subseteq B_i \quad \text{or} \quad B_k \cap B_i \subseteq B_j$$

(Nakano [1973b]).

(b)   The matrix $\mathbf{M}$ is *closed* if $B_i \cap B_j \neq \varnothing$ implies $B_i \cup B_j = B_k$ for some column set B_k of $\mathbf{M}$. The closure $cl(\mathbf{M})$ of $\mathbf{M}$ is defined by adding columns to $\mathbf{M}$ inductively: $\mathbf{M}^{(0)} = \mathbf{M}$; $\mathbf{M}^{(n)}$ has column sets $B_i^{(n-1)} \cup B_j^{(n-1)}$ for all $B_i^{(n-1)}$ and $B_j^{(n-1)}$ of $\mathbf{M}^{(n-1)}$ satisfying $B_i^{(n-1)} \cap B_j^{(n-1)} \neq \varnothing$.

Prove that $\mathbf{M}$ has the consecutive 1's property for columns if and only if $cl(\mathbf{M})$ has the column intersection property (Nakano [1973a]).

13.   Let $\mathscr{A} = \{A_x\}_{x \in V}$ be a finite collection of closed arcs of a circle. Show that there exists another collection $\mathscr{A}' = \{A'_x\}_{x \in V}$ of open arcs such that

(i) $A_x \cap A_y = \varnothing \Leftrightarrow A'_x \cap A'_y = \varnothing$,

(ii) $A_x \subseteq A_y \Leftrightarrow A'_x \subseteq A'_y$ $(x, y \in V)$.

14.   Let $\mathscr{A} = \{A_x\}_{x \in V}$ be a finite collection of open arcs of a circle. Show that there exists another collection $\mathscr{A}' = \{A'_x\}_{x \in V}$ of open arcs of a circle satisfying the following for all $x, y \in V$:

(i) $A_x \cap A_y = \varnothing \Leftrightarrow A'_x \cap A'_y = \varnothing$;

(ii) $A_x \subseteq A_y \Leftrightarrow A'_x \subseteq A'_y$;

(iii)   no two arcs of $\mathscr{A}'$ have a common endpoint.

15.   Let $\mathbf{A}$ be the $n \times m$ incidence matrix (vertices-versus-edges) of an undirected graph G.

(a)   Prove that G has a Hamiltonian circuit if and only if $\mathbf{A}$ has an $n \times n$ submatrix satisfying the circular 1's property.

(b)   Prove that G has a Hamiltonian path if and only if $\mathbf{A}$ has an $n \times (n-1)$ submatrix satisfying the consecutive 1's property (Booth [1975]).

16.   Give an example of a circular-arc graph whose augmented adjacency matrix does not have the circular 1's property.

17.   A graph G is a *unit circular-arc graph* if there exists a circular-arc representation for G in which every arc is of unit length. (The diameter of the circle is variable.) Verify that the graph in Figure 8.15 is a proper circular-arc graph but is not a unit circular-arc graph (Tucker [1974]). (Here we assume that all arcs are open or all arcs are closed.)

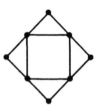

Figure 8.15.

18. Let G be a circular-arc graph. Show that $\chi(G) \leq 2\omega(G)$ (Tucker [1974]).
19. A matrix $\mathbf{M}$ is said to have the *unimodular* property if every square submatrix of $\mathbf{M}$ has determinant equal to 0, $+1$, or -1. (Every entry of such a matrix is necessarily 0, $+1$, or -1.) Show that any $(0, 1)$-valued matrix satisfying the consecutive 1's property is unimodular.

Research Problem. Let $\mathscr{I}$ be a collection of intervals whose intersection graph is G, and let $IC(\mathscr{I})$ denote the number of different sized intervals in $\mathscr{I}$. Define the *interval count* of G, denoted $IC(G)$, to be $\min\{IC(\mathscr{I}) | \mathscr{I}$ is an interval representation of $G\}$. Clearly, $IC(G) = 1$ iff G is a unit interval graph. Also, $IC(K_{1,3}) = 2$.
   (i)   For any $k \geq 2$, characterize all graphs G with $IC(G) = k$.
   (ii) Find good upper and lower bounds for $IC(G)$.
Leibowitz [1978] has constructed graphs of interval count k for all integers k. She has also found three classes of graphs with interval count 2, namely, trees that are interval graphs, interval graphs with a vertex whose removal leaves a unit interval graph, and threshold graphs.

Research Problem. Define the *interval number* of $G = (V, E)$, denoted $IN(G)$, to be the minimum number t for which there exists a collection of sets $\mathscr{U} = \{U_v\}_{v \in V}$, where U_v is the union of t (not necessarily disjoint) intervals on the real line, such that G is the intersection graph of $\mathscr{U}$, i.e., $xy \in E$ iff $U_x \cap U_y \neq \varnothing$. Clearly, $IN(G) = 1$ iff G is an interval graph. Also, any circular-arc graph has interval number at most 2.
   (i)   For any $k \geq 2$, characterize the graphs G with $IN(G) = 2$.
   (ii) Calculate the interval numbers of some special classes of graphs.
   (iii) What are the best bounds for $IN(G)$?
Trotter and Harary [1979] and Griggs and West [1979] have shown that the interval number of a tree is at most 2 and that

$$IN(K_{m,n}) = \lceil (mn + 1)/(m + n) \rceil.$$

Griggs and West have also shown that $IN(G) \leq \lceil (\delta + 1)/2 \rceil$, where δ is the maximum degree of a vertex, with equality holding for triangle-free regular graphs. Griggs [1979] has proven that $IN(G) \leq \lceil (n + 1)/4 \rceil$ for all n-vertex graphs.

Bibliography

Abbott, Harvey, and Katchalski, Meir
   [1979] A Turán type problem for interval graphs, *Discrete Math.* **25**, 85–88.
Armstrong, W. E.
   [1950] A note on the theory of consumer's behavior, *Oxford Econ. Papers* **2**, 119–122.

Bartholdi, John J., III, Orlin, James B., and Ratliff, H. Donald
  [1977] Circular ones and cyclic staffing, Tech. Report No. 21, Dept. of Oper. Res., Stanford
          Univ.
Berge, Claude
  [1973] "Graphs and Hypergraphs," Chapter 16. North-Holland, Amsterdam.
          MR50 #9640.
Benzer, S.
  [1959] On the topology of the genetic fine structure, *Proc. Nat. Acad. Sci. U.S.A.* **45**,
          1607–1620.
  [1962] The fine structure of the gene, *Sci. Amer.* **206**, 70–84.
Booth, Kellogg S.
  [1975] PQ-tree algorithms, Ph.D. thesis, Univ. of California. (Also available as UCRL-
          51953, Lawrence Livermore Lab., Livermore, California, 1975.)
Booth, Kellogg, S., and Leuker, George S.
  [1975] Linear algorithms to recognize interval graphs and test for the consecutive ones
          property, *Proc. 7th ACM Symp. Theory of Computing*, 255–265.
  [1976] Testing for the consecutive ones property, interval graphs, and graph planarity using
          PQ-tree algorithms, *J. Comput. Syst. Sci.* **13**, 335–379.
Cohen, Joel E.
  [1968] Interval graphs and food webs: a finding and a problem, RAND Document 17696-PR.
  [1978] "Food Webs and Niche Space." Princeton Univ. Press, Princeton, New Jersey.
Cohen, Joel E., Komlós, János, and Mueller, Thómas
  [1979] The probability of an interval graph, and why it matters, "Proc. Symp. on Relations
          Between Combinatorics and Other Parts of Mathematics" (D. K. Ray-Chaudhuri,
          ed.). Amer. Math. Soc., Providence, Rhode Island.
Dean, Richard A., and Keller, Gordon
  [1968] Natural partial orders, *Canad. J. Math.* **20**, 535–554. MR37 #1279.
Debreu, Gerard
  [1954] Representation of a preference ordering by a numerical function, *in* "Decision
          Processes" (R. M. Thrall, C. H. Coombs, and R. L. Davis, eds.), pp. 159–165.
          Wiley, New York. MR16, p. 606.
          Proves that there does *not* always exist a real-value order preserving function for an
          uncountable totally ordered set.
Eswaran, Kapali P.
  [1975] Faithful representation of a family of sets by a set of intervals, *SIAM J. Comput.* **4**,
          56–68. MR51 #14677.
          Representation is not the intersection model. It is another use of the consecutive 1's
          property.
Fishburn, Peter C.
  [1970a] An interval graph is not a comparability graph, *J. Combin. Theory* **8**, 442–443.
          MR42 #7541.
  [1970b] Intransitive indifference with equal indifference intervals, *J. Math. Psych.* **7**, 144–149.
          MR40 #7155.
  [1970c] "Utility Theory for Decision Making." Wiley, New York. MR41 #9401.
  [1970d] Intransitive indifference in preference theory: A survey, *Oper. Res.* **18**, 207–228.
          MR41 #8050.
  [1971] Betweenness, orders and interval graphs, *J. Pure Appl. Algebra* **1**, 159–178.
          MR47 #1689.
  [1973] Interval representations for interval orders and semi-orders, *J. Math. Psych.* **10**,
          91–105. MR47 #4870.
  [1975] Semiorders and choice functions, *Econometrica* **43**, 975–977. MR55 #14082.

Fulkerson, D. R., and Gross, O. A.
[1964] Incidence matrices with the consecutive 1's property, *Bull. Amer. Math. Soc.* **70**, 681–684. MR32 #7444.
[1965] Incidence matrices and interval graphs, *Pacific J. Math.* **15**, 835–855. MR32 #3881.
Gavril, Fanica
[1974] Algorithms on circular-arc graphs, *Networks* **4**, 357–369. MR51 #12614.
Ghosh, Sakti P.
[1972] File organization: the consecutive retrieval property, *Comm. Assoc. Comput. Mach.* **15**, 802–808. Zbl246 #68004.
[1973] On the theory of consecutive storage of relevant records, *J. Inform. Sci.* **6**, 1–9. MR48 #1551.
[1974] File organization: consecutive storage of relevant records on a drum-type storage, *Inform. Control* **25**, 145–165. MR49 #6706.
[1975] Consecutive storage of relevant records with redundancy, *Comm. Assoc. Comput. Mach.* **18**, 464–471. MR52 #4743.
Ghouilà- Houri, Alain
[1962] Caractérisation des graphes non orientés dont on peut orienter les arrêtes de manière à obtenir le graphe d'une relation d'ordre, *C.R. Acad. Sci. Paris* **254**, 1370–1371. MR30 #2495.
Gilmore, Paul C., and Hoffman, Alan J.
[1964] A characterization of comparability graphs and of interval graphs, *Canad. J. Math.* **16**, 539–548; abstract in *Int. Congr. Math.* (Stockholm) (1962), 29 (A) MR31 #87.
Griggs, Jerrold R.
[1979] Extremal values of the interval number of a graph, II, *Discrete Math.* **28**, 37–47.
Griggs, Jerrold R., and West, Douglas B.
[1979] Extremal values of the interval number of a graph, *SIAM J. Algebraic Discrete Meth.*, to be published.
Gupta, U.
[1979] Bounds on storage for consecutive retrieval, *J. Assoc. Comput. Mach.* **26**, 28–36.
Hadwiger, H., Debrunner, H., and Klee, V.
[1964] "Combinatorial Geometry in the Plane," p. 54. Holt, New York. M22 #11310; MR29 #1577.
    The problem of characterizing circular-arc graphs was posed here.
Hajös, G.
[1957] Über eine Art von Graphen, *Intern. Math. Nachr.* **11**, Problem 65.
    First posed the problem of characterizing interval graphs.
Hanlon, Phil
[1979a] Counting interval graphs, submitted for publication.
[1979b] The asymptotic number of unit interval graphs, submitted for publication.
Hodson, F. R., Kendall, D. G., and Tăutu, P., eds.
[1971] "Mathematics in the Archaeological and Historical Sciences." Edinburgh Univ. Press, Edinburgh.
Hubert, Lawrence
[1974] Some applications of graph theory and related non-metric techniques to problems of approximate seriation: The case of symmetric proximity measures, *British J. Math. Statist. Psych.* **27**, 133–153.
Jamison, Dean T., and Lau, Lawrence J.
[1973] Semiorders and the theory of choice, *Econometrica* **41**, 901–912; corrections **43** (1975), 979–980. MR55 #14081.
    See also Fishburn [1975].

Jean, Michel
[1969] An interval graph is a comparability graph, *J. Combin. Theory*, 189–190. MR39 #4036.
Result is false; see Fishburn [1970a].

Kendall, D. G.
[1969a] Incidence matrices, interval graphs, and seriation in archaeology, *Pacific J. Math.* **28**, 565–570. MR39 #1344.
[1969b] Some problems and methods in statistical archaeology, *World Archaeol.* **1**, 68–76. The consecutive 1's property is applied to sequence dating.

Klee, Victor
[1969] What are the intersection graphs of arcs in a circle? *Amer. Math. Monthly* **76**, 810–813.

Kolata, Gina Bari
[1977] Overlapping genes: more than anomalies? *Science* **176**, 1187–1188.

Kotzig, Anton
[1963] Paare Hajóssche Graphen, *Časopis Pěst. Mat.* **88**, 236–241. MR27 #2971.
Studies bipartite interval graphs.

Kou, L. T.
[1977] Polynomial complete consecutive information retrieval problems, *SIAM J. Comput.* **6**, 67–75. MR55 #7006.

Krantz, D. H., Luce, R. D., Suppes, P., and Tversky, A.
[1971] "Foundation of Measurement," Vol. I. Academic Press, New York.

Leibowitz, Rochelle
[1978] Interval counts and threshold graphs, Ph.D. thesis, Rutgers Univ.

Lekkerkerker, C. G., and Boland, J. Ch.
[1962] Representation of a finite graph by a set of intervals on the real line, *Fund. Math.* **51**, 45–64. MR25 #2596.

Leuker, G. S.
[1975] Interval graph algorithms, Ph.D. thesis, Princeton Univ.

Lipski, W., Jr., and Nakano, T.
[1976/1977] A note on the consecutive 1's property (infinite case), *Comment. Math. Univ. St. Paul.* **25**, 149–152. MR55 #12531.

Luce, R. Duncan
[1956] Semiorders and a theory of utility discrimination, *Econometrica* **24**, 178–191. MR17, p. 1222.
[1971] Periodic extensive measurement, *Compositio Math.* **23**, 189–198.

Mirkin, B. G.
[1972] Description of some relations on the set of real-line intervals, *J. Math. Psych.* **9**, 243–252. MR47 #4893.

Nakano, Takeo
[1973a] A characterization of intervals; the consecutive (one's or retrieval) property, *Comment. Math. Univ. St. Paul.* **22**, 49–59. MR48 #5871.
[1973b] A remark on the consecutivity of incidence matrices, *Comment. Math. Univ. St. Paul.* **22**, 61–62. MR49 #86.

Orlin, James B.
[1979a] Circular ones and cyclic capacity scheduling, Res. Report, Dept. of Operations Research, Stanford Univ.
[1979b] Coloring periodic interval graphs, Res. Report, Dept. of Operations Research, Stanford Univ.

Patrinos, A. N., and Hakimi, S. L.
[1976] File organization with consecutive retrieval and related properties, *in* "Large Scale Dynamical Systems" (R. Sacks, ed.). Point Lobos, North Hollywood, California.

Propp, James
   [1978]   A greedy solution for linear programs with circular ones, I.B.M. Res. Report
            RC 7421.

Rabinovitch, Issie
   [1977]   The Scott–Suppes theorem on semiorders, *J. Math. Psych.* **15**, 209–212.
            MR55 #10334.
   [1978]   The dimension of semiorders, *J. Combin. Theory A* **25**, 50–61.

Renz, Peter L.
   [1970]   Intersection representations of graphs by arcs, *Pacific J. Math.* **34**, 501–510.
            MR42 #5839.

Roberts, Fred S.
   [1968]   Representations of indifference relations, Ph. D. thesis, Stanford Univ.
   [1969]   Indifference graphs, *in* "Proof Techniques in Graph Theory" (F. Harary, ed.),
            pp. 139–146. Academic Press, New York. MR40 #5488.
   [1971]   On the compatibility between a graph and a simple order. *J. Combin. Theory* **11**,
            28–38. MR43 #7362.
   [1976]   "Discrete Mathematical Models, with Applications to Social, Biological and
            Environmental Problems." Prentice-Hall, Englewood Cliffs, New Jersey.
   [1978]   "Graph Theory and Its Applications to Problems of Society," NFS-CBMS Mono-
            graph No. 29. SIAM Publications, Philadelphia, Pennsylvania.
   [1979a] Indifference and seriation, *Ann. N.Y. Acad. Sci.* **328**, 173–182.
   [1979b] On the mobile radio frequency assignment problem and the traffic light phasing
            problem, *Ann. N.Y. Acad. Sci.* **319**, 466–483.
   [1979c] "Measurement Theory, with Applications to Decision-Making, Utility, and the
            Social Sciences." Addison-Wesley, Reading, Massachusetts.

Ryser, H. J.
   [1969]   Combinatorial configurations, *SIAM J. Appl. Math.* **17**, 593–602. MR41 #1559.

Scott, Dana S.
   [1964]   Measurement structures and linear inequalities, *J. Math. Psych.* **1**, 233–247.

Scott, Dana S., and Suppes, Patrick
   [1958]   Foundation aspects of theories of measurement, *J. Symbolic Logic* **23**, 113–128.
            MR22 #6716.

Stahl, F. W.
   [1967]   Circular genetic maps, *J. Cell. Physiol.* Suppl. **70**, 1–12.

Stoffers, K. E.
   [1968]   Scheduling of traffic lights—a new approach, *Transport. Res.* **2**, 199–234.

Suppes, Patrick, and Zinnes, J.
   [1963]   Basic measurement theory, *in* "Handbook of Mathematical Psychology," Vol. I
            (R. D. Luce, R. R. Bush, and E. Galanter, eds.), pp. 1–76. Wiley, New York.

Trotter, William T., Jr., and Harary, Frank
   [1979]   On double and multiple interval graphs, *J. Graph Theory* **3**, 205–211.

Trotter, William T., Jr. and Moore, John, I., Jr.
   [1979]   Characterization problems for graph partially ordered sets, lattices and families of
            sets (to be published).

Tucker, Alan C.
   [1970a] Characterizing the consecutive 1's property, *Proc. 2nd Chapel Hill Conf. on Com-
            binatorial Mathematics and its Applications*, Univ. North Carolina, Chapel Hill.
            472–477. MR42 #1681.

[1970b] Characterizing circular-arc graphs, *Bull. Amer. Math. Soc.* **76**, 1257–1260. MR43 #1877.

Superseded by Tucker [1971].

[1971] Matrix characterizations of circular-arc graphs, *Pacific J. Math.* **39**, 535–545. MR46 #8915.

[1972] A structure theorem for the consecutive 1's property, *J. Combin. Theory* **12**, 153–162. MR45 #4999.

[1974] Structure theorems for some circular-arc graphs, *Discrete Math.* **7**, 167–195. MR52 #203.

[1975] Coloring a family of circular-arc graphs, *SIAM J. Appl. Math.* **29**, 493–502. MR55 #10309.

[1978] Circular arc graphs: new uses and a new algorithm, *in* "Theory and Application of Graphs," Lecture Notes in Math., Vol. 642, pp. 580–589. Springer-Verlag, Berlin.

[1979] An efficient test for circular arc graphs, *SIAM J. Comput.*, to be published.

Waksman, Abraham, and Green, Milton W.

[1974] On the consecutive retrieval property in file organization, *IEEE Trans. Comput.* **C-23**, 173–174. MR50 #11886.

Wegner, G.

[1967] Eigenschaften der Nerven Homologische-einfacher Familien in R^n, Ph.D. thesis, Göttingen.

CHAPTER 9

Superperfect Graphs

1. Coloring Weighted Graphs

In this chapter we turn our attention to a notion of perfection in weighted graphs. In the process, a more general type of coloring the vertices of a graph will be introduced, suggesting many interesting applications. The concept of superperfection, introduced in Section 2, is due to Alan Hoffman and Ellis Johnson. They were motivated by the shipbuilding problem (Application 9.1), and most of the early results are theirs.

To each vertex x of a graph $G = (V, E)$ we associate a non-negative number $w(x)$, and we define the *weight* of a subset $S \subseteq V$ to be the quantity

$$w(S) = \sum_{x \in S} w(x).$$

(Without loss of generality we may assume throughout that all weights are integral.) The pair $(G; w)$ is called a *weighted graph*. The subset S will often be the vertices of a simple cycle or a clique or a stable set.

An *interval coloring* of a weighted graph $(G; w)$ maps each vertex x onto an (open) interval I_x of the real line of width (or measure) $w(x)$ such that adjacent vertices are mapped to disjoint intervals, that is, $xy \in E$ implies $I_x \cap I_y = \emptyset$. Figure 9.1 shows two colorings of a weighted graph. The *number of hues* of a coloring (i.e., its total *width*) is defined to be $|\bigcup_x I_x|$. The *interval chromatic number* $\chi(G; w)$ is the least number of hues needed to color the vertices with intervals. For the graph in Figure 9.1, $\chi(G; w) = 10$.

Example 1. If $w(x) = 1$ for every vertex $x \in V$, then $\chi(G; w) = \chi(G)$. That is, the notion of interval coloring reduces to the usual definition of coloring when all weights are equal.

203

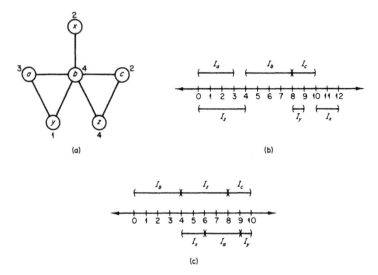

Figure 9.1. Two interval colorings of a weighted graph: (a) $(G; w)$; (b) a coloring of $(G; w)$ using 12 hues; and (c) a coloring of $(G; w)$ using 10 hues.

Application 9.1 (The Shipbuilding Problem).   In certain shipyards the sections of a ship are constructed on a dry dock, called the welding plane, according to a rigid time schedule. Each section s requires a certain width $w(s)$ on the dock during construction. Can the sections be assigned space on a welding plane of total width k so that no spot is reserved for two sections at the same time?

Let the sections be represented by the vertices of a graph G and connect two vertices if their corresponding sections have intersecting time intervals. Thus $(G; w)$ is a weighted interval graph. An interval coloring of $(G; w)$ will provide the assignment of the sections to spaces, of appropriate size, on the welding plane. This assignment will be consistent with the intersecting time restrictions. (The reader must be careful to distinguish between the *time intervals* which produced the edges of G and the *coloring intervals* which provide a solution to the assignment of space on the dock.)

Remark 1.   Larry Stockmeyer has shown that determining whether $\chi(G; w) \leq k$ is an NP-complete problem, even if w is restricted to the values 1 and 2 and G is an interval graph. It follows that the shipbuilding problem is NP-complete.

Application 9.2 (The Banquet Problem).   The menu for a banquet includes a number of cooked dishes which must be prepared in advance. A dish

d must be baked for $m(d)$ minutes at a temperature (not necessarily constant) between $t_1(d)$ and $t_2(d)$. Unfortunately, there is only one oven. How can we schedule the dishes so that the total cooking time is minimized?

Let G be the graph whose vertices represent dishes, with two vertices being connected if their corresponding dishes have disjoint temperature intervals (and therefore can never be in the oven at the same time). A $\chi(G; m)$-coloring* of $(G; m)$ will provide a solution to the scheduling problem by assigning an appropriate time interval to each dish during which it is to be in the oven. The Helly property for (the temperature) intervals insures that there is always a common acceptable temperature for all dishes being simultaneously baked. (Our solution does not take into account the size limitation of the oven.)

Remark 2. The banquet problem *can* be solved in polynomial time. This is particularly interesting in the context of Remark 1. The reason for the tractability here is that the graph obtained in the banquet problem is the *complement* of an interval graph. We will show in the next section that $\chi(G; w)$ can be calculated in polynomial time whenever G is a comparability graph.

Application 9.3 (Computer Storage Optimization). Most compilers maintain a one-to-one mapping between the variables in a program and their locations in storage. Therefore, in a tight storage situation, the programmer may have to overlay storage by deliberately using the same variable for more than one purpose, much to the detriment of clarity and reliability of the program. Using the notion of interval coloring, Fabri [1979] has investigated freeing the programmer from the task of overlaying by having the processor perform all storage allocation decisions. Thus, we want an automatic construction of a many-to-one correspondence between the variables and storage which guarantees the integrity of the variables. It is assumed that the variables have differing size requirements (as with arrays).

Let G be an undirected graph whose vertices correspond to the variables of a program. We connect two vertices v and u by an edge if and only if there is some node in the program flow graph at which v and u are simultaneously live† and thus enjoined from sharing storage. Associated with each vertex of G is a weight corresponding to the size of the variable. Since nonconflicting variables may overlay one another in storage, an interval coloring of G corresponds to a linear storage layout, and the interval chromatic number corresponds to the size of the optimum (i.e., smallest) such storage layout.

---

* From this point on, we will use the term *coloring* to mean *interval coloring* whenever the context allows.

†This can be determined by global data flow analysis.

We may regard an interval coloring in another manner. Associated with any such coloring of a weighted undirected graph $G = (V, E)$ is an implicit acyclic orientation F of G. This orientation is obtained by directing an edge toward the vertex whose coloring interval is to the right of the other, on the real line, that is,

$$xy \in F \Leftrightarrow I_x < I_y \qquad \text{(for all} \quad xy \in E).$$

This suggests the following alternative definition of $\chi(G; w)$.

Proposition 9.1.   Let $(G; w)$ be a weighted undirected graph. Then

$$\chi(G; w) = \min_F \left(\max_\mu w(\mu) \right)$$

where F is an acyclic orientation of G and μ is a path in F.

Proof.   Given F we define a coloring h of $(G; w)$ in the same way that one usually constructs a height function in a partial order. For a sink x, let $h(x) = (0, w(x))$. Proceeding inductively, for a vertex y let t be the largest endpoint of the intervals corresponding to the sons of y, and define $h(y) = (t, t + w(y))$. Thus, h is a coloring and its number of hues equals $\max_\mu w(\mu)$. This proves that $\chi(G; w) \leq \min_F(\max_\mu w(\mu))$.

Conversely, a minimum coloring gives us an acyclic orientation F' as mentioned above, and clearly $\chi(G; w) \geq w(\mu)$ for any path μ in F'. This proves the reverse inequality, and hence equality holds.   ∎

2.   Superperfection

The *clique number* of a weighted graph $(G; w)$ is defined as

$$\omega(T; w) = \max\{w(K) | K \text{ is a clique of } G\}.$$

As one might expect, $\omega(G; w) \leq \chi(G; w)$, which follows from Proposition 9.1. An undirected graph G is *superperfect* if for every non-negative weighting

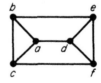

Figure 9.2.   A superperfect graph.

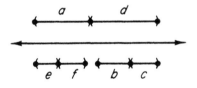

Figure 9.3.

w of the vertices $\omega(G;w) = \chi(G;w)$. Equivalently, G is superperfect if for every weighting w there exists an acyclic orientation F of G such that $w(\mu) \leq w(K)$ for every path μ in F and some clique K of G. In particular, the weight of the "heaviest" clique in G will equal the weight of the "heaviest" path in F. Thus, we have two basic methods for demonstrating superperfection: providing a suitable coloring or giving a suitable acyclic orientation. We shall illustrate these techniques on a few examples.

Example 2.   The graph in Figure 9.2 is superperfect. The heaviest clique is either (i) one of the two triangles or (ii) one of the three edges not contained in a triangle. Suppose (ii) is the case for some weighting w, and assume, without loss of generality, that $\{a, d\}$ is the heaviest. Then $w(b) + w(c) \leq w(d)$ and $w(e) + w(f) \leq w(a)$, so that the coloring in Figure 9.3 will do. Otherwise, suppose (ii) is *not* the case, and assume that $\{d, e, f\}$ is the heaviest clique. Therefore, $w(a) + w(b) + w(c) \leq w(d) + w(e) + w(f)$ and, since (ii) has been ruled out,

$$w(a) < w(e) + w(f), \qquad w(b) < w(d) + w(f), \qquad w(c) < w(d) + w(e).$$

By cyclically permuting the vertices of each triangle if necessary, we may also assume that $w(b) \leq w(d)$. If $w(a) \geq w(f)$, then the coloring in Figure 9.4a gives a solution; otherwise Figure 9.4b works. Therefore, in every case, we have exhibited a coloring whose number of hues equals the weight of the heaviest clique. We conclude that the graph is superperfect.

Example 3.   An undirected graph is perfect if and only if for every $(0, 1)$-valued weighting w of the vertices $\omega(G;w) = \chi(G;w)$. Thus *every superperfect graph is a perfect graph.*

(a)   (b)

Figure 9.4.

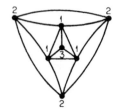

Figure 9.5.   *G*: not superperfect.

Example 4.   The graph *G* in Figure 9.5 is not superperfect since any acyclic orientation with the weighting shown would have a path of weight > 6. Its complement $\bar{G}$ (Figure 9.6), however is superperfect.

Example 5.   By extending a weighting *w* of $X \subseteq V$ to all of *V*, defining $w(v) = 0$ for all $v \in V - X$, it follows that *each induced subgraph of a superperfect graph is itself superperfect.*

Let *F* be a transitive orientation of a comparability graph *G*. By transitivity, every path in *F* is contained in a clique of *G*. So, in particular, for any weighting of the vertices of *G*, the weight of heaviest path in *F* equals the weight of the heaviest clique in *G*. This argument proves the following theorem.

Theorem 9.2.   A comparability graph is superperfect.

Theorem 9.2 was first noted by Alan Hoffman, and he raised the question of the existence of superperfect graphs which are not comparability graphs. Such a graph was found by the author in 1974; it is the graph in Figure 9.2. We shall explore this question further in Sections 3 and 4.

Theorem 9.2 has an algorithmic aspect as well. The interval chromatic number $\chi(G; w)$ of a weighted comparability graph can be calculated in polynomial time. One must simply obtain a transitive orientation *F*, for which Algorithm 5.2 may be used, and then apply Algorithm 5.4 to find a

Figure 9.6. $\bar{G}$: superperfect.

maximum weighted clique. In fact, the optimal coloring may be calculated efficiently by a depth-first search procedure utilizing the method described in the proof of Proposition 9.1.

3. An Infinite Class of Superperfect Noncomparability Graphs

Before describing our class of graphs, we will prove the following useful lemma.

Lemma 9.3. Let $a_0, \ldots, a_{n-1}$ and $b_0, \ldots, b_{n-1}$ be sequences of real numbers such that

$$\sum_{i=0}^{n-1} a_i \leq \sum_{i=0}^{n-1} b_i.$$

There exists a cyclic permutation π of $\{0, 1, \ldots, n-1\}$ such that

$$\sum_{i=0}^{m} a_{\pi_i} \leq \sum_{i=0}^{m} b_{\pi_i} \qquad (m = 0, 1, \ldots, n-1).$$

Proof. Let $c_i = b_i - a_i$. If each of the partial sums $\sum_{i=0}^{m} c_i \geq 0$ ($m = 0, 1, \ldots, n-1$), then the result holds. Otherwise, let $\sum_{i=0}^{j} c_i$ be the smallest of these partial sums (i.e, the most negative).

Consider the permutation $\pi_i = i + j + 1 \pmod{n}$. For $m = j + 1, \ldots, n-1$ we have

$$0 \leq \sum_{i=0}^{m} c_i - \sum_{i=0}^{j} c_i = \sum_{i=j+1}^{m} c_i,$$

where, for $m = 1, \ldots, j$,

$$0 \leq \sum_{i=0}^{n-1} c_i = \sum_{i=0}^{j} c_i + \sum_{i=j+1}^{n-1} c_i \leq \sum_{i=0}^{m} c_i + \sum_{i=j+1}^{n-1} c_i,$$

thus proving the lemma. ∎

Let n and k be arbitrary positive integers, $n \geq k$. Consider the undirected graph $G_{n,k} = (A + B, E)$, where

(i) $A = \{a_0, a_1, \ldots, a_{n-1}\}$ is a clique,
(ii) $B = \{b_0, b_1, \ldots, b_{n-1}\}$ is a clique, and
(iii) a_i is adjacent to $b_{i+j \pmod{n}}$ for $j = 1, 2, \ldots, k$.

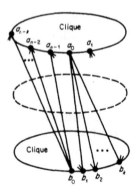

Figure 9.7. $G_{n,k}$.

Figure 9.7 illustrates these adjacencies. We remark here that $G_{n, n-2}$ is the same graph as $\bar{C}_{2n}$, the complement of a chordless cycle of length $2n$. The vertices of the cycle going around in order are $[a_0, b_0, a_1, b_1, \ldots, a_{n-1}, b_{n-1}]$.

Theorem 9.4 (Golumbic [1974]).   For arbitrary integers $n \geq k \geq 0$, the graph $G_{n, k}$ is superperfect.

Proof.   Assume that $n > k > 0$, for the other cases are trivial. Our method of showing superperfection has three steps.

Step I.   Assign an arbitrary weight to each vertex.
Step II.   Describe a particular acyclic orientation F.
Step III.   Show that every maximal path in F is either (i) contained in some clique, or (ii) has weight less than or equal to a path (already shown to be) in class (i).

We call F a *superperfect orientation* with respect to this weighting.

Step I.   We assign an arbitrary weight to each vertex. For simplicity, denote the weight of a_i and b_i by $\hat{a}_i$ and $\hat{b}_i$, respectively. We may assume that $\hat{a}_0 + \hat{a}_1 + \cdots + \hat{a}_{n-1} \leq \hat{b}_0 + \hat{b}_1 + \cdots + \hat{b}_{n-1}$ by interchanging the sets A and B, if necessary. Furthermore, applying Lemma 9.3, we may assume that the vertices have been indexed so that the partial sums satisfy

$$\hat{a}_0 + \cdots + \hat{a}_m \leq \hat{b}_0 + \cdots + \hat{b}_m \qquad (m = 0, 1, \ldots, n-1).$$

Step II.   Let us assign the acyclic orientation F of $G_{n, k}$ as follows:

$$\begin{aligned}
a_i a_j, \ b_i b_j \in F \qquad & (0 \leq i < j \leq n-1), \\
a_i b_j \in F \qquad & (0 \leq i < j \leq \min(n-1, i+k)), \\
b_i a_j \in F \qquad & (0 \leq i \leq k-1, \ n-k+i \leq j \leq n-1)
\end{aligned}$$

(see Figure 9.8).

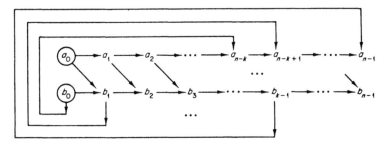

Figure 9.8. Maximal paths in F.

Step III. Any maximal path in F will start in a source node a_0 or b_0. (i) Consider a maximal path μ starting in b_0. Either $\mu = [b_0, b_1, \ldots, b_{n-1}]$, in which case it is contained in a clique, or, for some indices i and p, $0 \le i \le k-1$, $1 \le p < k - i$,

$$\mu = [b_0, \ldots, b_i, a_{n-k+i}, \ldots, a_{n-p}, b_{n-p+1}, \ldots, b_{n-1}].$$

(Obviously if $p = 1$, then there are no b's at the end.) Now $b_t a_j \in E$ for $t = 0, \ldots, i$ and $j = n - k + i, \ldots, n - 1$ and $a_{n-q} b_{n-q+j} \in E$ for $q \le k - i$ and $j < q + 1$. Thus, μ is contained in a clique.

(ii) Consider a maximal path v starting in a_0. Now v is of the form $v = [a_0, \ldots, a_r, b_{r+1}, \text{remainder}]$ where b_{r+1} is the first b_j in v. Since

$$\hat{a}_0 + \cdots + \hat{a}_r \le \hat{b}_0 + \cdots + \hat{b}_r,$$

the weight of v is no more than the weight of the path

$$[b_0, \ldots, b_r, b_{r+1}, \text{remainder}],$$

which is contained in a clique, concluding the proof of the theorem. ∎

Corollary 9.5. The complement of an even-length cycle with no chords is superperfect.

The next result shows that the graphs $G_{n,k}$ constitute a class of superperfect graphs distinct from the comparability graphs.

Theorem 9.6. $G_{n,k}$ is not a comparability graph, for $1 \le k \le n - 2$.

Proof. Recall that an undirected graph is a comparability graph if and only if every closed path with no triangular chords has even length (see Theorem 5.27). However,

$$[a_0, a_1, b_{k+1}, b_0, a_{k+1}, a_0, b_k, a_0, a_k, a_0, \ldots, a_0, b_1, a_0]$$

is such a closed path in $G_{n,k}$ and has odd length. ∎

4.   When Does Superperfect Equal Comparability?

Figure 9.9 illustrates part of the world of superperfect graphs. We have shown in Section 3 that the superperfect graphs properly contain the comparability graphs. This leads us to ask *under what conditions these two classes coincide.* In this section we shall give one answer to this question and we shall discuss some open problems.

Földes and Hammer [1977] have proved the following:

Theorem 9.7.   If G is a split graph, then G is a comparability graph if and only if G contains no induced subgraph isomorphic to H_1, H_2, or H_3 of Figure 9.10.

Proof.   The forward implication is immediate since none of the graphs in Figure 9.10 is a comparability graph. We shall show the reverse implication. Let G be a split graph whose vertices are partitioned into a stable set X and a complete set Y. An edge of G is called *pure* if both its endpoints are in Y and called *mixed* otherwise. A vertex from X (resp. Y) is denoted by a subscripted lower-case x (resp. y). The key to the proof is the observation that a minimal Γ-chain (i.e., one that does not properly contain another Γ-chain) will alternate between mixed and pure edges and will involve only two vertices of X. Assume that G contains no induced copy of $H_1, H_2,$ or H_3.

Let γ be a minimal Γ-chain. Since no two pure edges are Γ-related, how many mixed edges may separate consecutive pure edges e_1 and e_2 in γ? All such mixed edges will share a common vertex from Y, and hence they are

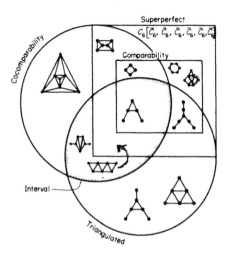

Figure 9.9.

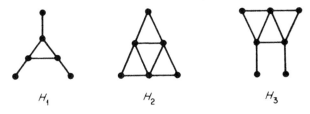

H_1 H_2 H_3

Figure 9.10.

Γ-related to one another. Thus, if there were more than two, the chain γ could be shortened, contradicting minimality. Suppose there are exactly two; then we have the following segment of γ,

$$\cdots f_0 \Gamma e_1 \Gamma f_1 \Gamma f_2 \Gamma e_2 \Gamma f_3 \cdots,$$

which corresponds to the diagram in Figure 9.11. By minimality of γ, y_1 and y_3 are adjacent, respectively, to x_2 and x_1. If $y_1 \in \mathrm{Adj}(x_3)$ or $y_3 \in \mathrm{Adj}(x_0)$, which includes the possibility of x_0 and x_3 coinciding, then G contains an induced copy of H_2; otherwise, G contains a copy of H_3. Therefore, γ *alternates between pure and mixed edges*, as claimed.

If G is not a comparability graph, then there exists a minimal Γ-chain γ from some mixed edge $x_0 y_1$ to its reversal, namely,

$$x_0 y_1 \Gamma y_2 y_1 \Gamma y_2 x_1 \Gamma y_2 y_3 \Gamma x_2 y_3 \Gamma y_4 y_3 \Gamma y_4 x_3 \Gamma \cdots \Gamma y_n x_{n+1} = y_1 x_0.$$

Now $x_0 \neq x_1$ and γ involves only these two vertices from X, since G has no induced copy of H_1. Thus, $x_0 = x_2 = x_4 = \cdots$ and $x_1 = x_3 = x_5 = \cdots$, and by the parity of the indices x_{n+1} equals x_1 and not x_0, a contradiction. This proves the theorem. ∎

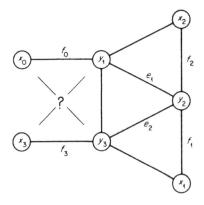

Figure 9.11.

One can easily verify that H_1, H_2, and H_3 are not superperfect (Exercise 1). From this observation we obtain a new result.

Corollary 9.8. For split graphs, G is a comparability graph if and only if G is superperfect.

Proof. Suppose G is not a comparability graph; then G contains one of the forbidden induced subgraphs of Figure 9.10. Since superperfection is a hereditary property, we deduce that G is not superperfect. The opposite implication is true for all graphs.   ∎

The class of split graphs is very restrictive. We wonder how much it is possible to weaken the hypothesis of Corollary 9.8 and yet obtain the same conclusion. For example, all the superperfect noncomparability graphs of Section 3 were neither triangulated nor cotriangulated. Is it true or false that for triangulated (or cotriangulated) graphs G is a comparability graph if and only if G is superperfect?

5. Composition of Superperfect Graphs

Recall from Section 5.2 the definition of the *composition* of graphs. In this section we investigate how this operation affects superperfection. Let $G_0, G_1, \ldots, G_n$ be undirected graphs, where G_0 has n vertices $v_1, v_2, \ldots, v_n$.

Theorem 9.9. If $G_0, G_1, \ldots, G_n$ are superperfect, then their composition $G = G_0[G_1, \ldots, G_n]$ is superperfect; i.e., superperfection is preserved under composition.

Proof. Let $G_i = (V_i, E_i)$ for $i = 0, 1, \ldots, n$ be disjoint superperfect graphs, and let w be a weighting of $V_1 + V_2 + \cdots + V_n$. (The vertices in V_0 are not weighted since they will be replaced.) Suppose further that F_i is a superperfect orientation of G_i with respect to w (restricted to G_i) for each $i = 0$, $1, \ldots, n$. We claim that $F = F_0[F_1, \ldots, F_n]$ is a superperfect orientation of G with respect to w.

Since each of the F_i ($i = 0, 1, \ldots, n$) are acyclic, so is F. Let K_i ($i = 1, \ldots, n$) be a clique of G_i whose weight $w(K_i)$ is greater than or equal to that of any path in F_i. Define $w'(v_i) = w(K_i)$ for all $v_i \in V_0$, and let K_0 be a clique of G_0 whose weight $w'(K_0)$ is greater than or equal to that of any path in F_0. Now any path μ in $F_0[F_1, \ldots, F_n]$ is of the form $\mu = [\mu_{i_1}, \mu_{i_2}, \ldots, \mu_{i_t}]$, where the

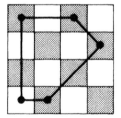

Figure 9.12.   A chordless 5-cycle in (V_4, Q).

μ_{i_j} are paths in distinct F_{i_j}, and the sequence of vertices $[v_{i_1}, v_{i_2}, \ldots, v_{i_t}]$ is a path in F_0. Hence, we have the following inequalities:

$$w(\mu) = w(\mu_{i_1}) + \cdots + w(\mu_{i_t}) \leq w(K_{i_1}) + \cdots + w(K_{i_t})$$
$$= w'(v_{i_1}) + \cdots + w'(v_{i_t})$$
$$\leq w'(K_0).$$

But the vertices of $\bigcup \{K_i | v_i \in K_0\}$ induce a clique K of $G_0[G_1, \ldots, G_n]$ whose weight $w(K)$ equals $w'(K_0)$. Thus, we have shown that G is superperfect. ∎

Example 6.   Let X_k be the set of positions of a $k \times k$ chessboard, and let Q be the binary relation defined on X_k as follows: $xy \in Q$ iff a queen can be moved from position x to position y in a single chess move.

Consider, for the moment, the graph (X_3, Q). Let x be the middle position and let $X = X_3 - x$. Notice that (X_3, Q) is the composition of the single vertex x and the induced subgraph (X, Q_X) with external factor K_2. However, (X, Q_X) is the complement of a chordless 8-cycle and is therefore a superperfect graph. Hence, (X_3, Q) *is also a superperfect noncomparability graph.*

Clearly (X_k, Q) is an induced subgraph of (X_{k+1}, Q), so (X_k, Q) is a noncomparability graph for all $k \geq 3$. Moreover, Figure 9.12 shows that (X_4, Q) is not perfect since it contains a chordless 5-cycle. Thus, (X_k, Q) is not perfect and hence not superperfect for $k \geq 4$.

6.   A Representation Using the Consecutive 1's Property

We now relate the concept of superperfection to some ideas of linear programming. The material presented here is due to Alan J. Hoffman and Ellis L. Johnson.

Recall that a $(0, 1)$-valued matrix is said to have the *consecutive 1's property* (for columns) if the rows can be permuted so that all the 1's in each column occur consecutively. Let $\mathbf{M}$ denote the stable sets-versus-vertices incidence matrix of an undirected graph G.

Theorem 9.10. *G is superperfect if and only if for every row vector $\mathbf{w} \geq \mathbf{0}$ the linear programming problem*

$$\mathbf{yM} \geq \mathbf{w}, \quad \mathbf{y} \geq \mathbf{0}, \tag{1a}$$

$$\text{minimize} \sum_i y_i, \tag{1b}$$

has an optimum (row vector) solution $\mathbf{y}$ such that

the submatrix of $\mathbf{M}$ consisting of those rows S_i with $y_i \neq 0$ has the consecutive 1's property. (2)

Assume that the vertex set V is indexed $v_1, v_2, \ldots, v_n$, and let us interpret what the theorem says.

Interpretation. Each stable set S_i is assigned a plot on the real line of width y_i for use only by its members. (Recall that no two members of S_i will need this communal plot simultaneously.)

Feasibility: By (1a) the sum of the widths of the plots available to a given vertex v_j must be at least w_j.
Minimality: By (1b) the combined width of the plots is smallest possible.
Consecutive 1's: By (2) the nonempty plots can be arranged so that those for each vertex are contiguous.

Proof. From the above interpretation, it is clear that any $\mathbf{y}$ satisfying both (1a) and (2) gives a coloring of $(G; \mathbf{w})$ of width $\sum y_i$. The following converse also holds:

For every coloring c of $(G; \mathbf{w})$ there exists a vector $\mathbf{y}$ (to be constructed below) satisfying (1a) and (2) such that $\sum y_i$ equals the width of c. (3)

Let c map V onto the interval from 0 to t. We may assume that c is *left justified*, that is, that no interval can be shifted to the left without disturbing the validity of c as an interval coloring.

Divide each interval $c(v_i)$ into subintervals labeled with exactly those vertices assigned to that subinterval (see Figure 9.12). Each of these labels is some stable set. Suppose there are two subintervals, I_1 and I_2, with the same label S_i. If they are adjacent, then combine them into one larger subinterval.

If they are nonadjacent, there is a vertex v such that $c(v)$ is wholly contained between I_1 and I_2 and whose left endpoint coincides with the right endpoint of I_1 (assume I_1 is to the left of I_2). However, shifting $c(v)$ to the left by the width of I_1 yields another coloring, contradicting left justification. Thus, we may assume that for each stable set S_i there is at most one subinterval with label S_i, and we define y_i to equal the width of that subinterval if it exists and zero otherwise. Clearly, **y** satisfies (1a) and (2) and $\sum y_i$ equals the width of the coloring c. This proves claim (3).

Consider the linear programming problem

$$\mathbf{Mx} \le 1, \qquad \mathbf{x} \ge 0,$$
$$\text{maximize } \sum_j w_j x_j. \tag{4}$$

By the Duality theorem, the optimum solutions of (1) and (4) are equal. Furthermore, if **x** is the characteristic vector of a clique of G, then **x** is a feasible solution to (4). Conversely, any integral feasible solution to (4) is the characteristic vector of a clique. Thus, an optimum solution $\bar{\mathbf{y}}$ to (1) satisfies

$$\sum \bar{y}_i \ge \omega(G; \mathbf{w}). \tag{5}$$

We do not necessarily have equality in (5) since (4) may not have an optimum solution which is integral. (For example, consider the graph C_5.)

We are now ready to prove the theorem in one direction. Suppose that G is superperfect, and let $\mathbf{w} \ge \mathbf{0}$ be given. Choose a coloring c of $(G; \mathbf{w})$ of width $\omega(G; \mathbf{w})$. By (3) we obtain a vector **y** satisfying (1a) and (2) with $\sum y_i = \omega(G; \mathbf{w})$; and by (5), **y** is optimum.

To prove the converse of the theorem, we need the following lemmas.

> If **A** is a (0, 1)-valued matrix whose columns have the consecutive
> 1's property, then **A** is totally unimodular (i.e., every subdetermi-   (6)
> nant is 0, 1 or -1).

Hence, if **w** is integral, then every optimum solution to (1) which satisfies (2) is integral. (See Hoffman and Kruskal [1956].)

> If for every integral $\mathbf{w}^* \ge \mathbf{0}$ (1) has an optimum solution which
> is integral, then for every $\mathbf{w} \ge \mathbf{0}$ (4) has an optimum solution   (7)
> which is integral. (See Hoffman [1974] for analogous theorem.)

Suppose that for all. $\mathbf{w} \ge \mathbf{0}$ (1) has an optimum solution $\bar{\mathbf{y}}$ satisfying (2). Then $\chi(G; \mathbf{w}) = \sum_i \bar{y}_i$. By (6) and (7), there is an optimum solution $\bar{\mathbf{x}}$ to (4) which is integral. But this optimum solution $\bar{\mathbf{x}}$ is the characteristic vector of a clique of G, so $\omega(G, \mathbf{w}) = \sum_i \bar{x}_i$. Finally, by the duality of (1) and (4) we obtain $\chi(G; \mathbf{w}) = \omega(G; \mathbf{w})$.   ∎

EXERCISES

1.   Using the technique of Example 5, prove that the graphs H_1, H_2, and H_3 of Figure 9.10 are not superperfect.

2.   Prove the following: If H is obtained from G by multiplication of vertices, then H is superperfect if and only if G is superperfect.

3.   Prove that the shipbuilding problem is NP-complete.

4.   Write a polynomial-time algorithm to solve the banquet problem. Analyze the complexity of your algorithm.

5.   Show that the bull's head graph (Figure 1.14) is an interval graph which is not superperfect.

Bibliography

Fabri, Janet
    [1979]   Automatic storage optimization, Ph.D. thesis, Courant Computer Science Report No. 14, New York Univ.

Földes, Stephane, and Hammer, Peter L.
    [1977]   Split graphs, *Proc. 8th Southeastern Conf. on Combinatorics, Graph Theory and Computing.* "Congressus Numerantium XIX," pp. 311–315. Utilities Math., Winnipeg.

Golumbic, Martin Charles
    [1974]   An infinite class of superperfect noncomparability graphs, I.B.M. Res. Rep. RC 5064.

Hoffman, Alan J.
    [1974]   A generalization of max flow – min cut, *Math. Programming* **6**, 352–359. MR50 #15906.

Hoffman, Alan J., and Kruskal, Joseph B.
    [1956]   Integral boundary points of convex polyhedra. "Linear Inequalities and Related Systems" (H. W. Kuhn and A. W. Tucker, eds.), Annals of Mathematics Studies, No. 38, pp. 223–246. Princeton Univ. Press, Princeton, New Jersey. MR18, p. 980.

CHAPTER 10

Threshold Graphs

In this chapter we discuss a particularly simple technique for distinguishing between stable and nonstable subsets of vertices in a special class of graphs. The graphs that admit this technique, which involves assigning certain weights to the vertices, are called *threshold graphs*. Threshold graphs were introduced by Chvátal and Hammer [1973]. Their results form the basis for much of the next two sections. We begin by introducing the more general notion of threshold dimension.

1. The Threshold Dimension

Let $V = \{v_1, v_2, \ldots, v_n\}$ be the vertex set of an undirected graph G. Any subset $X \subseteq V$ can be represented by its *characteristic vector* $\mathbf{x} = (x_1, x_2, \ldots, x_n)$, where for all i

$$x_i = \begin{cases} 1 & \text{if } v_i \in X, \\ 0 & \text{if } v_i \notin X. \end{cases}$$

Thus, the subsets of vertices are in one-to-one correspondence with the corners of the unit hypercube in $\mathbb{R}^n$ according to the coordinates of their characteristic vectors.

Let us consider the collection of all stable sets of G. We ask the following: Is there a hyperplane that cuts n-space in half in such a way that on one side all corners of the hypercube (characteristic vectors) correspond to stable sets of G and on the other side all corners correspond to nonstable sets? Equivalently, can we distinguish which subsets of V are stable sets using a single

Figure 10.1.

linear inequality? If the answer is affirmative, then the graph under consideration is a *threshold graph*. If not, we shall want to know how many inequalities are needed to distinguish between stable and nonstable sets.

   Example.   Consider the graph in Figure 10.1. Its stable sets correspond to the solid corners of the unit 3-cube in Figure 10.2. The inequality $x_1 + 2x_2 + x_3 \leq 2$ is satisfied only by the characteristic vectors of the stable sets. Thus, a separating plane does exist, namely, $x + 2y + z = 2$.
   The *threshold dimension* $\theta(G)$ of the graph $G = (V, E)$ is defined to be the minimum number k of linear inequalities

$$a_{11}x_1 + a_{12}x_2 + \cdots + a_{1n}x_n \leq t_1,$$
$$\vdots \tag{1}$$
$$a_{k1}x_1 + a_{k2}x_2 + \cdots + a_{kn}x_n \leq t_k,$$

such that X is a stable set if and only if its characteristic vector $\mathbf{x} = (x_1, x_2, \ldots, x_n)$ satisfies (1). Regarding each inequality of (1) as a hyperplane in n-space, X is stable iff $\mathbf{x}$ lies on or within the "good" side of each of those k hyperplanes. Since G is finite, $\theta(G)$ is finite and well defined.

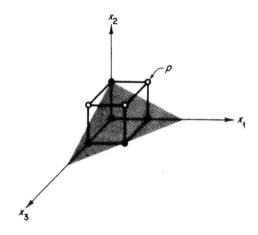

Figure 10.2.   The point $p = (1, 1, 0)$ corresponds to the set $\{v_1, v_2\}$, which is not stable.

Remark.   The only graphs G for which $\theta(G) = 0$ are those having no edges. In this case, the empty set of constraints suffices.

Let us first notice that *without loss of generality, we may assume that all the numbers a_{ij} and t_i in* (1) *are non-negative integers.* Suppose we are given a set of linear inequalities (1). Since the zero vector represents a stable set, we must have each $t_i \geq 0$. Furthermore, any negative coefficient a_{ij} can be changed to zero because for sets X not containing v_j the sum $\text{COUNT}_i(X)$, defined by

$$\text{COUNT}_i(X) = \sum_{v_p \in X} a_{ip} = \sum_{p=1}^{n} a_{ip} x_p,$$

would remain unchanged, whereas for sets X containing v_j this sum would be increased to $\text{COUNT}_i(X - \{v_j\})$ which is $\leq t_i$ if and only if X is stable. Finally, since the graph is finite and the x_i are integral, we can perturb the system by a small ε here and there to make all the numbers non-negative rationals. Then we multiply by the least common divisor in order to obtain integers.

An undirected graph $G = (V, E)$ whose threshold dimension $\theta(G)$ is ≤ 1 is a *threshold graph.* Equivalently, $G = (V, E)$ is *threshold* if there exists a threshold assignment $[a; t]$ consisting of a labeling a of the vertices by non-negative integers and an integer threshold t such that

$$X \text{ is stable} \Leftrightarrow \sum_{x \in X} a(x) \leq t \qquad (X \subseteq V). \qquad (2)$$

Examples.   The star graph $K_{1,n}$ is easily seen to be a threshold graph by assigning $a(v)$ to be the degree of v and $t = n$, (Figure 10.3). Labeling by degree, however, does not always work. The labeling in Figure 10.4a fails to satisfy (2) for any value of t since there is a stable set of weight 7 and a non-stable set of weight 6. It is not a threshold assignment. On the other hand, the

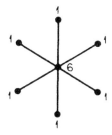

Figure 10.3.   The graph $K_{1,6}$ and a threshold assignment with $t = 6$.

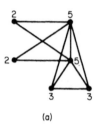
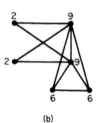

(a)   (b)

Figure 10.4.   (a) Degree labeling. (b) A threshold assignment for $t = 10$.

labeling in Figure 10.4b of the same graph is a threshold assignment for $t = 10$. There are graphs which are not threshold graphs. For example, the chordless cycle C_n with $n \geq 4$ is not threshold and neither is the path P_n for $n \geq 4$ (Figure 10.5).

It should also be noted that an induced subgraph of a threshold graph is a threshold graph. Therefore, any graph which contains an induced subgraph isomorphic to one of those in Figure 10.5 is not threshold.

The threshold dimension $\theta(G)$ of an arbitrary graph G can be defined in an alternate but equivalent manner using threshold graphs. Take $\theta(G)$ to be the *minimum number of threshold graphs needed to cover the edges of G*, i.e., partial subgraphs of G, which are themselves threshold, and include every edge at least once. For example, $\theta(C_4) = 2$ since C_4 can be covered by two copies of $K_{1,2}$. The formalities of proving the definitions equivalent are left to the reader (Exercise 13). However, one can easily see that each inequality of (1) corresponds to one threshold graph and vice versa, and taken together they determine the adjacencies of the graph. This idea of covering by threshold graphs can be used to prove the following theorem. Let $\alpha(G)$ denote the size of the largest stable set of G.

Theorem 10.1 (Chvátal and Hammer [1973]).   If G is an undirected graph with n vertices, then $\theta(G) \leq n - \alpha(G)$. Moreover, equality holds if (but not only if) G contains no triangle.

Proof.   Let X be a stable set of cardinality $\alpha(G)$. For each vertex $v \notin X$ let S_v be the star graph with v at the center and having edges vv' for each v'

Figure 10.5.   The graphs C_4, P_4, and $2K_2$ are not threshold since any assignment would require the inequalities $w + y \leq t$, $w + z > t$, $x + z \leq t$, and $x + y > t$, which are inconsistent.

adjacent to v in G. Thus, $\{S_v | v \notin X\}$ forms an edge covering of cardinality $n - \alpha(G)$, proving the first assertion.

Since C_m and P_m are not threshold for $m \geq 4$ and since being threshold is a property inherited by induced subgraphs, it follows that any minimum cover $\{G_i | i = 1, \ldots, \theta(G)\}$ of a *triangle-free* graph G consists only of stars G_i, whose center vertices we denote by u_i. Moreover, any edge of G has at least one endpoint in $U = \{u_i | i = 1, \ldots, \theta(G)\}$, implying that $V - U$ is stable and $\alpha(G) \geq |V - U| = n - \theta(G)$. Combining this with the first assertion, we obtain $\alpha(G) = n - \theta(G)$ for triangle-free graphs. ∎

Corollary 10.2.   For the following graphs we have

(i) $\theta(C_n) = \lceil n/2 \rceil$ $(n > 3)$,
(ii) $\theta(K_{m,n}) = \mathrm{MIN}\{m, n\}$,
(iii) $\theta(P_n) = \lfloor n/2 \rfloor$.

Proof. Each of these graphs is triangle free, so the theorem provides the equalities. ∎

As pointed out in Chvátal and Hammer [1977], *the problem of computing* $\theta(G)$ *is NP-complete* in view of Poljak's proof (Theorem 2.1) that computing $\alpha(G)$ for triangle-free graphs is NP-complete. We shall see, however, that deciding whether or not $\theta(G) = 1$ can be done in linear time.

Unfortunately, since $\theta(K_n) = 1$, the bound of the theorem is sometimes useless. However, the next result shows that it is the best possible.

Corollary 10.3 (Chvátal and Hammer [1973]).   For every $\varepsilon > 0$ there exists a graph G with n vertices such that $(1 - \varepsilon)n < \theta(G)$.

Proof. Erdös [1961] has proved that for any positive integer N there is a triangle-free graph G_N with $\alpha(G_N) < N$ and $n > c(N/\log N)^2$ vertices. (Here c is a positive constant independent of N.) Given $\varepsilon > 0$, choose N large enough so that $\varepsilon c N \geq (\log N)^2$ and consider the Erdös graph G_N. Since $\varepsilon \geq N(\log N)^2/cN^2 > N/n$, it follows that $(1 - \varepsilon)n < n - N < n - \alpha(G_N) = \theta(G_N)$. ∎

2.   Degree Partition of Threshold Graphs

In this section we present a number of characterizations of threshold graphs. Let $G = (V, E)$ be a threshold graph with threshold assignment $[a; t]$. The following properties are immediate:

$$a(x) \leq t \quad (x \in V), \tag{3}$$

$$xy \in E \Leftrightarrow a(x) + a(y) > t \quad (x, y \in V, x \neq y). \tag{4}$$

A labeling satisfying (3) and (4) is not in general a threshold assignment since the sets being tested for stability are restricted to those of cardinality ≤ 2. However, condition (4) does imply the existence of a different labeling and threshold satisfying (2), as we shall prove in Theorem 10.4. For example, the labeling given in Figure 10.4a does satisfy (3) and (4) with $t = 5$, but it is not a threshold assignment. On the other hand, the labeling in Figure 10.4b is a threshold assignment for $t = 10$.

We begin by defining the degree partition of an undirected graph $G = (V, E)$ in which we associate vertices having the same degree. Let $0 < \delta_1 < \delta_2 < \cdots < \delta_m < |V|$ be the degrees of the nonisolated vertices; the δ_i are distinct and there may be many vertices of degree δ_i. Define $\delta_0 = 0$ and $\delta_{m+1} = |V| - 1$. The *degree partition* of V is given by

$$V = D_0 + D_1 + \cdots + D_m,$$

where D_i is the set of all vertices of degree δ_i. Only D_0 is possibly empty.

The following theorem is due to Chvátal and Hammer [1973]. The equivalence of (i) and (ii) was discovered independently by Henderson and Zalcstein [1977].

Theorem 10.4.   Let $G = (V, E)$ be an undirected graph with degree partition $V = D_0 + D_1 + \cdots + D_m$. The following statements are equivalent:

(i) G is a threshold graph;
(ii)   there exists an integer labeling c of V and an integer (threshold) t such that for distinct vertices x and y,

$$xy \in E \Leftrightarrow c(x) + c(y) > t;$$

(iii)   for all distinct vertices $x \in D_i$ and $y \in D_j$,

$$xy \in E \Leftrightarrow i + j > m;$$

(iv)   the recursions below are satisfied:

$$\delta_{i+1} = \delta_i + |D_{m-i}| \qquad (i = 0, 1, \ldots, \lfloor m/2 \rfloor - 1),$$
$$\delta_i = \delta_{i+1} - |D_{m-i}| \qquad (i = m, m, -1, \ldots, \lfloor m/2 \rfloor + 1).$$

Before proving the theorem let us understand its significance. Statement (iii) says that *the structure of the graph is entirely determined by the indices of the degree partition*. The vertices contained in the first half of the partition cells form a stable set, while those contained in the later half of the partition cells from a complete set. Furthermore, the adjacencies possess a natural containment, as illustrated in Figure 10.6. Statement (iv) is most important computationally for it allows us to verify that we have a threshold graph by using purely arithmetic operations without making reference to edges or

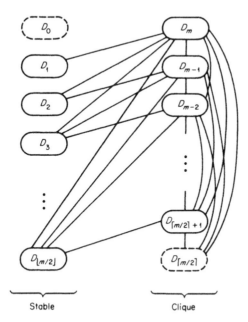

Stable   Clique

Figure 10.6.   The typical structure of a threshold graph. A line between cells D_i and D_j indicates that each vertex in D_i is adjacent to each vertex of D_j. Cell D_0 contains all isolated vertices and may be empty. Cell $D_{\lceil m/2 \rceil}$ only exists if m is odd.

adjacency sets—a very unusual situation in graph theory.* Since these recursive relations can be verified within $O(n)$ computational steps for a graph with n vertices, we obtain the following.

Corollary 10.5.   Given only the degrees of the vertices of an undirected graph G, there is an algorithm which decides whether or not G is a threshold graph and which runs in time proportional to the number of vertices of G.

Proof of this corollary is given as Exercise 7.

Proof of Theorem 10.4.   (i) $\Rightarrow$ (ii)   This is just Property (4).
   (ii) $\Rightarrow$ (iii)   The proof is by induction on the length of the degree partition. We may assume that $0 < i \le j \le m$.
   Let y be a vertex of largest label $c(y)$. For any other vertex x, if x is adjacent to some vertex w (i.e., $x \notin D_0$), then $t < c(x) + c(w) \le c(x) + c(y)$, implying that x is adjacent to y. Hence, $y \in D_m$, $\delta_m = |V| - |D_0| - 1$ and each vertex

---

* The reader will notice that the two sets of recursions actually use the same equation. They are stated separately to emphasize the method of calculation (δ_0 and δ_{m+1} are known), and to indicate how they may be proved inductively.

in D_m is adjacent to all nonisolated vertices. This proves (iii) in the case $j = m$.

Furthermore, *the vertices of D_1 are adjacent only to those in D_m.* For suppose $\delta_1 > |D_m|$, then *each* vertex $x \in V - D_0$ would be adjacent to some vertex in $V - D_m$; hence x would also be adjacent to z where z has the largest label in $V - D_m$. This forces z to be in D_m, a contradiction. Thus, $\delta_1 = |D_m|$.

Finally, let $V' = V - D_0 - D_m$ and consider the induced subgraph $G_{V'} = (V', E_{V'})$ which also satisfies (ii). Since its degree partition $V' = D_1 + \cdots + D_{m-1}$ is shorter by 2 than our original, the induction hypothesis proves the claim for $j < m$.

(iii) $\Rightarrow$ (iv) After some reflection (iv) is seen simply as a restatement of (iii).

(iv) $\Rightarrow$ (i) We shall assign an integer label a_i to each $x \in D_i$ such that the sum of the labels of the vertices in $X \subseteq V$ is less than or equal to a designated integer t if and only if X is a stable set. Now $D_0 + \cdots + D_{\lfloor m/2 \rfloor}$ is stable, and if X is a stable set containing a vertex $y \in D_j$ with $j > \lfloor m/2 \rfloor$, then $X - y \subseteq D_0 + \cdots + D_{m-j}$.

The reader may verify that the following labeling is a threshold assignment. (He should do the arithmetic base $|V|$.)

$$a_i = |V|^i \qquad\qquad (i = 0, 1, \ldots, \lfloor m/2 \rfloor),$$
$$t = 2|V|^{\lfloor m/2 \rfloor + 1},$$
$$a_j = t + 1 - |V|^{m-j+1} \qquad (j = \lfloor m/2 \rfloor + 1, \ldots, m). \qquad\blacksquare$$

Remark. Orlin [1977] has given a construction of the unique *integral* threshold assignment which minimizes the threshold t.

Notice that almost the mirror image of Figure 10.6 will appear if we replace edges by nonedges in that illustration. This is not surprising in light of the following corollary.

Corollary 10.6. The complement of a threshold graph is a threshold graph.

Proof. Assume that a labeling satisfying condition (ii) is given. The labeling $\bar{c}(x) = t - c(x)$ (for all $x \in V$) with the threshold $\bar{t} = t - 1$ satisfies (ii) for the complement since

$$xy \notin E \Leftrightarrow 0 \le t - c(x) - c(y)$$
$$\Leftrightarrow t \le \bar{c}(x) + \bar{c}(y)$$
$$\Leftrightarrow \bar{t} < \bar{c}(x) + \bar{c}(y). \qquad\blacksquare$$

From this corollary we conclude that a graph with n vertices is threshold if and only if there exists a hyperplane in $\mathbb{R}^n$ separating the characteristic vectors of the complete sets of vertices from the characteristic vectors of the noncomplete sets. An alternative proof of Corollary 10.6 follows from the next result.

Theorem 10.7 (Chvátal and Hammer [1973]).   A graph is threshold if and only if it has no induced subgraph isomorphic to $2K_2$, P_4, or C_4.

Proof.   The graphs $2K_2$, P_4, and C_4 are not threshold graphs (Figure 10.4), hence no threshold graph can contain one of them. Conversely, suppose $G = (V, E)$ is not threshold, then there exists a subset $X \subseteq V$ with $|X| \geq 4$ such that $\varnothing \neq \text{Adj}(x) \cap X \neq X - \{x\}$ for each $x \in X$. (This is a straightforward consequence of Theorem 10.4; see Exercise 3.) Choose $x_1 \in X$ to have the smallest degree in G_X and pick $x_2, x_3 \in X$ so that $x_1 x_2 \in E$ but $x_2 x_3 \notin E$. Since $|\text{Adj}(x_1) \cap X| \leq |\text{Adj}(x_3) \cap X|$ there exists an $x_4 \in X$ such that $x_3 x_4 \in E$ but $x_1 x_4 \notin E$. Thus, the set $\{x_1, x_2, x_3, x_4\}$ induces one of the three forbidden graphs $2K_2$, P_4, or C_4, which proves the theorem. ∎

Benzaken and Hammer [1978] have studied an analogous threshold problem for absorbent (or dominating) sets. A subset X of vertices is *absorbent* if every vertex not in X is adjacent to some member of X. The class of graphs obtained properly contains the threshold graphs. They give a number of characterizations of this class.

3.   A Characterization Using Permutations

Where does a threshold graph G fit into the world of perfect graphs? First of all, G is a split graph since its vertices can be partitioned into a stable set and a complete set; the edges between these sets are structured in a manner that has already been described. Secondly, the edges of G can be transitively oriented; let the vertices of G be numbered according to ascending degree and orient each edge toward its larger numbered endpoint. By Corollary 10.6, the complement $\bar{G}$ can also be transitively oriented, so G is a special kind of permutation graph. In the nomenclature of Section 6.1, every threshold graph is a triangulated–cotriangulated–comparability–cocomparability graph, or symbolically,

$$\text{THRESHOLD} \subset T \cap \bar{T} \cap C \cap \bar{C}.$$

This inclusion is proper as demonstrated by the graph P_4.

Let us characterize threshold graphs in the context of permutation graphs.

Let π be a permutation of the numbers $\{1, 2, \ldots, n\}$. In Chapter 7 we defined the graph of π, denoted by $G[\pi]$, to have vertices numbered v_1,

$v_2, \ldots, v_n$, with v_i and v_j adjacent if and only if $(i - j)(\pi_i^{-1} - \pi_j^{-1}) < 0$. For example, writing π as the sequence $\pi_1 \pi_2 \cdots \pi_n$ we see that $G[1, 2, \ldots, n]$ has no edges whereas $G[n, n - 1, \ldots, 1]$ is the complete graph. Recall that the graphs $G[\pi]$ and $G[\pi^\rho]$ are complementary, where π^ρ denotes π written in reversed sequential order. Finally, an undirected graph G is a *permutation graph* if it is isomorphic to $G[\pi]$ for some π.

Let σ and τ be two sequences over some alphabet. The *shuffle product* is defined as follows:

$$\sigma \sqcup \tau = \{\sigma_1 \tau_1 \cdots \sigma_k \tau_k | \sigma = \sigma_1 \cdots \sigma_k \quad \text{and} \quad \tau = \tau_1 \cdots \tau_k\}.$$

Here the σ_i and τ_i are subsequences, k ranges over all integers, and juxtaposition means concatenation. The notion of shuffle product appears in automata theory (see Eilenberg [1974]).

Theorem 10.8 (Golumbic [1978a]).   The threshold graphs are precisely those permutation graphs corresponding to sequences contained in

$$[1, 2, \ldots, p] \sqcup [n, n - 1, \ldots, p + 1], \tag{5}$$

where p and n are positive integers and $\sqcup$ denotes shuffle product.

Proof.   Let $G = (V, E)$ be a given threshold graph with degree partition $V = D_0 + D_1 + \cdots + D_m$. Let $s_i = \sum_{q=0}^{i} |D_q|$ and rename the vertices $v_1, v_2, \ldots, v_n$ such that $\deg v_i < \deg v_j$ implies $i < j$. We define a permutation π as follows:

$$\gamma_0 = \begin{cases} [1, \ldots, D_0] & \text{if } |D_0| > 0, \\ \text{empty sequence} & \text{otherwise}; \end{cases}$$

$$\gamma_i = \begin{cases} [1 + s_{i-1}, \ldots, s_i] & \text{for } 1 \leq i \leq \lfloor m/2 \rfloor, \\ [s_i, \ldots, 1 + s_{i+1}] & \text{for } \lfloor m/2 \rfloor < i \leq m; \end{cases}$$

$$\pi = \gamma_0 \gamma_m \gamma_1 \gamma_{m-1} \gamma_2 \gamma_{m-2} \cdots \gamma_{\lfloor m/2 \rfloor}.$$

Note that π is of the form (5), and that

$$v_z \in D_k \Leftrightarrow s_{k-1} < z \leq s_k \Leftrightarrow z \in \gamma_k. \tag{6}$$

We will show that $G = G[\pi]$.

Choose vertices $v_x \in D_i$ and $v_y \in D_j$; we may assume that $x < y$ and hence, by construction, $i \leq j$. By (6), v_x and v_y are adjacent in $G[\pi]$ if and only if y appears to the left of x in π. This will occur if and only if either (i) $i \leq \lfloor m/2 \rfloor$ and γ_j is strictly to the left of γ_i or (ii) $\lfloor m/2 \rfloor < i \leq j$. But conditions (i) and (ii) together are equivalent with $i + j > m$. Hence, by Theorem 10.4(iii), v_x and v_y are adjacent in $G[\pi]$ if and only if $v_x v_y \in E$, proving that $G = G[\pi]$.

Conversely, any permutation of the form (5) yields a threshold graph.   ∎

Teng and Liu [1978] use the shuffle product for the integration of several logically independent and concurrently operating transmission grammars. Transition grammars describe the rules, or *protocols*, which regulate the interactions between the attached entities in a computer network to ensure that they proceed in an orderly fashion.

4. An Application to Synchronizing Parallel Processes

Threshold graphs were rediscovered and studied by others, including Henderson and Zalcstein [1977]; they are responsible for the application presented here. See also Vantilborgh and van Lamsweede [1972].

A hypergraph $H = (S, \mathscr{E})$ consisting of a vertex set S and a hyperedge collection $\mathscr{E}$ of subsets of S, is called a *threshold hypergraph* if there exists a non-negative integer labeling c of S and an integer threshold t such that for all $X \subseteq S$,

$$X \text{ contains no hyperedge} \Leftrightarrow \sum_{x \in X} c(x) \leq t.$$

As before, we call the pair $[c; t]$ a *threshold assignment* for H.

Unlike the special case when H is a graph for which many results are known, the problem of characterizing threshold hypergraphs is unsolved and appears to be quite difficult. Nevertheless, threshold graphs and hypergraphs can be useful in an application to computing which we will now present.

Consider a set of computer programs $\mathscr{P} = \{P_i\}$ to be run in parallel. (Some of the P_i might actually be subroutines of larger programs.) Because of overall memory constraints or common memory location requirements some conflict may arise when a certain subset $\mathscr{P}'$ of $\mathscr{P}$ is not able to run simultaneously. Let $\mathscr{E}$ denote the collection of all such forbidden $\mathscr{P}'$. Hence, the programs in $X \subseteq \mathscr{P}$ can be run together without conflict if and only if X contains no member of $\mathscr{E}$.

When $(\mathscr{P}, \mathscr{E})$ is a threshold hypergraph a particularly simple programming technique can be applied to let the computer prevent conflicts automatically and control the traffic of programs running and waiting. Let $[c; t]$ be a threshold assignment for $(\mathscr{P}, \mathscr{E})$ and denote $c_i = c(P_i)$. The technique is as follows.

(1) Precede each program P_i with a call to procedure $P(s, c_i)$.
(2) Follow each program P_i with a call to procedure $V(s, c_i)$.
(3) Initialize a new global variable s with the value t.

```
procedure P(s, c):                procedure V(s, c):
  if s ≥ c then                       s ← s + c
     s ← s − c                     return
     enter P_i
  else
     call again
  return
```

Figure 10.7.   Subroutine P requests permission to begin and subroutine V informs the traffic controller that the program is finished. Variable s records how much "room" is currently available.

(See Figure 10.7.) The variable s, called a *semaphore*, never allows the sum of the c_i for those programs currently running to exceed t. The number c_i resembles the space required to do P_i. Every time we wish to execute a routine P_i, the procedure P checks whether or not there is sufficient space (i.e., is $s \geq c_i$). If so, we reduce s by c_i and begin; if not, then we wait (in a queue) until there is enough space. When we finish P_i the procedure V releases c_i units of space.

Example 1.   Given a set of programs $\{P_i\}$ such that at most 12 of them can be executed simultaneously, assign $t = 12$ and $c(P_i) = 1$ for each i.

Example 2.   Let $\mathscr{P}$ consist of three types of processes: the readers $R_1, \ldots,$ R_r; the writers $W_1, \ldots, W_w$; and the mathematicians $M_1, \ldots, M_m$. Assume that we may execute simultaneously either at most one mathematician plus an unlimited number of readers or at most one writer.* This problem has the threshold assignment

$$
\begin{aligned}
c(R_i) &= 1 & (i = 1, \ldots, r), \\
c(M_j) &= r + 1 & (j = 1, \ldots, m), \\
c(W_k) &= 2r + 1 & (k = 1, \ldots, w), \\
t &= 2r + 1.
\end{aligned}
$$

Although there is no accurate graph theoretic formulation for Example 1, Example 2 can be viewed as a graph G with edges connecting the readers with the writers, the mathematicians with each other and the writers, and the writers with everyone. In this case the stable sets of G correspond to the subsets which can be executed simultaneously.

Example 3.   If we add some bureaucrats $B_1, \ldots, B_b$ to Example 2 who can work with writers but cannot work with mathematicians, then the system no longer has a threshold assignment.

---

* If someone is writing on the system, no one else may have access since changes are being made. Otherwise, as many readers can work as want, but only one mathematician can work because there is only one calculator and he needs a calculator.

Finally, suppose we have a system with threshold dimension k. We can proceed similarly using P's and V's except that k semaphores will be needed. One semaphore handles each inequality (or labeling), and a segment P_i can be entered if and only if there is sufficient resource according to each of the k semaphores.

EXERCISES

1. Prove that $\theta(G) \le n + 1 - \omega(G)$, where $\theta(G)$ and $\omega(G)$ denote the threshold dimension of G and the size of the largest clique of G, respectively.
2. Show that $\theta(H) \le \theta(G)$ for any induced subgraph H of G.
3. Prove the following: $G = (V, E)$ is a threshold graph if and only if for each subset $X \subseteq V$ there exists a vertex $x \in X$ such that $\mathrm{Adj}(x) \cap X = \emptyset$ or $\mathrm{Adj}(x) \cap X = X - \{x\}$ (i.e., x is adjacent to all the vertices of $X - \{x\}$ or to none of them; Chvátal and Hammer [1973].)
4. Show that the following procedure will recognize threshold graphs. What is its complexity?

```
Boolean procedure THRESHOLD(G):
begin
    while the edge set is nonempty do
        begin
            delete all isolated vertices;
            if there is a vertex x adjacent to all remaining vertices then delete x;
            else
                return false;
        end
    return true;
end
```

5. Prove the following: A graph $G = (V, E)$ is threshold if and only if its vertices can be ordered and partitioned into a stable set $X = \{x_1, x_2, \ldots, x_s\}$ and a complete set $Y = \{y_1, \ldots, y_t\}$ such that

$$x_i y_j \in E \Rightarrow x_{i'} y_{j'} \in E \qquad (i' \ge i, \quad j' \ge j).$$

6. Prove that a threshold graph G with degree partition $V = D_0 + D_1 + \cdots + D_m$ has a Hamiltonian circuit if and only if the following relations are satisfied:

$$|D_0| = 0,$$

$$\sum_{i=1}^{k} |D_i| < \sum_{j=m+1-k}^{m} |D_j| \qquad (k = 1, 2, \ldots, \lfloor (m-1)/2 \rfloor),$$

$$\sum_{i=1}^{m/2} |D_i| \le \sum_{j=m/2+1}^{m} |D_j| \qquad \text{(if } m \text{ is even).}$$

Show how one may obtain the Hamiltonian circuit.

7. Let H have vertices $1, 2, \ldots, n$ and let DEG (i) equal the degree of vertex i. Write an algorithm which verifies the recurrence relations in Theorem 10.4(iv). Prove that your algorithm runs in $O(n)$ time.

8. Calculate the threshold dimension for the graphs in Figure 10.8 thus showing that $\theta(G)$ is not in general equal to $\theta(\bar{G})$ for nonthreshold graphs.

Figure 10.8.

9. Find necessary and sufficient conditions for a sequence $[a_1, a_2, \ldots, a_n; t]$ to be a threshold assignment for some threshold graph.

10. Prove that the number of mutually nonisomorphic n-vertex threshold graphs is 2^{n-1}.

11. Prove that G is a threshold graph if and only if equality holds in *each* of the Erdös–Gallai inequalities (see Section 6.3) (Hammer, Ibaraki, and Simeone [1978]).

12. Verify that the labeling given at the end of the proof of Theorem 10.4 is a threshold assignment.

13. Let $\theta'(G)$ denote the smallest integer k for which there exist partial subgraphs $(V_1, E_1), (V_2, E_2), \ldots, (V_k, E_k)$ of $G = (V, E)$ satisfying $E = E_1 \cup E_2 \cup \cdots \cup E_k$, where each (V_i, E_i) is a threshold graph. Prove that $\theta'(G)$ equals the threshold dimension $\theta(G)$ of G. (Note: You may assume $V_i = V$ for each i. Why?)

14. Let G be a threshold graph whose vertices are numbered according to increasing degree. Prove that the orientation obtained by directing each edge of G toward its larger numbered endpoint is transitive.

15. Let X be a set of propositions and let Y be a set of subjects in a psychological experiment. A subject either agrees or disagrees with a proposition. A *Guttman scale* is a linear ordering of $X \cup Y$ such that a subject agrees with all items following it and disagrees with all items preceding it. Let G be an undirected graph with vertex set $X \cup Y$ constructed as follows: X forms a stable set; Y forms a clique; subject y is adjacent to proposition x if and only if subject y agrees with proposition x. The following are from Leibowitz [1978]:

(i) Prove that there exists a Guttman scale for $X \cup Y$ if and only if G is a threshold graph.

(ii)   Give an algorithm to construct a Guttman scale. For a discussion of Guttman scales, see Coombs [1964].

16.   Prove that every threshold graph has an interval representation using intervals of at most two different lengths (Leibowitz [1978]).

17.   Let $m(G)$ denote the number of maximal cliques of an undirected graph G and let $\alpha(G)$ be the *stability number*. Clearly,

$$\alpha(G) \le m(G),$$

since there must be $\alpha(G)$ distinct cliques containing the members of a maximum stable set.

An undirected graph $G = (V, E)$ is said to be *trivially perfect* if for each $A \subseteq V$, the induced subgraph G_A of G satisfies $\alpha(G_A) = m(G_A)$. This name was chosen since it is trivial to show that such a graph is perfect. Prove the following (Golumbic [1978b]):

(i)   A graph $G = (V, E)$ is trivially perfect if and only if it contains no induced subgraph isomorphic to C_4 or P_4.

(ii)   A connected graph is trivially perfect if and only if it is a comparability graph whose Hasse diagram is a rooted tree.

(iii) G and $\bar{G}$ are both trivially perfect iff G is a threshold graph.

Research Problem.   Characterize the graphs of threshold dimension 2.

Research Problem.   Let S be a finite set and let $\mathscr{E}$ be a collection of subsets of S each of size r. The pair $H = (S, \mathscr{E})$ is usually called an r-regular hypergraph. If $r = 2$, then H is just an undirected graph. Consider the following properties:

(T_1)   There exists a (positive integer) labeling c of S and an (integer) threshold t such that, for all subsets $X \subseteq S$,

$$X \text{ contains no member of } \mathscr{E} \Leftrightarrow \sum_{x \in X} c(x) \le t.$$

(T_2)   There exists a (positive integer) labeling c' of S and an (integer) threshold t' such that for all subsets $A \subseteq S$ of size r,

$$A \in \mathscr{E} \Leftrightarrow \sum_{x \in A} c'(x) > t'.$$

(T_3)   For $x, y \in S$ define $x \geqslant y$ if x can replace y in any hyperedge (member) of $\mathscr{E}$. That is, $x \geqslant y$ if $[y \in A \in \mathscr{E}$ and $x \notin A]$ imply $A - \{y\} + \{x\} \in \mathscr{E}$. Then, for all $x, y \in S$, either $x \geqslant y$ or $y \geqslant x$ or both.

Clearly $(T_1) \Rightarrow (T_2) \Rightarrow (T_3)$. Either prove or disprove the reverse implications. [We know they are both true when $r = 2$. Perhaps a proof for $r = 3$ would generalize to arbitrary r.]

Bibliography

Benzaken, Claude, and Hammer, Peter L.
[1978]   Linear separation of dominating sets in graphs, *Ann. Discrete Math.* 3, 1–10.
Chvátal, Václáv, and Hammer, Peter L.
[1973]   Set-packing and threshold graphs, Univ. Waterloo Res. Report, CORR 73-21.
[1977]   Aggregation of inequalities in integer programming, *Ann. Discrete Math.* 1, 145–162.
Coombs, C. H.
[1964]   "A Theory of Data." Wiley, New York.
Ecker, K., and Zaks, S.
[1977]   On a graph labeling problem, Gesellschaft für Mathematik und Datenverarbeitung mbH, Seminarbericht No. 99.
Eilenberg, Samuel
[1974]   "Automata, Languages and Machines," Vol. A. Academic Press, New York.
Erdös, Paul
[1961]   Graph theory and probability, II, *Canad. J. Math.* 13, 346–352. MR22 #10925.
Erdös, Paul, and Gallai, Tibor
[1960]   Graphen mit Punkten vorgeschriebenen Grades, *Mat. Lapok* 11, 264–274.
Golumbic, Martin Charles
[1978a] Threshold graphs and synchronizing parallel processes, *in* "Combinatorics" (A. Hajnal and V. T. Sós, eds.), Colloq. Math. Soc. Janos Bolyai, Vol. 18, pp. 419–428. North-Holland, Budapest.
[1978b] Trivially perfect graphs, *Discrete Math.* 24, 105–107.
Hammer, Peter L., Ibaraki, T., and Simeone, B.
[1978]   Degree sequences of threshold graphs, Univ. of Waterloo, Dept. of Combinatorics and Optimization, Res. Report CORR 78-10.
Henderson, Peter B., and Zalcstein, Y.
[1977]   A graph-theoretic characterization of the PV chunk class of synchronizing primatives, *SIAM J. Comput.* 6, 88–108.
Leibowitz, Rochelle
[1978]   Interval counts and threshold graphs, Ph.D. thesis, Rutgers Univ.
Orlin, J.
[1977]   The minimal integral separator of a threshold graph, *Ann. Discrete Math.* 1, 415–419.
Peled, Uri
[1977]   Matroidal graphs, *Discrete Math.* 20, 263–286.
Poljak, S.
[1974]   A note on stable sets and coloring of graphs, *Comm. Math. Univ. Carolinae* 15, 307–309.
Teng, Albert Y., and Liu, Ming T.
[1978]   A formal approach to the design and implementation of network communication protocol, *Proc. COMPSAC 78*, Chicago, Illinois, 722–727.
Vantilborgh, H., and van Lamsweede, A.
[1972]   On an extension of Dijkstra's semaphore primitives, *Inform. Process. Lett.* 1, 181–186.

CHAPTER 11

Not So Perfect Graphs

1. Sorting a Permutation Using Stacks in Parallel

Let π be a permutation of the numbers $\{1, 2, \ldots, n\}$, which we will regard as the sequence $\pi = [\pi_1, \pi_2, \ldots, \pi_n]$. We would like to sort π into natural order using a system of stacks arranged in parallel, illustrated in Figure 11.1. Initially, the permutation sits on the input queue. Two types of moves are allowed: (i) moving the number at the head of the input queue onto the top of one of the stacks or (ii) moving a number from the top of a stack to the tail of the output queue. A successful sorting is accomplished by transferring all numbers to the output queue in the order $[1, 2, \ldots, n]$ by repeatedly applying (i) and/or (ii).

Given a sufficient number of stacks, any permutation can be sorted in this manner. But when can a permutation π be sorted using a system of only m stacks in parallel? For example, the sequence $\pi = [3, 5, 4, 1, 6, 2]$ requires three stacks, since the numbers 3, 5, and 6 must be stored on different stacks until 2 has reached the output queue. The observation that 3, 5, and 6 occur in their natural order but are followed by the smaller number 2 is the key to converting this sorting problem into a graph coloring problem.

Let $H[\pi]$ be the undirected graph having vertices $\{1, 2, \ldots, n\}$ with j and k adjacent if there exists an i such that

$$i < j < k \qquad \text{and} \qquad \pi_j^{-1} < \pi_k^{-1} < \pi_i^{-1}.$$

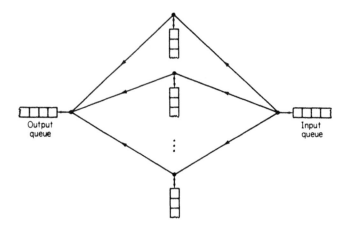

Figure 11.1.   A system of stacks in parallel.

We can read $\pi_i^{-1} = (\pi^{-1})_i$ as "the position of i in π." An example of this construction is given in Figure 11.2. We call $H[\pi]$ the *stack sorting graph* of π. It is a straightforward exercise to show the following:

π can be sorted in a system of m stacks in parallel if and only if the chromatic number of $H[\pi]$ is at most m.

One possible application of this sorting technique is in rearranging the railroad cars of a train in a switching yard (see Knuth [1969, Section 2.2.1; 1973, pp. 169–170], Even and Itai [1971], and Tarjan [1972]).

Let $\mathscr{H}$ be the collection of all graphs G such that G is isomorphic to $H[\pi]$ for some permutation π. Very little is known about the class $\mathscr{H}$. There have been no efficient recognition or coloring algorithms produced for the graphs in $\mathscr{H}$ which are in general *not perfect graphs*. Neither is there a good graph theoretic characterization. Our reason for introducing the class $\mathscr{H}$ is to show an equivalence between $\mathscr{H}$ and another class of graphs which has frustrated mathematicians for some years, namely, the circle graphs.

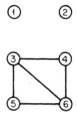

Figure 11.2.   The graph $H[3, 5, 4, 1, 6, 2]$.

2.   Intersecting Chords of a Circle

An undirected graph G is called a *circle graph* if it is isomorphic to the intersection graph of a finite collection of chords of a circle (see Figure 11.3). Without loss of generality, we may assume that no two chords share a common endpoint.

Theorem 11.1 (Even and Itai [1971]).   An undirected graph G is a graph of intersecting chords of a circle if and only

$$G - \{\text{all isolated vertices}\} \cong H[\pi] - \{\text{all isolated vertices}\}$$

for some permutation π.

This theorem will be proved constructively by demonstrating two techniques, Algorithms 11.1 and 11.2, whose correctness will be shown in Propositions 11.2 and 11.3, respectively. The algorithms transform one representation into the other.

Remark 1.   The subtraction of isolated vertices in the theorem is required. Two intersecting chords would give the complete graph on two vertices, whereas any graph $H[\pi]$ which has an edge must have at least three vertices.

Remark 2.   From the point of view of coloring, covering by cliques, and finding a maximum stable set or maximum clique, isolated vertices neither add to nor subtract from the essential complexity of the problem.

Remark 3.   A circle with intersecting chords enables us to generalize the notion of a matching diagram which we encountered in Section 7.4. Furthermore, sorting a permutation using stacks in parallel is very much like the problem of sorting a permutation using parallel queues discussed in Section

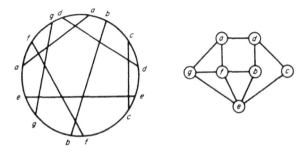

Figure 11.3.   A set of chords and its intersection graph.

7.5. The remarkable feature of Theorem 11.1 is that the equivalence established in Chapter 7 between permutation graphs and sorting in parallel queues extends to an equivalence between circle graphs and sorting in parallel stacks.

Algorithm 11.1.   Constructing a circle with chords from a permutation.

Input: A permutation π of the numbers $1, 2, \ldots, n$.
Output: A circle $\mathscr{C}$ with n chords.
Method: The algorithm is as follows:

1.   DRAW A CIRCLE. Label nodes $\pi_1, \pi_2, \ldots, \pi_n$ in a clockwise manner.
2.   We go once around the circle clockwise starting just prior to π_1.
3.   **for** $i \leftarrow 1$ **to** n **do**
4.   **if** you have not passed by node i
5.   **then** SKIP clockwise to i;
6.   Draw another node i immediately clockwise (but before the next node);
7.   **next** i;
8.   DRAW chords matching the pairs of numbers.

Example 11.1.   We apply Algorithm 11.1 to the permutation $\pi = [2, 9, 4, 6, 7, 1, 3, 8, 5]$. The instructions executed by the algorithm are given in Figure 11.4 along with the stack sorting graph $H[\pi]$ and the initial and final configurations for the circle $\mathscr{C}$ of chords.

Proposition 11.2.   Given a permutation π, Algorithm 11.1 constructs a set of chords of a circle whose intersection graph is isomorphic to $H[\pi]$.

Proof.   Suppose j and k are adjacent in $H[\pi]$ and assume $j < k$. Then there is an i such that $i < j < k$ and $\pi_j^{-1} < \pi_k^{-1} < \pi_i^{-1}$, which implies that after the ith iteration of Algorithm 11.1 j and k have already been passed

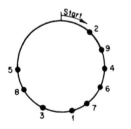

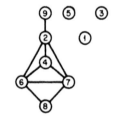

Initial configuration   Final configuration $H[2, 9, 4, 6, 7, 1, 3, 8, 5]$

Figure 11.4.   Algorithm 11.1 applied to the permutation $\pi = [2, 9, 4, 6, 7, 1, 3, 8, 5]$. The instructions executed are as follows.
Skip to 1; draw a 1. Draw a 2. Skip to 3; draw a 3. Draw a 4. Skip to 5; draw a 5. Draw 6–9.

over. Hence we shall write the new j before the new k, and their chords will therefore intersect.

Conversely, suppose two chords intersect and, reading clockwise around the circle from the starting point, their endpoints are labeled $j, k, j, k.$* Thus $j < k$ and $\pi_j^{-1} < \pi_k^{-1}$. Since in the jth iteration we had already passed the first k, there must be an $i, i < j$, such that during the ith iteration we skipped over k to the first occurrence of i. Thus, $\pi_k^{-1} < \pi_i^{-1}$, so j and k are adjacent in $H[\pi]$. ∎

Algorithm 11.2. Constructing a permutation from a circle with chords.

Input: A circle $\mathscr{C}$ with chords.
Output: A permutation π of the numbers $1, 2, \ldots, n$.
Method: The algorithm is as follows:

1. Pick a number n (a lucky choice will eliminate renumbering later);
2. Initialize: $i \leftarrow n$; pick a starting point (not an endpoint of a chord);
3. **for** once around the circle going counterclockwise **do**
   begin
4.   **if** next endpoint p is unlabeled
       then
5.   label it i; label its opposite endpoint i';
6.   decrement: $i \leftarrow i - 1$;
       else
7.   create a dummy endpoint on the circle just preceding p;
8.   label the dummy i';
9.   decrement: $i \leftarrow i - 1$;
10.   **skip** to just prior to the next unlabeled endpoint;
    end
11. **renumber** everything so that the smallest label is 1 (by subtracting the final value of i from each);
12. **print** the sequence of primed numbers running clockwise from the starting point and call them $\pi_1, \pi_2, \ldots,$ respectively;

Example 11.2. Applying Algorithm 11.2 to the circle $\mathscr{C}$ in Figure 11.5, we obtain the permutation $\pi = [7, 4, 2, 10, 6, 1, 8, 3, 9, 5]$. The instructions executed by the algorithm and the final (labeled) configuration for $\mathscr{C}$ are also given.

Proposition 11.3. Given a set of chords of a circle $\mathscr{C}$, Algorithm 11.2 finds a permutation π such that the intersection graph of $\mathscr{C}$ is isomorphic to $H[\pi] - \{\text{some isolated vertices}\}$.

---

* The second occurrences of j and k were created from the index of the loop, which is increasing.

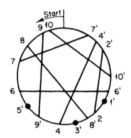

C: Initial configuration   Final configuration

Figure 11.5.   Algorithm 11.2 applied to the circle $\mathscr{C}$. The instructions executed are the following.
   We cheat and pick $n = 10$. Label chords 10, 9, 8, 7, and 6. Create dummy 5′ and skip over 9′. Label chord 4. Create dummy 3′ and skip over 8′. Label chord 2. Create dummy 1′ and skip over 6′, 10′, 2′, 4′, and 7′. The permutation is $\pi = [7, 4, 2, 10, 6, 1, 8, 3, 9, 5]$.

Proof.   Suppose j and k are adjacent in $H[\pi]$ and assume $j < k$. Then there is an i such that $i < j < k$ and their primed versions appear in the clockwise order $j′, k′, i′$. This implies that $j′$ and $k′$ are not dummy endpoints.* Since unprimed numbers occur in decreasing order going counterclockwise, it follows that the jth and kth chords intersect.
   Conversely, if the jth and kth chords intersect with $j < k$, then $k′$ was skipped over after labeling some dummy endpoint $i′$, where $i < j$. So $\pi_j^{-1} < \pi_k^{-1} < \pi_i^{-1}$ and j and k are adjacent in $H[\pi]$.   ∎

For small examples these algorithms are easy to do by hand. We would like to suggest a data structure suitable for performing the algorithms on a computer. A circle $\mathscr{C}$ with chords may be represented by either a list or an array consisting of the endpoints of the chords given in the counterclockwise order, beginning with a fixed but arbitrary starting point. There will be pointers providing direct access from one endpoint of a chord to the opposite endpoint. An example of this data structure is given in Figure 11.6. Notice that ARRAY(i) = ARRAY(OPPOSITE(i)) for all i in the example.
   To implement Algorithm 11.2 we scan the data structure corresponding to $\mathscr{C}$ once from left to right, labeling endpoints appropriately The property of an endpoint being primed can be coded into the label. On the other hand, Algorithm 11.1 would receive its input π as the reversed list $[\pi_n, \ldots, \pi_2, \pi_1]$ into which the new nodes are inserted. As the list is scanned from right to left, we keep track of which numbers have been passed by using an auxiliary bit vector. Both of these implementations can be carried out in time and space proportional to the size of the input.

---

* Because any dummy endpoint following $i′$ counterclockwise would have smaller value.

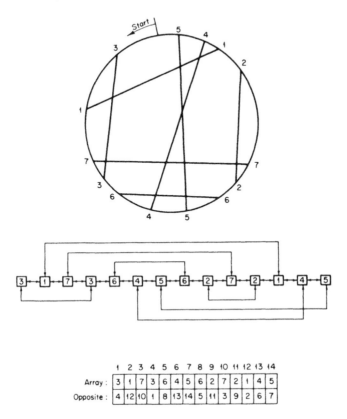

Figure 11.6. Data structures for a circle with chords.

Mark Buckingham has suggested the following algorithm to construct the adjacency sets of the intersection graph obtained from a circle with chords.

Algorithm 11.3.
Input: The data structure DS, as described above, representing a collection of *n* chords of a circle.
Output: The adjacency sets of the intersection graph.
Method: The algorithm is given in Figure 11.7. We traverse DS (i.e., the circle counterclockwise) exactly once. Chords j and k intersect if and only if their endpoints occur in the order $[k, j, k, j]$ or $[j, k, j, k]$. Each chord k is added to the end of a list called LIST when its first endpoint is encountered (line 4). It remains there until the second endpoint is reached at which time all chords j on the LIST following k are discovered to intersect chord k (lines 5–7). Then k is removed from the LIST (line 8). An array POINTER(k),

```
begin
1.      initialize: LIST ← ∅; for i ← 1 to n do Adj(i) ← ∅;;
2.      for each entry k of DS do in order
3.          if k is not on LIST
               then
4.                  append k to the tail LIST;
               else
5.                  for each j to the right of k on LIST do
6.                      add j to Adj(k);
7.                      add k to Adj (j);
                   next j;
8.                  delete k from LIST;
           next k;
end
```

Figure 11.7.   Algorithm 11.3.

initially undefined, may be used to execute efficiently the test in line 3 and the access to the starting point in line 5. A proof of correctness is left as Exercise 8.

Remark.   Touchard [1952], Riordan [1975], and Read [1979] investigate a generating function for the number of ways of drawing n chords of a circle so as to obtain k intersections.

3.   Overlap Graphs

The circle graphs are equivalent to yet another popular class of graphs, namely, the *overlap graphs*. Given a collection of intervals on a line, each pair of intervals will satisfy exactly one of the following properties.

Overlap.   The two intervals intersect but neither properly contains the other.

Containment.   One of the two intervals properly contains the other.

Disjointness.   The two intervals have empty intersection.

A graph G is called an *overlap graph* if its vertices may be put into one-to-one correspondence with a collection of intervals on a line such that two vertices are adjacent in G if and only if their corresponding intervals overlap (not just intersect). Without loss of generality we may assume that the intervals are either open or closed and that no two intervals have a common endpoint.

Let $\mathscr{I} = \{I_x\}_{x \in V}$ be a collection of intervals on a line, and assume that no two intervals have a common endpoint. The pairs of distinct indices are partitioned into three mutually disjoint sets A, B, D as follows: For distinct $x, y \in V$,

$$xy \in A \quad \text{if} \quad \varnothing \neq I_x \cap I_y \neq I_x, I_y$$

(i.e., the intervals overlap);

$$xy \in B \quad \text{if} \quad \text{either } I_x \subset I_y \quad \text{or} \quad I_y \subset I_x$$

(i.e., one interval properly contains the other);

$$xy \in D \quad \text{if} \quad I_x \cap I_y = \varnothing$$

(i.e., the intervals are disjoint).

Clearly, $A, B,$ and D are symmetric relations partitioning all pairs. Thus we have that (V, A) is the overlap graph represented by $\mathscr{I}$, $(V, A + B)$ is the interval graph represented by $\mathscr{I}$, and (V, D) is a comparability graph since its complement is an interval graph. Furthermore, defining

$$xy \in C \quad \text{if} \quad I_x \subset I_y$$

and

$$xy \in F \quad \text{if} \quad I_x \text{ lies entirely to the left of } I_y,$$

it follows that (V, C) and (V, F) are transitive orientations of (V, B) and (V, D), respectively. An example of these graphs is illustrated in Figure 11.8.

(a)

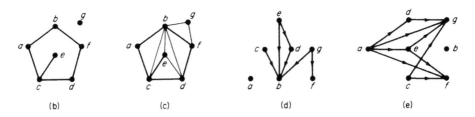

(b)　　　　　(c)　　　　　(d)　　　　　(e)

Figure 11.8. (a) A collection $\mathscr{I}$ of intervals. (b) The overlap graph (V, A) of $\mathscr{I}$. (c) The interval graph $(V, A + B)$ of $\mathscr{I}$. (d) The transitive orientation (V, C) representing proper containment. (e) The transitive orientation (V, F) representing disjointness.

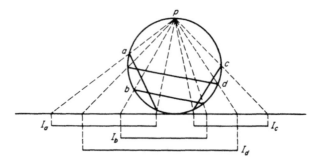

Figure 11.9. Intersecting chords of the circle correspond to overlapping intervals on the line. This projection method is suggested by Gavril [1973]. Relative to this choice of p, we have I_a and I_b overlapping, I_b and I_c overlapping, I_a and I_c disjoint, and I_d containing I_b.

We have the following easy result.

Proposition 11.4. An undirected graph G is a circle graph if and only if G is an overlap graph.

Proof. Given a circle with chords, choose a point p on the circle which is not an endpoint of a chord. To each chord with endpoints c and c' we associate the arc along the circle from c to c' which does not contain p. If the circle is cut at point p, then we will obtain a line with a collection of intervals having the desired property.

This process may be reversed by wrapping a collection of intervals of a line around a circle and then drawing a chord between the two endpoints of each interval. ∎

An alternate way to visualize this equivalence is by placing the circle tangent to a line, with p as the north pole, and the point of tangency as the south pole. Projecting the chords down to the line, as in Figure 11.9, we obtain the correspondence. Our data structure for representing a circle with chords is simply a discrete version of this linearization.

4. Fast Algorithms for Maximum Stable Set and Maximum Clique of These Not So Perfect Graphs

In the preceding sections we demonstrated the equivalence of the stack sorting graphs, the circle graphs, and the overlap graphs. We are therefore free to choose whichever model suits us best in order to prove properties about the class.

As we mentioned earlier, there are a number of open problems concerning this class of graphs.

(i)   Find an algorithm which recognizes circle graphs and constructs a representation for the graph as intersecting chords of a circle.

(ii)*   Are the coloring and clique-cover problems NP-complete for circle graphs?

(iii)   Is the strong perfect graph conjecture true for circle graphs?

The stable set problem and the clique problem are tractable when restricted to our not so perfect graphs. In the context of open problem (i), it is essential that we be given, *a priori*, a representation of the graph as overlapping intervals, intersecting chords, or a permutation to be sorted. We choose to use the overlap graph model. We first present a solution to the stable set problem, due to Gavril [1973], which can be implemented to run in polynomial time.

Let $\mathscr{I} = \{I_x\}_{x \in V}$ be a collection of intervals, let $G = (V, A)$ be the overlap graph of $\mathscr{I}$, and let C and F be the oriented containment and disjointness relations as defined in the preceding section. For all $x \in V$, let

$$U(x) = \{v \in V \mid vx \in C\}$$

be the set of indices whose corresponding intervals are (properly) contained in I_x. The algorithm is as follows.

Algorithm 11.4.   Maximum stable set of an overlap graph.

Input: The (transitively) oriented containment relation (V, C) and the (transitively) oriented disjointness relation (V, F) of a collection $\mathscr{I}$ of intervals whose overlap graph is $G = (V, A)$.
Output: A maximum stable set S of G.
Method: We assign, to each vertex $x \in V$, a weight $w(x)$ and a maximum stable set $S(x)$ of $G_{\{x\} \cup U(x)}$, where $w(x) = |S(x)|$. This is carried out recursively in such a way that vertices are assigned weights in a topologically sorted order with respect to C. At the heart of the algorithm is the subroutine (from Chapter 5) MAXWEIGHT CLIQUE, which finds a set of pairwise disjoint intervals $\{I_v\}_{v \in T}$ that generates, in line 6 of MAXSTABLE, the best possible stable set. The entire algorithm consists of the single call,

begin
  $S \leftarrow$ MAXSTABLE(V);
end

and uses the recursive procedure in Figure 11.10. The assumption that F is transitive is crucial since it allows line 5 to be executed efficiently.

---

* Garey, Johnson, Miller and Papadimitrious [1979] report that the coloring problem for circle graphs *is* NP-complete.

```
procedure MAXSTABLE(X):
   begin
1.    if X = ∅ then return ∅;;
2.    while there exists x ∈ X with w(x) undefined do
3.        S(x) ← {x} ∪ MAXSTABLE(U(x));
4.        w(x) ← |S(x)|;;
5.    T ← MAXWEIGHT CLIQUE(X, Fₓ);
6.    return ⋃ᵥ∈ₜ S(v);
   end
```

Figure 11.10

Theorem 11.5 (Gavril [1973]).   Algorithm 11.4 correctly finds a maximum stable set of an overlap graph.

An example of Algorithm 11.4 applied to the intervals in Figure 11.8 follows the proof of the theorem.

Proof.   We shall show that the procedure returns a maximum stable set of the subgraph $G_X = (X, A_X)$ for any subset $X \subseteq V$ satisfying $U(x) \subset X$ for all $x \in X$. The claim is certainly true if $|X| = 0$. Assume that it is true for all subsets smaller than X. In particular, by induction, $S(x) - \{x\}$ is a maximum stable set of $G_{U(x)}$, so $S(x)$ is a maximum stable set of $G_{\{x\} \cup U(x)}$ for each $x \in X$.

Let T be as defined in line 5. Since the intervals $\{I_v\}_{v \in T}$ are pairwise disjoint, and since $I_x \subset I_v$ for all $x \in S(v) - \{v\}$ and $v \in T$, it follows that $J = \bigcup_{v \in T} S(v)$ is a stable set of G_X. Thus

$$\alpha(G_X) \geq \sum_{v \in T} w(v) = |J|. \tag{1}$$

We must show that J is maximum.

Let J' be a maximum stable set of G_X, and let T' be the set of sinks of $(J', C_{J'})$; i.e.,

$$T' = \{y \in J' \mid I_y \subset I_z \text{ implies } z \notin J'\}.$$

Note that the intervals represented by T' are also pairwise disjoint, so, by the correctness of MAXWEIGHT CLIQUE,

$$\sum_{y \in T'} w(y) \leq \sum_{v \in T} w(v). \tag{2}$$

Now, clearly, if $S'(y) = \{x \in J' \mid I_x \subset I_y\}$, then $|S'(y)| = w(y)$ for all $y \in T'$, for otherwise we could replace $S'(y)$ with $S(y)$ and obtain a larger stable set. Hence,

$$\alpha(G_X) = |J'| = \sum_{y \in T'} w(y). \tag{3}$$

Combining (1)–(3), we obtain

$$\alpha(G_X) = |J|,$$

which concludes the required proof.   ∎

Example 11.3. We apply Algorithm 11.4 to the intervals of Figure 11.8 in order to find a maximum stable set of their overlap graph $G = (V, A)$. The vertices can be assigned weights in any topologically sorted order with respect to C. We arbitrarily choose the order a,c,e,g,d,b,f. Clearly, $w(a) = w(c) = w(e) = w(g) = 1$ and $S(a) = \{a\}$, $S(c) = \{c\}$, $S(e) = \{e\}$, and $S(g) = \{g\}$ since they are all sources of C. Next, $S(d) = \{d, e\}$ and $w(d) = 2$ since MAX-WEIGHT CLIQUE $(U(d), F_{U(d)}) = U(d) = \{e\}$. Now $U(b) = \{c, d, e, g\}$ and the heaviest clique in $F_{U(b)}$ is $\{d, g\}$; thus $S(b) = \{b, d, g\}$ and $w(b) = 3$. Similarly, $S(f) = \{f, g\}$ and $w(f) = 2$. Finally, MAXWEIGHT CLIQUE (V, F) could be either $\{a, d, g\}$ or $\{a, e, f\}$, both having weight 4. They give, respectively, the stable sets $\{a, d, e, g\}$ and $\{a, e, f, g\}$.

Next we provide an algorithm due to Gavril [1973] which solves the clique problem for circle graphs in polynomial time. The notions of matching diagram and permutation graph from Chapter 7 will be used.

Let $G = (V, E)$ be a circle graph with representing family $\{C_v\}_{v \in V}$ of chords of a circle $\mathscr{C}$, and let $N(v) = \{v\} + \text{Adj}(v)$.

Lemma 1. For every vertex $v \in V$, the induced subgraph $G_{N(v)}$ is a permutation graph.

Proof. We may assume that no two chords have a common endpoint. Thus, the chord C_v cuts $\mathscr{C}$ into two pieces such that for $x \in \text{Adj}(v)$ the chord C_x has one endpoint in each piece. Therefore, the subset $D = \{C_x\}_{x \in \text{Adj}(v)}$ is a matching diagram whose permutation graph is $G_{\text{Adj}(v)}$. Since connecting a new vertex to every vertex of a permutation graph results in another permutation graph, it follows that $G_{N(v)}$ is a permutation graph.   ∎

Lemma 2. If K is a clique of G, then K is a clique of $G_{N(v)}$ for each $v \in K$.

Proof. Trivial.   ∎

Algorithm 11.5. Maximum clique of a circle graph. The algorithm is as follows:

```
begin
1.   for v ∈ V do K_v ← MAXCLIQUE(G_{N(v)});
2.   return the largest K_v;
end
```

By Lemma 1, statement 1 can be executed efficiently. By Lemma 2, the algorithm is correct. Details are left to the reader as Exercise 11.

5.   A Graph Theoretic Characterization of Overlap Graphs

Although we have shown the equivalence of circle graphs, stack sorting graphs, and overlap graphs, we have not really characterized them from a traditional graph theoretic point of view. In this section we shall present such a characterization. The solution, however, will fall short of providing us with an efficient recognition algorithm.

Theorem 11.6 (Fournier [1978]).   An undirected graph $G = (V, E)$ is an overlap graph if and only if there exists an acyclic orientation P of G and two linear extensions L_1 and L_2 of P such that the relation $F = (L_1 \cap L_2) - P$ satisfies

$$xy \in F, \; yz \in L_1 \Rightarrow xz \in F \tag{4}$$

and

$$xy \in L_2, \; yz \in F \Rightarrow xz \in F. \tag{5}$$

Remark.   Such a relation F is transitive.

Proof. ($\Rightarrow$)   Let $\mathscr{I} = \{I_x\}_{x \in V}$ be a collection of closed intervals on the real line, no two intervals sharing an endpoint, such that xx' is an edge of G if and only if I_x and $I_{x'}$ overlap (i.e., they intersect but neither contains the other). We denote $I_x = [a, b]$ and $I_{x'} = [a', b']$. Consider the binary relations defined on V as follows:

$$xx' \in P \iff a < a' < b < b',$$
$$xx' \in L_1 \iff a < a',$$
$$xx' \in L_2 \iff b < b'.$$

Clearly P is an acyclic orientation of G, and L_1 and L_2 are linear extensions of P. The relation $F = (L_1 \cap L_2) - P$ satisfies

$$xx' \in F \iff a < b < a' < b'$$

and represents I_x being entirely to the left of $I_{x'}$. It is easy to see that (4) and (5) are satisfied.
    ($\Leftarrow$)   In the remainder of the proof, for any binary relation R we denote

$$R(x) = \{y \mid xy \in R\}.$$

Let $G = (V, E)$ be an undirected graph on n vertices and let P, L_1, L_2, and F satisfy the conditions of the theorem. To a vertex x of G we associate the interval $I_x = [a, b]$ as follows:

$$a = 1 + |L_1^{-1}(x)| + |P^{-1}(x)|,$$
$$b = 2n - |L_2(x)| - |P(x)|.$$

We shall show that $\mathscr{I} = \{I_x\}_{x \in V}$ is an overlap representation of G.

Claim 1. $1 \le a < b \le 2n$.

We shall prove the inequality $a < b$, the others being trivial. By the definitions of a and b, it is enough to prove the inequality

$$|L_1^{-1}(x)| + |F^{-1}(x)| + |L_2(x)| + |F(x)| \le 2(n-1). \tag{6}$$

If $x' \in F^{-1}(x)$, then $x' \notin L_2(x)$, because $F \subset L_2$ and L_2 is antisymmetric; thus $F^{-1}(x) \cap L_2(x) = \varnothing$. Similarly, $F(x) \cap L_1^{-1}(x) = \varnothing$. Thus, each member x' of V is counted at most twice on the left side of (6) except for x itself, which is not counted at all. This proves Claim 1.

For vertices x and x' $(x \ne x')$, where $I_x = [a, b]$ and $I_{x'} = [a', b']$ are defined as above, we shall show the following three implications:

Claim 2.
(i) $xx' \in L_1 - (P \cup F) \Rightarrow a < a' < b' < b,$
(ii) $xx' \in F \Rightarrow b < a',$
(iii) $xx' \in P \Rightarrow a < a' < b < b'.$

Notice that $xx' \in L_1 - (P \cup F)$ if and only if $xx' \in L_1 - L_2$, and since L_2 is a total order, implication (i) would follow directly from

(i_1) $xx' \in L_1 \Rightarrow a < a'$ and (i_2) $xx' \in L_2 \Rightarrow b < b'.$

Let $xx' \in L_1$. Since L_1 is a total order we have $L_1^{-1}(x) \subset L_1^{-1}(x')$, and thus $|L_1^{-1}(x)| < |L_1^{-1}(x')|$. (The strict inequality is due to $x \in L_1^{-1}(x')$ and $x' \notin L_1^{-1}(x)$.) Also, from property (i_1), we have $F^{-1}(x) \subset F^{-1}(x')$, so $|F^{-1}(x)| < |F^{-1}(x')|$. Combining these inequalities we obtain

$$|L_1^{-1}(x)| + |F^{-1}(x)| < |L_1^{-1}(x')| + |F^{-1}(x')|,$$

which yields $a < a'$ and proves (i_1). Implication (i_2) is proved in the same fashion. This proves (i).

For implication (ii) we shall show that if $xx' \in F$, then

$$|L_1^{-1}(x')| + |F^{-1}(x')| + |L_2(x)| + |F(x)| \ge 2n. \tag{7}$$

Let $x'' \in V$. If $x'' \notin L_1^{-1}(x')$ (i.e., $x'x'' \notin L_1^{-1}$), then $x'x'' \in L_1$ since L_1 is a total order; moreover, (4) implies that $xx'' \in F$ (i.e., $x'' \in F(x)$), and hence $x'' \in L_2(x)$. In an analogous manner, if $x'' \notin L_2(x)$, then $x'' \in F^{-1}(x')$ and

$x'' \in L_1^{-1}(x')$. Thus, each vertex of V, including x and x', is counted at least twice on the left side of (7). This proves (ii).

For implication (iii), since (i_1) and (i_2) hold, it is sufficient to show that if $xx' \in P$ then $a' < b$, or equivalently, if $xx' \in P$, then

$$|L_1^{-1}(x')| + |F^{-1}(x')| + |L_2(x)| + |F(x)| \leq 2(n-1). \tag{8}$$

It is easy to verify that if $x'' \in F^{-1}(x')$ then $x'' \notin L_2(x)$ and $x'' \notin F(x)$ (since F is a transitive relation). Similarly, if $x'' \in F(x)$, then $x'' \notin L_1^{-1}(x')$ and $x'' \notin F^{-1}(x')$. Thus, each element of V is counted at most twice on the left side of (8), except for x and x', which are counted exactly once each ($x \in L_1^{-1}(x')$ and $x' \in L_2(x)$). This proves (iii) and concludes Claim 2.

Finally, the three conditions of (i)–(iii) are mutually exclusive and cover all possibilities. Therefore, the opposite implications also hold in (i)–(iii). In particular,

$$xx' \in P \Leftrightarrow a < a' < b < b'.$$

Since P is an orientation of G, we conclude that

$$xx' \in G \Leftrightarrow I_x \text{ and } I_{x'} \text{ overlap,}$$

and $\mathscr{I} = \{I_x\}_{x \in V}$ is the desired overlap model of G.   ∎

Remark.   If the relation F in Theorem 11.6 is empty, then G is a permutation graph; conversely, for every permutation graph there exist relations P, L_1, and L_2 as in the theorem with $P = L_1 \cap L_2$ (as in the proof of Theorem 7.1.) However, even when G is a permutation graph there may very well exist other relations P, L_1, and L_2 satisfying the conditions of the theorem for which $P \neq L_1 \cap L_2$. For example, letting $G = \bar{K}_3$ with $P = \varnothing$, $L_1 = [x < y < z]$, and $L_2 = [x < z < y]$, we obtain $L_1 \cap L_2 \neq \varnothing$.

Historical Note

We began this chapter by discussing a problem using stacks. We conclude with an historical note on one of the oldest written references to the notion of "last-in, first-out."* The reference occurs in a commentary by Rashi (Rabbi Solomon ben Isaac) on the Biblical verse

> Then his brother emerged, his hand seizing Esau's heel; so they named him Jacob.† Isaac was sixty years old when they were born [Genesis XXV, 26].

---

* We are indebted to Gideon Ehrlich for pointing out this reference in a communication with Edward M. Reingold, who then passed it on to the author. The translation of the Rashi quotation is due to E. M. Reingold.

† In Hebrew, Ya'akov, play on the word 'aqev meaning "heel."

Rashi lived from 1040 to 1105 A.D., residing primarily in Troyes, France, his birthplace, where he founded one of the leading schools of the time. He is the most famous biblical and talmudic commentator in all of Jewish history. His commentary on this verse is as follows:

> I heard a Midrashic legend expounding on the meaning [of the phrase "Then his brother emerged ..."]. It was his [Jacob's] right, the grasping of his [Esau's] heel: Jacob was conceived from the first drop and Esau from the second. Consider a tube with narrow mouth. Put two stones into it, one after the other—the first to enter exits last and the last to enter exists first. It is [thus] found [that] Esau, conceived last, exited first, and Jacob, conceived first, exited last. [Thus] Jacob went to delay him [Esau] so that he [Jacob] would be first born [just] as he was first produced and to be a first fruit of the womb and to take the birthright [as he deserved] according to the law.

Rashi is clearly describing a stack mechanism.

EXERCISES

1. Find a permutation π whose graph $H[\pi]$ is a chordless pentagon plus some isolated vertices.

2. Using the data structure suggested in the text, write computer programs to implement Algorithms 11.1 and 11.2 and test them on the examples given in this chapter.

3. Does $H[\pi]$ always have some isolated vertices? Prove that if $H[\pi]$ has exactly one isolated vertex then $H[\pi] = \overline{G[\pi]}$ but not conversely (see Chapter 7).

4. Show that if the output of Algorithm 11.2 is used as the input of Algorithm 11.1, then the resulting composition may change the set of chords. Modify Algorithm 11.2 so that this does not happen.

5. By an arbitrary convention we have discussed sorting a permutation π in a parallel system of stacks from right to left. The problem of sorting π from left to right* is equivalent to forming π from $[1, 2, \ldots, n]$ from right to left. In this case one should study the coloring problem on the undirected graph $H^p[\pi]$ having vertices $\{1, 2, \ldots, n\}$ with i and j adjacent if there is a k such that $i < j < k$ and $\pi_k^{-1} < \pi_i^{-1} < \pi_j^{-1}$. In general the chromatic numbers of $H[\pi]$ and $H^p[\pi]$ are different.

---

* n would emerge first and 1 last.

Let the function $\rho: i \mapsto n + 1 - i$ act on the labeled vertices of a graph*
and be composed with other permutations in the obvious way. Also let
$\mathcal{H}^\rho = \{G \mid G \cong H^\rho[\pi] \text{ for some } \pi\}$. Prove the following:

    (i) $\rho \circ H^\rho[\pi] = H[\rho \circ \pi \circ \rho]$,

    (ii) $\rho \circ H[\pi] = H^\rho[\rho \circ \pi \circ \rho]$,

    (iii) $\mathcal{H} = \mathcal{H}^\rho$.

6. Let the graph $H[\pi]$ be properly colored using t colors. Show that the
following algorithm correctly sorts π using a network of t stacks in parallel.
Input: The permutation π with the numbers properly colored.
Output: The permutation $[1, 2, \ldots, n]$ sorted.
Method: The t stacks are in one-to-one correspondence with the colors.
The algorithm is as follows:

```
begin
   k ← 1;
   while k ≤ n do
      if k can be moved onto the output queue then
         move k onto the output queue;
      else
         move the next number of the input queue onto the stack of its color;;
end
```

7. Show that a permutation π can be sorted in a network of k stack in
parallel under the restriction that all numbers *must* be loaded into the stacks
before any unloading can begin if and only if its reversal π^ρ can be sorted in a
network of k queues in parallel. Give some additional equivalent conditions.

8. Prove that Algorithm 11.3 correctly calculates the adjacency sets of the
graph $G = (V, E)$ of intersecting chords of a circle. Show that the algorithm
can be implemented to run in $O(|V| + |E|)$.

9. Let $f(n)$ be the length of the shortest string of numbers from $\{1, 2, \ldots, n\}$
which contains all $n!$ permutations as subsequences. Prove that $f(n) \leq n^2 -
2n + 4 \, (n \geq 3)$ (Koutas and Hu [1975]).

10. Determine the computational complexity of Algorithm 11.4.

11. Determine the computational complexity of Algorithm 11.5 taking into
consideration that each subgraph $G_{N(v)}$ must be calculated.

12. A circle $\mathscr{C}$ with chords *admits an equator* if an additional chord may be
added to $\mathscr{C}$ which will intersect every other chord.

    (i) Prove that G is a permutation graph if and only if G is the intersection
graph of a circle of chords which admits an equator.

    (ii) Give an example of a circle $\mathscr{C}$ which does *not* admit an equator but
whose intersection graph is a permutation graph.

13. The sequence of operations used in sorting the permutation $[3,5,4,1,6,2]$

---

*denoted $\rho \cdot G$.

as in Section 11.1 can be abbreviated by the code

$$S_1 S_2 S_2 S_1 X_1 S_3 S_1 X_1 X_1 X_2 X_2 X_3,$$

where S_i stands for "move the next number from the input queue onto stack i" and X_i stands for "move the number at the top of stack i to the output queue." Some sequences of S_i's and X_i's specify meaningless operations; for example, the sequence $S_1 S_2 S_1 X_2 X_1 S_3 X_2 X_1 S_2 X_3$ cannot be carried out.

We call a sequence of S_i's and X_i's *admissible* if it contains the same number of S_i's and X_i's for each integer i, and if it specifies no operation that cannot be performed. Formulate a rule which distinguishes between admissible and inadmissible sequences (Knuth [1969, Exercise 2.2.1, No. 3]).

Bibliography

Even, Shimon, and Itai, Alon
  [1971] Queues, stacks and graphs, *in* "Theory of Machines and Computations," pp. 71–86. Academic Press, New York.
Fournier, Jean-Claude
  [1978] Une caractérization des graphes de cordes, *C.R. Acad. Sci. Paris* **286A**, 811–813.
Garey, M. R., Johnson, D. S., Miller, G. L., and Papadimitrious, C. H.
  [1979] The complexity of coloring circular arcs and chords, submitted for publication.
Gavril, Fanica
  [1973] Algorithms for a maximum clique and a minimum independent set of a circle graph, *Networks* **3**, 261–273. MR49 #4862.
Knuth, Donald E.
  [1969] "The Art of Computer Programming," Vol. 1. Addison-Wesley, Reading, Massachusetts.
  [1973] "The Art of Computer Programming," Vol. 3. Addison-Wesley, Reading, Massachusetts.
Koutas, P. J., and Hu, T. C.
  [1975] Shortest string containing all permutations, *Discrete Math.* **11**, 125–132. MR50 #12740.
Read, Ronald C.
  [1979] The chord intersection problem. *Ann. N.Y. Acad. Sci.* **319**, 444–454.
Riordan, John
  [1975] The distribution of crossings of chords joining pairs of $2n$ points on a circle. *Math. Comp.* **29**, 215–222. MR51 #2933.
Tarjan, Robert Endre
  [1972] Sorting using networks of queues and stacks, *J. Assoc. Comput. Mach.* **19**, 341–346. MR45 #7852.   ·
Touchard, Jacques
  [1952] Sur un problème de configurations et sur les fractions continues, *Canad. J. Math.* **4**, 2–25. MR13, p. 716.
Zelinka, Bohdan
  [1965] The graph of the system of chords of a given circle, *Mat.-Fyz. Casopis Sloven. Akad. Vied* **15**, 273–279. MR33 #2575.

CHAPTER 12

Perfect Gaussian Elimination

1. Perfect Elimination Matrices

Let $\mathbf{M}$ be a nonsingular $n \times n$ matrix with entries m_{ij} in some field (like the real numbers). We *reduce* $\mathbf{M}$ to the identity matrix $\mathbf{I}$ by repeatedly

(a) choosing a nonzero entry m_{ij} to act as *pivot* and
(b) updating the matrix by using elementary row and column operations to change m_{ij} to 1 and to make all other entries in the ith row and jth column equal to 0.

This familiar technique is called Gaussian elimination* and can be found in most books on linear algebra.

When performing Gaussian elimination on a sparse matrix, an arbitrary choice of pivots may result in the filling in of some zero positions with nonzeros. One may ask, when is there a sequence of pivots which induces no fill-in? A *perfect elimination scheme* for $\mathbf{M}$ is a sequence of pivots which reduces $\mathbf{M}$ to $\mathbf{I}$ without ever changing a zero entry (even temporarily) to a nonzero. Such a sequence does not exist for every matrix. When $\mathbf{M}$ has a perfect elimination scheme and $\mathbf{M}$ is also sparse, then the sparseness can be preserved throughout the reduction. This is important for the storage requirements of $\mathbf{M}$ since a sparse matrix is most efficiently represented in a computer by listing its nonzero entries.

---

* In practice one usually "zeros out" just the jth column postponing calculations on the ith row until the end, at which time back substitution is used. For our purposes the methods are the same.

Bad Choice:
$$\begin{pmatrix} ④ & 1 & 1 & 1 \\ 1 & 1 & 0 & 0 \\ 1 & 0 & 1 & 0 \\ 1 & 0 & 0 & 1 \end{pmatrix} \rightarrow \begin{pmatrix} 1 & 0 & 0 & 0 \\ 0 & ③ & -1 & -1 \\ 0 & -1 & 3 & -1 \\ 0 & -1 & -1 & 3 \end{pmatrix} \rightarrow \begin{pmatrix} 1 & 0 & 0 & 0 \\ 0 & 1 & 0 & 0 \\ 0 & 0 & ⑧ & -4 \\ 0 & 0 & -4 & 8 \end{pmatrix} \rightarrow \begin{pmatrix} 1 & 0 & 0 & 0 \\ 0 & 1 & 0 & 0 \\ 0 & 0 & 1 & 0 \\ 0 & 0 & 0 & ⑫ \end{pmatrix}$$

$\downarrow$

Good Choice:
$$\begin{pmatrix} 4 & 1 & 1 & 1 \\ 1 & 1 & 0 & 0 \\ 1 & 0 & 1 & 0 \\ 1 & 0 & 0 & ① \end{pmatrix} \rightarrow \begin{pmatrix} 3 & 1 & 1 & 0 \\ 1 & 1 & 0 & 0 \\ 1 & 0 & ① & 0 \\ 0 & 0 & 0 & 1 \end{pmatrix} \rightarrow \begin{pmatrix} 2 & 1 & 0 & 0 \\ 1 & ① & 0 & 0 \\ 0 & 0 & 1 & 0 \\ 0 & 0 & 0 & 1 \end{pmatrix} \rightarrow \begin{pmatrix} 1 & 0 & 0 & 0 \\ 0 & 1 & 0 & 0 \\ 0 & 0 & 1 & 0 \\ 0 & 0 & 0 & 1 \end{pmatrix}$$

Figure 12.1. Two reductions of a matrix.

Examples. Two reductions of the same matrix are given in Figure 12.1; the first introduces new nonzeros but the second does not. What causes the fill-in? It is easy to see that pivoting on a nonzero entry m_{ij} will introduce some fill-in if $m_{sj} \neq 0$ and $m_{it} \neq 0$ but $m_{st} = 0$ for some s and t (see Figure 12.2). The matrix in Figure 12.3 is nonsingular but has no perfect elimination scheme. Any entry chosen as the first pivot will cause some fill-in. A tri-diagonal matrix, as indicated in Figure 12.4, has a number of perfect elimination schemes, one being the positions on the main diagonal ordered from top to bottom.

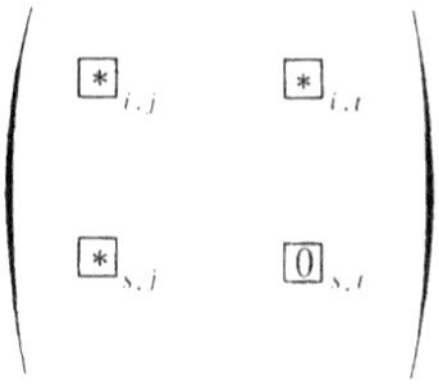

$$\begin{pmatrix} \boxed{*}_{i,j} & & \boxed{*}_{i,t} \\ & & \\ \boxed{*}_{s,j} & & \boxed{0}_{s,t} \end{pmatrix}$$

Figure 12.2. Choosing position (i, j) as pivot results in filling in position (s, t). Thus the choice is unacceptable. The asterisks indicate nonzeros.

$$\begin{pmatrix} 1 & 1 & 0 & 0 & 1 \\ 0 & 1 & 1 & 0 & 1 \\ 0 & 1 & 1 & 1 & 0 \\ 1 & 0 & 1 & 1 & 0 \\ 1 & 0 & 0 & 1 & 1 \end{pmatrix}$$

Figure 12.3. A matrix with no perfect elimination scheme.

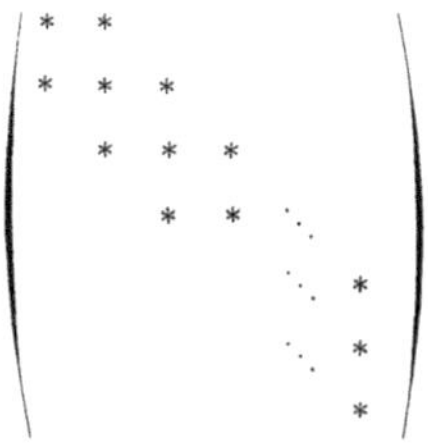

Figure 12.4. A tridiagonal matrix. The asterisks indicate the nonzero entities.

Throughout this chapter we shall assume that M is nonsingular and that arithmetic coincidence does not cause new zeros to occur during the reductions.*

Let us look at the problem graph theoretically.

The *graph* $G(M)$ *of* M has vertices $v_1, \ldots, v_n$, with $v_i v_j$ being an edge if and only if $i \neq j$ and entry $m_{ij} \neq 0$. The *bipartite graph* $B(M)$ of M has vertices $x_1, \ldots, x_n$ and $y_1, \ldots, y_n$, corresponding to the rows and columns, respectively, where x_i is adjacent to y_j if and only if $m_{ij} \neq 0$. We call x_i and y_i partners; their correspondence with vertex v_i of $G(M)$ is obvious.

2.   Symmetric Matrices

If M is a symmetric matrix, then $G(M)$ is an undirected graph. In this case a nonzero diagonal entry m_{ii} is acceptable as a pivot if and only if v_i is a simplicial vertex of $G(M)$. (Why?) In fact, pivoting on m_{ii} is equivalent to making $\text{Adj}(v_i)$ into a complete subgraph by adding any missing edges and deleting v_i. Therefore, the perfect elimination schemes for M under the restriction that

(R)   all pivots are chosen along the main diagonal whose entries are each nonzero

correspond precisely to the perfect vertex elimination schemes of $G(M)$. By Theorem 4.1 we obtain the equivalence of statements (ii) and (iii), first obtained by Rose [1970], in the following theorem. We present a generalization due to Golumbic [1978].

Theorem 12.1.   Let M be a symmetric matrix with nonzero diagonal entries. The following conditions are equivalent:

(i) M has a perfect elimination scheme;
(ii) M has a perfect elimination scheme under restriction (R);
(iii) $G(M)$ is a triangulated graph.

Before proving the theorem, we must introduce a bipartite graph model of the elimination process. This model will be used here and throughout the chapter.

An edge $e = xy$ of a bipartite graph $H = (U, E)$ is *bisimplicial* if $\text{Adj}(x) + \text{Adj}(y)$ induces a complete bipartite subgraph of H. Take note that *the*

---

* Actually it is sufficient to assume that no coincidental new zero will be found in a position which we want to choose as pivot (expecting it to be nonzero) at the time when we want to choose it.

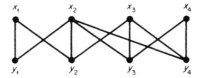

Figure 12.5.

bisimpliciality of an edge is retained as a hereditary property in an induced subgraph. Let $\sigma = [e_1, e_2, \ldots, e_k]$ be a sequence of pairwise nonadjacent edges of H. Denote by S_i the set of endpoints of the edges $e_1, \ldots, e_i$, and let $S_0 = \varnothing$. We say that σ is a *perfect edge elimination scheme* for H if each edge e_i is bisimplicial in the remaining induced subgraph $H_{U-S_{i-1}}$ and H_{U-S_n} has no edge. Thus, we regard the elimination of an edge as the removal of all edges adjacent to e. For example, the graph in Figure 12.5 has the perfect edge elimination scheme $[x_1 y_1, x_2 y_2, x_3 y_3, x_4 y_4]$. Notice that initially $x_2 y_2$ is not bisimplicial. A subsequence $\sigma' = [e_1, e_2, \ldots, e_k]$ of σ ($k \le n$) is called a *partial scheme*. The notations $H - \sigma'$ and H_{U-S_k} will be used to indicate the same subgraph.

Consider the bipartite matrix $B(\mathbf{M})$ of $\mathbf{M}$. It is evident that bisimplicial edges of $B(\mathbf{M})$ correspond precisely to acceptable pivots of $\mathbf{M}$ and that perfect edge elimination schemes for $B(\mathbf{M})$ correspond to perfect elimination schemes for $\mathbf{M}$. (See Figure 12.2 again in conjunction with Figure 12.6.)

Proof of Theorem 12.1. We have already remarked that (ii) and (iii) are equivalent, and since (ii) trivially implies (i), it suffices to prove that (i) implies (iii). Let us assume that $\mathbf{M}$ is symmetric with nonzero diagonal entries, and let σ be a perfect edge elimination scheme for $B(\mathbf{M})$.

Suppose $G(\mathbf{M})$ has a chordless cycle $[v_{\alpha_1}, v_{\alpha_2}, \ldots, v_{\alpha_m}, v_{\alpha_1}]$. This corresponds in $B(\mathbf{M})$ to the configuration B_C (Figure 12.7) induced by $C = \{x_{\alpha_1}, y_{\alpha_1}, \ldots, x_{\alpha_m}, y_{\alpha_m}\}$. Consider the first edge e of σ involving a vertex of C. Clearly, e involves only *one* vertex of C since none of the edges of B_C is bisimplicial. Assume without loss of generality that $e = x_{\alpha_1} y_s$ for some vertex $y_s \notin C$, and let x_s be the partner of y_s.

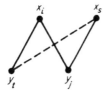

Figure 12.6. The edge $x_i y_j$ is not simplicial. A broken line indicates a nonedge.

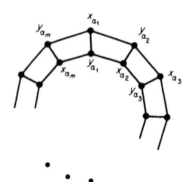

Figure 12.7.   The indicated subgraph B_C of $B(\mathbf{M})$.

Since in B_C, $\mathrm{Adj}(x_{\alpha_1}) = \{y_{\alpha_1}, y_{\alpha_2}, y_{\alpha_m}\}$ and $\mathrm{Adj}(y_{\alpha_1}) \cap \mathrm{Adj}(y_{\alpha_2}) \cap \mathrm{Adj}(y_{\alpha_m})$ $= \{x_{\alpha_1}\}$, the simpliciality of $x_{\alpha_1} y_s$ implies that $C \cap \mathrm{Adj}(y_s) = \{x_{\alpha_1}\}$, and by symmetry $C \cap \mathrm{Adj}(x_s) = \{y_{\alpha_1}\}$. Thus, we have that $x_{\alpha_1} y_{\alpha_2}$ and $x_s y_s$ are edges but $x_s y_{\alpha_2}$ is not an edge, which contradicts the bisimpliciality of $x_{\alpha_1} y_s$. Therefore, $G(\mathbf{M})$ is triangulated.   ∎

Corollary 12.2.   A symmetric matrix with nonzero diagonal entries can be tested for possession of a perfect elimination scheme in time proportional to the number of nonzero entries.

Proof.   Let m denote the number of nonzero entries of matrix $\mathbf{M}$. If $\mathbf{M}$ is stored in $O(m)$ space, then the data structures needed for applying Algorithms 4.1 and 4.2 to $G(\mathbf{M})$ can be initialized in $O(m)$ time. The result follows from Corollary 4.6.   ∎

Theorem 12.1 characterized perfect elimination for symmetric matrices. Moreover, it says that it suffices to consider only the diagonal entries. Haskins and Rose [1973] treat the nonsymmetric case under (R), and Kleitman [1974] settles some questions left open by Haskins and Rose. The unrestricted case was finally solved by Golumbic and Goss [1978], who introduced perfect elimination bipartite graphs. These graphs will be dis-cussed in the next section. Additional background on these and other matrix elimination problems can be found in the following survey articles and their references: Tarjan [1976], George [1977], and Reid [1977]. A discussion of the complexity of algorithms which calculate minimal and minimum fill-in under (R) can be found in Ohtsuki [1976], Ohtsuki, Cheung, and Fujisawa [1976], Rose, Tarjan, and Lueker [1976], and Rose and Tarjan [1978].

3. Perfect Elimination Bipartite Graphs

The general problem of deciding if a given (nonsymmetric) matrix $\mathbf{M}$ has a perfect elimination scheme can be answered using the bipartite graph model introduced in the preceding section. A perfect elimination bipartite graph is one for which there exists *some* perfect edge elimination scheme. But how do we construct such a scheme? Is it possible to choose *any* bisimplicial edge, eliminate it, and then continue from there to construct the remainder of a scheme? The next theorem answers this question in the affirmative. (Throughout this section the term *scheme* will mean perfect edge elimination scheme.)

Theorem 12.3 (Golumbic and Goss [1978]). If $e = xy$ is a bisimplicial edge of a perfect elimination bipartite graph $H = (X, Y, E)$, then $H_{X-\{x\}+Y-\{y\}}$ is also a perfect elimination bipartite graph.

Proof. We wish to show that if H has a scheme $[e_1, e_2, \ldots, e_n]$ then it also has a scheme beginning with e. Let $e_i = x_i y_i$ with $x_i \in X$ and $y_i \in Y$ ($i = 1, 2, \ldots, n$), and define H_i to be the subgraph of H induced by $X - \{x_1, \ldots, x_{i-1}\} + Y - \{y_1, \ldots, y_{i-1}\}$.

Case 1. $x = x_i$ and $y = y_i$ for some i. Since bisimpliciality of an edge is preserved in induced subgraphs, it follows that $[e, e_1, \ldots, e_{i-1}, e_{i+1}, \ldots, e_n]$ is a scheme.

Case 2. $x = x_i$ and $y = y_j$ for some $i \neq j$. We may assume that $i < j$ by interchanging X and Y if necessary; hence $[e, e_1, \ldots, e_{i-1}]$ is a partial scheme.

Consider an edge $x_h y_h$ for some $i < h < j$. Suppose there exists an $m > h$ such that $x_m y_h$ and $x_h y_i$ are edges in H. We would then have the following implications:

$$x_i y_i \text{ bisimplicial in } H_i \text{ implies } x_h y_j \in E,$$
$$x_h y_h \text{ bisimplicial in } H_h \text{ implies } x_m y_j \in E,$$
$$x_i y_j \text{ bisimplicial in } H \text{ implies } x_m y_i \in E.$$

This shows that $\sigma = [e, e_1, \ldots, e_{i-1}, e_{i+1}, \ldots, e_{j-1}]$ is a partial scheme.

Similarly, the following argument shows that $e' = x_j y_i$ is in H and is simplicial in $H - \sigma$. If $x_j y_t$ and $x_s y_i$ are in E for $s, t > j$, then

$$x_i y_j \text{ bisimplicial in } H \text{ implies } x_j y_i \in E,$$
$$x_i y_i \text{ bisimplicial in } H_i \text{ implies } x_s y_j \in E,$$
$$x_j y_j \text{ bisimplicial in } H_j \text{ implies } x_s y_t \in E.$$

Since $(H - \sigma) - [e'] = H_{j+1}$, we conclude that

$$[e, e_1, \ldots, e_{i-1}, e_{i+1}, \ldots, e_{j-1}, e', e_{j+1}, \ldots, e_n]$$

is a scheme.

Case 3.   One of x and y is not among the x_i and y_j. Assume that $x = x_i$ and $y \neq y_j$ for some i and for all j. By an argument similar to, but shorter than, Case 2, $[e, e_1, \ldots, e_{i-1}, e_{i+1}, \ldots, e_n]$ is a scheme.   ∎

Corollary 12.4.   The greedy algorithm of repeatedly eliminating a bisimplicial edge of the remaining graph (thus also removing all edges adjacent to it) until no edge remains will succeed if and only if H is a perfect elimination bipartite graph.

Proof.   Assume the corollary is true for all subgraphs of H. If H is not perfect elimination, then the algorithm will surely fail (i.e., at some time before all edges are removed, there will be no bisimplicial edge). If H is perfect elimination, then eliminate some bisimplicial edge e. Since $H - [e]$ is also perfect elimination, we have, by induction, that the algorithm will succeed.   ∎

A pair of edges ab and cd of $H = (U, E)$ is *separable* if the subgraph induced by them is isomorphic to $2K_2$. The graph H is said to be *separable* if it contains a pair of separable edges; otherwise H is *nonseparable*. Clearly a nonseparable graph has at most one nontrivial connected component. Furthermore, any induced subgraph of a nonseparable graph is nonseparable.

Theorem 12.5 (Golumbic and Goss [1978]).   If $H = (X, Y, E)$ is a nonseparable bipartite graph, then each *nonisolated vertex z is the endpoint of some bisimplicial edge of H.*

Suppose that z is a nonisolated vertex which is not the endpoint of any bisimplicial edge. We may assume that $z \in Y$; let $x_0 z$ be any edge. We shall construct an infinite chain of subsets of X

$$X_0 \subset X_1 \subset \cdots X_k \subset \cdots$$

which will contradict the finiteness of X. Assume we are given subsets

$$X_k = \{x_0, x_1, \ldots, x_k\} \subseteq X$$

and

$$Y_k = \{z, y_1, \ldots, y_k\} \subseteq Y$$

such that

$$x_i y_i \in E \Leftrightarrow i < j \qquad \text{for all } 0 \leq i, j \leq k$$

and

$$x_i z \in E \qquad \text{for all } 0 \le i \le k.$$

(The arbitrary edge $x_0 z$ will start the induction when $k = 0$.)

Since $x_k z$ is not bisimplicial, there exist vertices x and y ($\ne z$) such that $x_k y$, $xz \in E$ but $xy \notin E$. Hence, $y \notin Y_k$. Moreover, for all $0 \le i < k$ the edges $x_i y_{i+1}$ and $x_k y$ are not separable, implying that $x_i y \in E$. But $xy \notin E$, so $x \notin X_k$. Therefore, by renaming $x = x_{k+1}$ and $y = y_{k+1}$ and setting $X_{k+1} = X_k \cup \{x_{k+1}\}$ and $Y_{k+1} = Y_k \cup \{y_{k+1}\}$, we are ready for the next iteration of our construction. This algorithm goes on indefinitely, but X and Y are finite, a contradiction. Thus H must have a bisimplicial edge with z as one of its endpoints. ∎

Corollary 12.6. Every nonseparable bipartite graph H is a perfect elimination bipartite graph.

Proof. By Theorem 12.3, it suffices to show that H has a bisimplicial edge. The corollary follows from Theorem 12.5. ∎

We have accomplished two things in regard to perfect elimination bipartite graphs: We have provided an algorithm for recognizing them, and we have proven a sufficient (but not necessary) condition for them. Being perfect elimination, however, cannot tell us much about the structure of a graph. Indeed, let H be any bipartite graph with vertices $u_1, u_2, \ldots, u_n$; add new vertices $w_1, w_2, \ldots, w_n$ and connect u_i with w_i for each $i = 1, \ldots, n$. This augmented graph is a perfect elimination bipartite graph and completely masks the structure of H. It follows from this negative result that there cannot exist a characterization of perfect elimination bipartite graphs in terms of some forbidden configurations or subgraphs.

4. Chordal Bipartite Graphs

In the preceding section we have successfully generalized the perfect elimination aspect of triangulated graphs. This raises the following question: Is there an appropriate notion of chordality for bipartite graphs? A triangulated graph may have 3-cycles, but any longer cycle must have a chord. In bipartite graphs the smallest allowable cycle has length 4, so we make the following definition. A bipartite graph is *chordal* if every cycle of length strictly greater than 4 has a chord.

Remark. Every nonseparable bipartite graph is chordal bipartite.

Separable edges can be equivalently defined as follows: A pair of edges ab and cd of $G = (V, E)$ is *separable* if there exists a set S of vertices whose removal from G causes ab and cd to lie in distinct connected components of the remaining subgraph G_{V-S}. The set S is called an *edge separator* for ab and cd; S is *minimal* if no proper subset of S is an edge separator for ab and cd. The next result is analogous to Theorem 4.1(iii).

Theorem 12.7 (Golumbic and Goss [1978]). A bipartite graph $H = (X, Y, E)$ is chordal bipartite if and only if every minimal edge separator induces a complete bipartite subgraph.

Proof. Let $C = [v_1, v_2, \ldots, v_k, v_1]$ be a cycle of H having even length $k \geq 6$. Consider the set $S = \mathrm{Adj}(v_2) + \mathrm{Adj}(v_3) - \{v_2, v_3\}$. Clearly S separates $v_2 v_3$ from $v_5 v_6$, and $S \cap C = \{v_1, v_4\}$. Let $S' \subseteq S$ be a minimal edge separator for $v_2 v_3$ and $v_5 v_6$. Thus $v_1 \in S'$ and $v_4 \in S'$. If S' is complete bipartite, then $v_1 v_4$ is a chord of C due to the opposite parity of the subscripts.

Conversely, let T be a minimal edge separator and let H_A and H_B be connected components of the graph remaining after removing T. Let x and y be any pair of vertices of T of opposite parity. Since H_A and H_B are connected, there exist minimum length paths $[x, a_1, a_2, \ldots, y]$ and $[y, b_1, b_2, \ldots, x]$ with $a_i \in A$ and $b_i \in B$. Because these paths are of odd length ≥ 3, they join to give a cycle of length ≥ 6. If this cycle has a chord, it must be the edge xy, since by construction no other pair may be adjacent. Hence, T will be a complete bipartite set. ∎

The next theorem generalizes Lemma 4.2 with separability in bipartite graphs corresponding to nonadjacency in undirected graphs.

Theorem 12.8 (Golumbic [1979]). Let H be a chordal bipartite graph. If H is separable, then it has at least two separable bisimplicial edges.

Proof. Assume that $H = (X, Y, E)$ has separable edges α and β and that the theorem is true for all graphs with fewer vertices than H. Let S be a minimal edge separator for α and β with H_A and H_B being the connected components of H_{X+Y-S} containing α and β, respectively. We claim H_{A+S} has a bisimplicial edge whose endpoints are both in A.

Case 1. H_{A+S} is separable. By induction H_{A+S} has two separable bisimplicial edges $x_1 y_1$ and $x_2 y_2$. Since S is complete, at most two of the four endpoints are in S, either those with the same parity or those with the same subscript. Suppose $x_1, x_2 \in S$ and $y_1, y_2 \in A$. Take a minimum length path $[y_1, a_1, \ldots, a_j, y_2]$ in H_A and a minimum length path $[x_2, b_1, \ldots, b_k, x_1]$ with the b_i in H_B. Gluing these together we obtain a cycle of length at least 6 which must have a chord. But minimality permits only the chords $x_1 y_2$ or

$x_2 y_1$, contradicting the separability of $x_1 y_1$ and $x_2 y_2$. Similarly y_1 and y_2 cannot both be in S. Therefore, H_{A+S} has a bisimplicial edge whose endpoints are both in A.

Case 2. H_{A+S} is nonseparable. Let $x_1 y_2$ be any edge of H_A. By Theorem 12.5 there exist vertices y_1 and x_2 such that $x_1 y_1$ and $x_2 y_2$ are bisimplicial in H_{A+S}. Suppose both y_1 and x_2 are in S, for otherwise the claim is true. Then x_2 and y_1 are adjacent since S is complete. The bisimpliciality of $x_1 y_1$ implies that $\mathrm{Adj}(x_1) \subseteq \mathrm{Adj}_{A+S}(x_2)$, and the bisimpliciality of $x_2 y_2$ implies that $\mathrm{Adj}_{A+S}(x_2) \subseteq \mathrm{Adj}(x_1)$. Hence, $\mathrm{Adj}(x_1) + \mathrm{Adj}(y_2) = \mathrm{Adj}_{A+S}(x_2) + \mathrm{Adj}(y_2)$, which we know induces a complete bipartite subgraph. Thus, $x_1 y_2$ is also bisimplicial in H_{A+S}, so the claim for this case is proven.

Finally, a bisimplicial edge of H_{A+S} whose endpoints lie in A is also bisimplicial in H since $\mathrm{Adj}(A) \subseteq A + S$. Therefore, by the claim, H has a bisimplicial edge α' whose endpoints lie in A and, similarly, a bisimplicial edge β' whose endpoints lie in B, and α' and β' are separable. ∎

The proof of Theorem 12.8 actually gives a slightly stronger result.

Corollary 12.9. Let $H = (X, Y, E)$ be a chordal bipartite graph. If S is a minimal edge separator for some pair of edges, then H has a simplicial edge in each nontrivial connected component of H_{X+Y-S}.

Theorem 12.10 (Golumbic and Goss [1978]). Every chordal bipartite graph is a perfect elimination bipartite graph.

Proof. Since chordal bipartiteness is a hereditary property, it is sufficient to show that a chordal bipartite graph H has a bisimplicial edge. Applying Theorem 12.5 for H nonseparable or Theorem 12.8 for H separable, we obtain the desired result. ∎

Unlike the case of triangulated graphs (Theorem 4.1), the converse of Theorem 12.10 is false. Each of the edges $u_i w_i$ in Figure 12.8 is bisimplicial, and the elimination of any one of them breaks the 6-cycle. Nevertheless, we do have a necessary and sufficient condition for chordality in terms of perfect elimination by adding a hereditary condition.

Corollary 12.11. *A graph is chordal bipartite if and only if every induced subgraph is perfect elimination bipartite.*

Proof. If H possesses a chordless cycle C of length strictly greater than 4, then C would be an induced subgraph which is not perfect elimination. Conversely, if H is chordal bipartite, then so is every induced subgraph H', and by Theorem 12.10, H' is perfect elimination. ∎

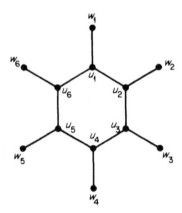

Figure 12.8.   A perfect elimination bipartite graph which is not chordal. The cycle is broken when one of the edges $u_i w_i$ is eliminated.

Summary

We have presented bipartite generalizations of triangulated graphs according to two important properties: perfect elimination and chordality. Although these notions do not coincide in the general setting, both do extend certain aspects of triangulated graphs. Perfect elimination bipartite graphs correctly model the application to Gaussian elimination with no fill-in. Chordal bipartite graphs satisfy the separation theorems analogous to those of Dirac. Alan Hoffman and Michel Sakarovich have recently discovered that the chordal bipartite graphs give a characterization of the matrices in an important class of linear programming problems for which the greedy heuristic approach gives an optimum solution.

EXERCISES

1.   Verify that the matrix below has a perfect elimination scheme but does *not* have one under restriction (R).

$$\begin{pmatrix} 1 & 1 & 0 & 0 & 0 \\ 2 & 1 & 1 & 0 & 0 \\ 3 & 0 & 1 & 1 & 0 \\ 4 & 1 & 0 & 1 & 1 \\ 5 & 1 & 1 & 0 & 1 \end{pmatrix}$$

2. Let $G = (V, E)$ be an undirected graph with vertices $v_1, v_2, \ldots, v_n$. The *bipartite graph* $B(G)$ of G has vertices $x_1, x_2, \ldots, x_n$ and $y_1, y_2, \ldots, y_n$, with x_i adjacent to y_j if and only if $i = j$ or $v_i v_j \in E$.

  (i) Show that if $B(G)$ is chordal bipartite, then G is a triangulated graph.

  (ii) For which of the triangulated graphs in Figure 12.9 is $B(G)$ *not* chordal bipartite? Find a minimal edge separator which does not induce a bipartite clique.

Figure 12.9.

3. Consider the graph H constructed as follows: $[a_1, \ldots, a_n, a_1]$ and $[b_1, \ldots, b_n, b_1]$ are vertex disjoint chordless cycles with n even, and $a_i b_j$ is an edge iff $i + j \equiv 1 \pmod 2$. Show that H is not a perfect elimination bipartite graph for $n \geq 6$.

4. Let $H = (X, Y, E)$ be a perfect elimination bipartite graph with perfect scheme σ. For $xy \in E$ define the *deficiency* of xy in G to be

$$D(xy) = \{ab \notin E \,|\, a, b \in \text{Adj}(x) + \text{Adj}(y)\}.$$

Show that σ is also a perfect scheme for the graph $H' = (X, Y, E + D(xy))$.

  Suppose you made a stupid pivot choice and caused some fill-in on your perfect elimination bipartite graph; is all hope lost? No, you can still continue perfectly, as the next exercise shows.

5. If $H = (X, Y, E)$ is a perfect elimination bipartite graph and xy is any edge, then the *xy-elimination graph*

$$H_{xy} = (X - \{x\}, Y - \{y\}, E_{X - \{x\} + Y - \{y\}} + D(xy))$$

is also perfect elimination. (Hint: Use Theorem 12.3 or modify its proof.) If σ is the perfect scheme for H which was misplaced when xy was stupidly eliminated, how can σ be cleverly modified to give a perfect scheme for H_{xy}?

6. Prove the claim in case 3 of Theorem 12.3.

7. Let $H = (X, Y, E)$ be a bipartite graph and let $H' = (X, Y, E')$ denote its bipartite complement; that is, for all $x \in X$ and $y \in Y$, $xy \in E'$ iff $xy \notin E$. Prove the following: The graphs H and H' are both chordal bipartite if and only if H contains no induced subgraph isomorphic to C_6, $3K_2$, or C_8 (Golumbic and Goss [1978]).

8. Let $G = (V, E)$ be an undirected graph and let $B(G)$ be its bipartite graph (see Exercise 2). For $S \subset V$, let $B(S) = \{x_i \,|\, v_i \in S\} \cup \{y_i \,|\, v_i \in S\}$. Prove that S is a minimal vertex separator of G if and only if $B(S)$ is a minimal edge separator of $B(G)$ (Golumbic [1979]).

9. The Venn diagram in Figure 12.10 has nine regions representing all possibilities for a bipartite graph to satisfy or not satisfy the following properties.

       P.E.B.   The graph is a perfect elimination bipartite graph.

        C.B.   The graph is chordal bipartite.

   (P.E.B.)′   The bipartite complement of the graph is perfect elimination bipartite.

    (C.B.)′   The bipartite complement of the graph is chordal bipartite.

For each region give an example of a graph which lives in that region. (One solution is shown in Appendix E, but try to find your own examples without referring to it.)

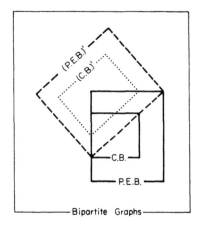

Figure 12.10.   Bipartite graphs.

10. Let $H = (X, Y, E)$ be a bipartite graph, and let G be the split graph obtained from H by connecting every pair of vertices in Y. Prove that H is nonseparable if and only if G is a threshold graph. For an application, see Chapter 10, Exercise 15.

Bibliography

George, J. Alan

   [1977]   Solution of linear systems of equations: Direct methods for finite element problems, *in* "Sparse Matrix Techniques," Lecture notes in math., Vol. 572, pp. 52–101. Springer-Verlag, Berlin. MR55 #13751.

Golumbic, Martin Charles
  [1978] A note on perfect Gaussian elimination, *J. Math. Anal. Appl.* **64**, 455–457.
  [1979] A generalization of Dirac's theorem on triangulated graphs, *Ann. N.Y. Acad. Sci.*
          319, 242–246.
          The "only if" portion of Theorem 4 is false. The correct version is given here in
          Exercise 2.
Golumbic, Martin Charles, and Goss, Clinton F.
  [1978] Perfect elimination and chordal bipartite graphs, *J. Graph Theory* **2**, 155–163.
Haskins, Loren, and Rose, Donald J.
  [1973] Toward a characterization of perfect elimination digraphs, *SIAM J. Comput.* **2**,
          217–224. MR49 #8895.
Kleitman, D. J.
  [1974] A note on perfect elimination digraphs, *SIAM J. Comput.* **3**, 280–282. MR56 #5351.
Ohtsuki, Tatsuo
  [1976] A fast algorithm for finding an optimal ordering for vertex elimination on a graph,
          SIAM J. Comput. **5**, 133–145. MR52 #13515.
Ohtsuki, Tatsuo, Cheung, L. K., and Fujisawa, T.
  [1976] Minimal triangulation of a graph and optimal pivoting order in a sparse matrix,
          J. Math. Anal. Appl. **54**, 622–633.
Parter, Seymour V.
  [1961] The use of linear graphs in Gaussian elimination, *SIAM Rev.* **3**, 119–130. MR26 #908.
          This is the first paper to introduce the notion of perfect elimination for preserving
          sparseness during Gaussian elimination.
Perl, Yehoshua
  [1973] A counterexample to a conjecture of D. J. Rose on minimum triangulation, *J. Math.
          Anal. Appl.* **42**, 594–595. MR53 #212.
Reid, John K.
  [1977] Solution of linear systems of equations: Direct methods (general), *in* "Sparse Matrix
          Techniques," Lecture notes in math., Vol. 572, pp. 102–129. Springer-Verlag, Berlin.
Rose, Donald J.
  [1970] Triangulated graphs and the elimination process, *J. Math. Anal. Appl.* **32**, 597–609.
          MR42 #5840.
  [1972] A graph-theoretic study of the numerical solution of sparse positive definite systems
          of linear equations. *in* "Graph Theory and Computing" (Ronald C. Read, ed.),
          pp. 183–217. Academic Press, New York. MR49 #6579.
Rose, Donald J., and Tarjan, Robert Endre
  [1975] Algorithmic aspects of vertex elimination, *Proc. 7th Annual ACM Symp. on Theory
          of Computing*, 245–254.
  [1978] Algorithmic aspects of vertex elimination of directed graphs, *SIAM J. Appl. Math.*
          34, 176–197.
Rose, Donald J., Tarjan, Robert Endre, and Leuker, George S.
  [1976] Algorithmic aspects of vertex elimination on graphs, *SIAM J. Comput.* **5**, 266–283.
          MR53 #12077.
Tarjan, Robert Endre
  [1976] Graph theory and Gaussian elimination, *in* "Sparse Matrix Computations" (J. R.
          Bunch and D. J. Rose, eds.), pp. 3–22. Academic Press, New York.

Appendix

A. A Small Collection of NP-Complete Problems

GRAPH COLORING (decision version)
Instance: An undirected graph G and an integer $k > 0$.
Question: Does there exist a proper k-coloring of G? Equivalently, is $\chi(G) \leq k$?

GRAPH COLORING (optimization version)
Instance: An undirected graph G.
Question: What is $\chi(G)$?

CLIQUE (decision version)
Instance: An undirected graph G and an integer $k > 0$.
Question: Does there exist a complete subset of vertices of G of size k? Equivalently, is $\omega(G) \geq k$?

CLIQUE (optimization version)
Instance: An undirected graph G.
Question: What is $\omega(G)$?

STABLE SET (decision version)
Instance: An undirected graph G and an integer $k > 0$?
Question: Does G have a stable set of size k? Equivalently, is $\alpha(G) \geq k$?

STABLE SET (optimization version)
Instance: An undirected graph G.
Question: What is $\alpha(G)$?

CLIQUE COVER (decision version)
Instance: An undirected graph G and an integer $k > 0$.
Question: Can the vertices of G be covered by k cliques of G? Equivalently, is $k(G) \le k$?

CLIQUE COVER (optimization version)
Instance: An undirected graph G.
Question: What is $k(G)$?

HAMILTONIAN PATH
Instance: An undirected graph G with vertices $v_1, v_2, \ldots, v_n$.
Question: Can the vertices be ordered $[v_{\pi_1}, v_{\pi_2}, \ldots, v_{\pi_n}]$ so that v_{π_i} and $v_{\pi_{i+1}}$ are adjacent in G for $i = 1, 2, \ldots, n - 1$?

HAMILTONIAN CIRCUIT
Instance: An undirected graph G with vertices $v_1, v_2, \ldots, v_n$.
Question: Can the vertices be ordered $[v_{\pi_1}, v_{\pi_2}, \ldots, v_{\pi_n}]$ so that v_{π_i} and $v_{\pi_{i+1}}$ are adjacent in G for $i = 1, 2, \ldots, n - 1$ and v_{π_n} and v_{π_1} are also adjacent in G?

STABLE SET ON TRIANGLE-FREE GRAPHS
Instance: An undirected graph G having no 3-cycle.
Question: What is $\alpha(G)$?

B. An Algorithm for Set Union, Intersection, Difference, and Symmetric Difference of Two Subsets

Input: Two subsets S and T of a universal set U whose members are numbered $u_1, u_2, \ldots, u_n$. All subsets are represented as lists of numbers (the indices of its members).
Output: The sets $S \cup T, S \cap T, S - T, T - S$, and $(S - T) \cup (T - S)$.
Method: An auxilliary Boolean n-vector $B = \langle b_1, b_2, \ldots, b_n \rangle$, initially containing only zeros, is used. As the list S is scanned, B is changed to the characteristic vector of S (line 3). In the loop 4–9, $S \cap T$ and $T - S$ are formed, $(S - T) \cup (T - S)$ is half formed, and B is changed to the characteristic vector of $S - T$. In the loop 10–15, $S \cup T$ and $(S - T) \cup (T - S)$ are completed and $S - T$ is formed. Also B is restored to the zero vector.

begin
1.　**remark**: $B = \langle 0, 0, \ldots, 0 \rangle$
2.　initialize: $S \cup T \leftarrow T$; $S \cap T \leftarrow S - T \leftarrow T - S \leftarrow (S - T) \cup (T - S) \leftarrow \varnothing$;
3.　**for all** $i \in S$ **do** $b_i \leftarrow 1$;;
4.　**for all** $j \in T$ **do**
5.　　**if** $B_j = 1$
　　　　then
6.　　　　Add j to $S \cap T$;
7.　　　　$b_j \leftarrow 0$;
　　　　else
8.　　　　Add j to $T - S$;
9.　　　　Add j to $(S - T) \cup (T - S)$;;
10.　**for all** $i \in S$ **do**
11.　　**if** $b_i = 1$
　　　　then
12.　　　　Add i to $S \cup T$;
13.　　　　Add i to $S - T$;
14.　　　　Add i to $(S - T) \cup (T - S)$;
15.　　　　$b_i \leftarrow 0$;;
16.　**remark**: $B = \langle 0, 0, \ldots, 0 \rangle$
end

Complexity.　Assuming no charge for initializing B (line 1), the complexity is dominated by the three loops. Thus, the algorithm runs in $O(|S| + |T|)$ steps.

C.　Topological Sorting: An Example of Algorithm 2.4

Let us assume that the graph in Figure C1 is stored as sorted adjacency lists. Initially, the DFSNUMBER and the TSNUMBER of each vertex is

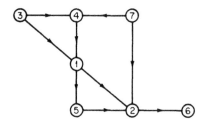

Figure C1

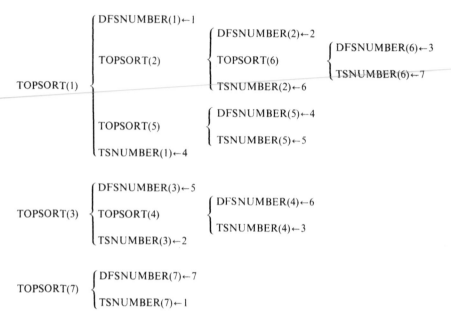

Figure C2

set to 0, j is set to 7, and i is set to 0. The search begins with vertex 1. TOP-SORT(1) will call TOPSORT(2), which will call TOPSORT(6); when control is eventually returned to TOPSORT(1) it will resume its scan of Adj(1) and will call TOPSORT(5). When TOPSORT(1) is finished, the main routine will call TOPSORT(3), etc. These recursive calls are illustrated in Figure C2. The final values of the depth-first search numbering and the topological sorting numbering are as follows:

Vertex	DFSNUMBER	TSNUMBER
1	1	4
2	2	6
3	5	2
4	6	3
5	4	5
6	3	7
7	7	1

D.   An Illustration of the Decomposition Algorithm

The decomposition algorithm in Section 5.4 as applied to a noncomparability graph is illustrated in Figure D1.

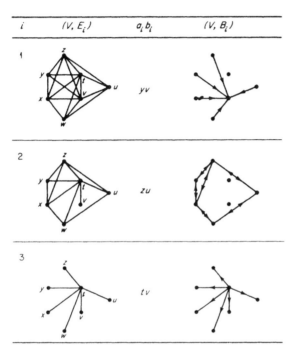

i	(V, E_i)	$a_i b_i$	(V, B_i)

Figure D1

E.   The Properties P.E.B., C.B., (P.E.B.)′, (C.B.)′ Illustrated

Figure E1 gives examples of graphs satisfying or not satisfying the following properties:

    P.E.B.:   the graph is a perfect elimination bipartite graph;

    C.B.:   the graph is chordal bipartite;

  (P.E.B.)′:   the bipartite complement of the graph is perfect elimination bipartite;

  (C.B.)′:   the bipartite complement of the graph is chordal bipartite.

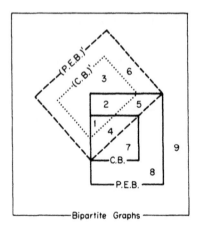

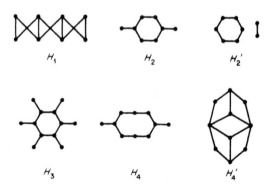

Figure E1

The regions in Figure E1 are illustrated by the given examples as follows:

Region	Example
1	H_1
2	H_2'
3	$3K_2$
4	H_2
5	H_3
6	H_4
7	C_6
8	H_4'
9	$C_n \ (n = 8, 10, 12, \cdots)$

F. The Properties $C, \bar{C}, T, \bar{T}$ Illustrated

Examples of graphs which are or whose complements are comparability graphs and/or triangulated graphs:

Property				
C	$\bar{C}$	T	$\bar{T}$	Examples*
+	+	+	+	Any threshold graph
+	+	+	−	$\bar{C}_4 = 2K_2$
+	+	−	+	C_4
+	+	−	−	G_6
+	−	+	+	G_5
+	−	+	−	G_4
+	−	−	+	$\bar{G}_2$ or $\bar{G}_3$
+	−	−	−	C_6, C_8, etc.
−	+	+	+	$\bar{G}_5$
−	+	+	−	G_2 or G_3
−	+	−	+	$\bar{G}_4$
−	+	−	−	$\bar{C}_6, \bar{C}_8$, etc.
−	−	+	+	G_7 or $\bar{G}_7$
−	−	+	−	$\bar{G}_1$
−	−	−	+	G_1
−	−	−	−	C_5, C_7, etc.

* See Figure F1 for the G_i.

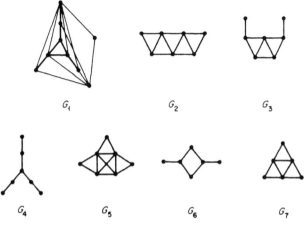

G_1 G_2 G_3

G_4 G_5 G_6 G_7

Figure F1

Epilogue 2004

1. Introduction – Foundations and Applications

In this Epilogue chapter, we will take a "short tour" of some of the new results in algorithmic graph theory and perfect graphs. So many new research directions have been the subject of investigation since this book was first published in 1980, that it is impossible to survey them all. The algorithms and applications associated with structured families of graphs have grown to maturity in these 24 years. The literature has increased tenfold, and the world of perfect graphs has grown to include over 200 special graph classes. Perfect graphs now have their own Mathematical Reviews Classification 05C17, as so do geometric and intersection representations 05C62.

We present here a sample of the many results of the Second Generation of Algorithmic Graph Theory from the author's biased view. Necessarily, it must be only a small fraction of what would otherwise require a large sequel volume. Fortunately, the availability of several new books, listed earlier in the Prologue of this edition, can also aid the reader eager to pursue further exploration in this area.

The sections of this Epilogue are numbered to correspond with the chapters of the book. Our intention is, as with the former chapters, to send the reader back to the literature, laboratory and library to continue research.

Intersection Graphs

We saw many of the early uses of the intersection graph model in the sneak preview Section 1.3, in the application sections on permutation graphs, interval graphs, and elsewhere in the book. But the volume and scope of research in this general area has expanded significantly both from the modeling and algorithmic points of view. Some of these applications include mobile frequency assignment (Osput and Roberts [1983]), pavement deterioration analysis (Gattass and

Nemhauser [1981]), relational databases (see Golumbic [1988]), evolutionary trees (see Waterman [1995]), physical mapping of DNA (Goldberg, et al. [1995], Golumbic, Kaplan and Shamir [1994]), container ship stowage (Avriel, Penn and Shpirer [2000]), and VLSI circuit design (Dagan, Golumbic and Pinter [1988]). Intersection graphs have become a necessary and important tool for solving real-world problems. McKee and McMorris [1999] is devoted to the topic of intersection graph models and their application. We will present several other examples in this new Epilogue chapter.

Temporal Reasoning

One of the "traditional" applications of interval graphs is reasoning about time intervals, which started with the original question of Hajös (Section 8.1, page 171). Temporal reasoning is an essential part of many applications in artificial intelligence. Given a set of explicit relationships between certain events, we would like to be able to infer additional relationships which are implicit in those given. For example, the transitivity of "before" and "contains" may allow us to derive information regarding the sequence of events. Seriation problems, like the example in Section 8.4, ask for a mapping of temporal events onto the time line such that all the given relations are satisfied, that is, a consistent scenario. Similarly, there are problems of scheduling, planning, and story understanding in which one is interested in constructing a time line where each particular event or phenomenon or task corresponds to an interval representing its duration.

Allen [1983] introduced a model for temporal reasoning using the thirteen primitive interval relations obtained by considering all possible orderings of their four endpoints. Several authors working in AI have studied and adapted Allen's model further, and have incorporated such models into reasoning systems. The paper by Golumbic and Shamir [1993] has provided a bridge linking some of these temporal reasoning notions from the AI community with those of the combinatorics community. Their approach has been to simplify Allen's model in order to study its complexity using graph theoretic techniques. We refer the reader to Golumbic [1999] which is a survey paper [1] on the topic, written in the same spirit as this book. It describes a number of directions of current work on reasoning about time, many of which employ graph algorithms.

2. The Design of Efficient Algorithms

Maximum Network Flow Problem

Progress on lowering the computational complexity of the maximum network flow (MAXFLOW) problem was presented in Table 2.1 as one of several illustrations

---

[1] This survey paper also includes some of the author's newest illustrative stories, "Goldie and the Four Bears", "Will Allan get to Judy's in time?", and "Five Autonomous Golumbic Women".

of closing the gap between the best known algorithms and the lower complexity bound. Further improvement for MAXFLOW have been given by Goldberg and Tarjan [1988] $O(ne \log(n^2/e))$, King, Rao and Tarjan [1994] $O(ne + n^{2+\varepsilon})$ and $O(ne \log_{e/(n \log n)} n)$, Philips and Westbrook [1993] $O(ne \log_{e/n} n + n^2(\log n)^{2+\varepsilon})$. We refer the interested reader to Cook, Cunningham, Pulleyblank and Schrijver [1998], Corman, Leiserson Rivest, second edition [2001], Goldberg [1998] and Johnson and McGeoch [1993].

Although it is not a central topic in studying perfect graphs, maximum network flow algorithms do play an important role in certain optimization problems on perfect graphs. For example, we invoked its complementary relative MINFLOW at the end of Section 5.7 when sketching a polynomial method for finding the stability number $\alpha(G)$ of a cocomparability graph. This method has subsequently been used as part of the algorithm by Narasimhan and Manber [1992] for the stability number of tolerance graphs.

Graph Sandwich Problems

In this book, we placed a major focus on the minimum coloring, maximum clique, and recognition problems for special families of graphs. Another important algorithmic direction has been the study of various completion problems. For example, the *minimum completion problem* requires adding a minimum number edges to an arbitrary graph G in order to obtain a new graph G' which satisfies the desired property Π, such as being an interval graph or a triangulated graph, see Garey and Johnson [1979], Yannakakis [1981]. One motivation for such completion problems is as a heuristic for coloring G, since $\chi(G) \leq \chi(G')$.

Another variation of the completion problem, called the *graph sandwich problem*, is defined by allowing only some of the nonedges to be eligible to be added to the original graph. Specifically, given a graph $G = (V, E)$ and a subset $E_0 \subseteq \overline{E}$ of (optional) nonedges, we ask whether there exists a completion $G' = (V, E')$ which satisfies the desired property Π, such that $E \subseteq E' \subseteq E \cup E_0$.

Sandwich problems arise in applications where only partial information about the graph is known. The interval graph sandwich problem was shown to be NP-complete by Golumbic and Shamir [1993]. It arises in molecular biology in problems of physical mapping of DNA and in problems of temporal reasoning, similar in spirit to the early work described in Applications 8.2 and 8.3 on pages 182–183. For example, see Atkins and Middendorf [1996] and Golumbic, Kaplan and Shamir [1994].

Golumbic, Kaplan and Shamir [1995] investigates graph sandwich problems for other special families of graphs. Specifically, the sandwich problem is polynomial for split graphs, cographs, and threshold graphs, but is NP-complete for chordal graphs, proper interval graphs, comparability graphs, permutation graphs, and others (see also Golumbic and Trenk [2004, Sections 4.7–4.8]).

There is also a body of literature on graph modification problems, where one may add edges and delete edges to transform a graph G into a modified graph G' satisfying a property Π. The problem is to minimize the total number of additions and deletions, thus changing the edge set as little as possible. For a good survey and introduction to this area, see Natanzon, Shamir and Sharan [2001].

3. Perfect Graphs

The Strong Perfect Graph Conjecture/Theorem

In May of 2002, the announcement was made that the Strong Perfect Graph Conjecture (SPGC) of Claude Berge had been proven by the team of researchers consisting of Maria Chudnovsky, Neil Robertson, Paul Seymour and Robin Thomas [2002]. This most important result, after 42 years, is currently submitted for publication.

The term *Berge graph* has been defined as an undirected graph which contains neither an odd chordless cycle C_{2k+1} (an odd hole) nor its complement $\overline{C}_{2k+1}$ (for $k \geq 2$) (an odd antihole) as an induced subgraph. Thus, what should now be called the *Strong Perfect Graph Theorem* states that an undirected graph is perfect if and only if it is a Berge graph, the "only if" direction being immediate.

During the decades preceding this solution, a large body of research developed involving the structure of minimally imperfect graphs, (see Brandstädt, Le and Spinrad [1999, Chapter 14]). The spin-off effect of these investigations has been the birth of many new children in the world of perfect graphs, both new problems and a generation of young researchers. The collections edited by Berge and Chvátal [1984] and Ramirez-Alfonsin and Reed [2001] provide a good cross-section of the work in this area.

Although proving the SPGC was a major mathematical challenge rather than an algorithmic one, it raised several related interesting algorithmic questions: Is there a polynomial time algorithm which recognizes whether or not an undirected graph G has an odd chordless cycle of length ≥ 5? Is there a polynomial time algorithm which recognizes Berge graphs? The second problem has been solved, by Chudnovsky and Seymour [2002], Chudnovsky, Cornuéjols, Liu, Seymour and Vuškovič [2002], Cornuéjols, Liu and Vuškovič [2002], and awaits publication. Combining this algorithm with the Strong Perfect Graph Theorem shows that *there is a polynomial time algorithm which recognizes whether or not a graph is perfect*. A solution to the first question has so far not been found (as of spring 2003), but when/if a solution is found, it will give an alternate solution to the second question by applying it to the graph and its complement.

Finally, Vashek Chvátal maintains a *perfect website*[2] which contains a long list of references and historical notes. A useful technical report by Hougardy [1998]

[2] http://www.cs.rutgers.edu/~chvatal/perfect/problems.html

lists, for each ordered pair (C_1, C_2) of 96 classes of perfect graphs, whether the class C_1 is a subset of the class C_2, and if this is not the case, provides and example of a graph which is in C_1 but not in C_2. An on-line tool for checking the class containment for a larger set of more than 200 graph classes has been developed at Rostock University[3]. It also contains the known complexity results for recognition of the classes, the maximum stable set and domination problems. Several unresolved containments and complexities remain as open problems. Figure 13.3, at the end of this Epilogue (p. 305), gives a complete hierarchy of some of the main classes of perfect graphs, ordered by inclusion (reprinted from Golumbic and Trenk [2004] where it appears together with separating examples).

Stable Sets

On the stable set problem for perfect graphs, there has also been progress. Grötschel, Lovász and Schrijver [1981] have shown that the ellipsoid method of solving linear programming problems can be applied to obtain a polynomial algorithm to find maximum stable sets and minimum colorings for perfect graphs. Also, since G is perfect if and only if its complement $\overline{G}$ is perfect, this same approach can be used to find maximum cliques and minimum clique covers.

The major importance of this result is that it generalizes what had been known for many classes of perfect graphs. Although the complexity of the algorithm is polynomial, it may not be practical to implement. As the authors point out, it is not intended to compete with the special purpose algorithms designed to solve these problems for interval graphs, cocomparability graphs, triangulated graphs, and other classes of perfect graphs which so often arise in applications. For further reading on algorithms for the stable set problem and the clique problem, see Hertz [1995] and Johnson and Trick [1996].

4. Triangulated Graphs – Chordal Graphs

Throughout this book, we have followed the French tradition by using the name *triangulated graph* for an undirected graph containing no chordless cycle C_k ($k \geq 4$). In this section, however, we will use its popular synonym *chordal graph*. Two important variations of chordal graphs will be briefly presented here, namely the strongly chordal and the weakly chordal graphs. As their names indicate, every strongly chordal graph is chordal, and every chordal graph is weakly chordal. Weakly chordal graphs are also perfect graphs.

---

[3] http://wwwteo.informatik.uni-rostock.de/isgci

Strongly Chordal Graphs

Strongly chordal graphs, introduced by Farber [1985], specialize chordal graphs in several ways. They are characterized by several equivalent definitions using chords of a cycle, forbidden subgraphs and elimination orderings. We will also encounter these graphs later in Section 12 of this Epilogue in relation to chordal bipartite graphs.

Let $C = [u_1, u_2, \ldots, u_{2k}, u_1]$ be a cycle of even length $2k \geq 6$. A chord $u_i u_j \in E$ is called an *odd chord* if one of i and j is even and the other is odd, that is, it divides C into two even length cycles. A graph $G = (V, E)$ is defined to be *strongly chordal* if it is chordal and every cycle of even length greater than or equal to 6 has an odd chord. For example, referring to Figure F1 on page 275, the graphs G_2, G_3, G_4 and G_5 are strongly chordal, however, the others are not strongly chordal, as follows: G_1 and G_6 are not chordal so they are not strongly chordal; the graph G_7 is chordal, but the 6-cycle going around the outside of the graph has no odd chord, so the graph is not strongly chordal.

The graph G_7 is often called the 3-*sun* and is one of a family of forbidden subgraphs characterizing strongly chordal graphs. The k-*sun* S_k ($k \geq 3$) consists of $2k$ vertices, a stable set $X = \{x_1, x_2, \ldots, x_k\}$ and a clique $Y = \{y_1, y_2, \ldots, y_k\}$ and edges $E_1 \cup E_2$ where $E_1 = \{x_1 y_1, y_1 x_2, x_2 y_2, y_2 x_3, \ldots, x_k y_k, y_k x_1\}$ forms the *outer cycle* and $E_2 = \{y_i y_j \mid i \neq j\}$ forms the inner clique. The suns are split graphs, so they are chordal by Theorem 6.3, but they are not strongly chordal since the outer cycle has no odd chord.

A *trampoline* of order k ($k \geq 3$) is a graph obtained from a k-cycle C by adding for each edge of C a new vertex adjacent only to the two endpoints of that edge, and then adding enough chords to C to make it chordal. A *complete trampoline* is one in which all the chords are added to the cycle C, making it a clique and identical to the k-sun S_k. It is not difficult to show that a chordal graph which contains an induced trampoline also contains a (smaller) complete trampoline.

A vertex x is called *simple* if for every pair of neighbors y and z of x, either $N(y) \subseteq N(z)$ or $N(z) \subseteq N(y)$. An ordering of the vertices $[v_1, v_2, \ldots, v_n]$ is called a *simple elimination ordering* for G if v_i is a simple vertex in the induced subgraph H_i, for all i, where $H_i = G_{\{v_i, \ldots, v_n\}}$ is the subgraph remaining after $v_1, \ldots, v_{i-1}$ have been eliminated. Note that the 3-sun S_3 (G_7 on page 275) has no simple vertex, so it does not have a simple elimination ordering.

A *strong elimination ordering* is defined to be an ordering of the vertices $[v_1, v_2, \ldots, v_n]$ where, for all $i < j < k < \ell$, if $v_i v_k, v_i v_\ell, v_j v_k \in E$ then $v_j v_\ell \in E$. It is an easy exercise to verify that simple elimination orderings and strong elimination orderings are special cases of perfect elimination orderings.

The next Theorem, due to Farber [1983], provides the following characterizations of strongly chordal graphs:

Theorem 13.1. The following conditions are equivalent for an undirected graph $G = (V, E)$:
  (i) G is strongly chordal.
 (ii) G has a simple elimination ordering.
(iii) G has a strong elimination ordering.
 (iv) G is chordal and sun-free.
  (v) G is chordal and trampoline-free.

Strongly chordal graphs are closely related to the class of chordal bipartite graphs (Section 12.4, page 261). We will present this connection below in Theorems 13.15 and 13.17. For further reading on strongly chordal graphs, and additional characterizations, see Brandstädt, Le and Spinrad [1999] and McKee and McMorris [1999].

Weakly Chordal Graphs

Hayward [1985] introduced the class of *weakly chordal* graphs (also called *weakly triangulated*) as those having no induced subgraph isomorphic to C_n or to $\overline{C}_n$ for $n \geq 5$. The class of weakly chordal graphs contains the class of chordal graphs, since $C_5 = \overline{C}_5$ and $\overline{C}_n$ contains induced copies of C_4 for $n \geq 6$. Also, the *weakly chordal graphs are perfect graphs.* This result now follows immediately from the Strong Perfect Graph Theorem, however, the first proof was obtained by combining a result by Chvátal [1985], that neither a minimally imperfect graph G nor its complement $\overline{G}$ can contain a "star-cutset", with a result by Hayward [1985], that if G is a weakly chordal graph (with at least 3 vertices) then either G or $\overline{G}$ must contain a "star-cutset".

We call vertices x and y a *two-pair* if every chordless path between x and y has exactly two edges. The weakly chordal graphs have been characterized using two-pairs as follows.

Theorem 13.2. The following are equivalent:
  (i) G is a weakly chordal graph.
 (ii) Every induced subgraph of G is either a clique or has a two-pair.
(iii) If edges are repeatedly added between two-pairs in G, the result is eventually a clique.

The implication (ii) $\Longrightarrow$ (i) follows from the observation that nonadjacent vertices in C_n or $\overline{C}_n$ ($n \geq 5$) are not a two-pair. The implication (i) $\Longrightarrow$ (ii) is due to Hayward, Hoang and Maffray [1990], and (i) $\Longleftrightarrow$ (iii) is due to Spinrad and Sritharan [1995]. The latter equivalence also leads to an $O(n^4)$ recognition algorithm for weakly chordal graphs.

LexBFS

In Section 4.3, we presented lexicographic breadth first search (LexBFS), using it to obtain the linear time algorithm for recognizing chordal graphs. We also mentioned that maximum cardinality search (MCS) provides a conceptually and computationally simpler method for getting a perfect elimination scheme by changing the *label* of a vertex x from a "list" of the marked neighbors to a simple "counter" of those marked neighbors. Tarjan's proof that MCS orderings correctly recognize chordal graphs can be found in Golumbic [1984].

Fans and friends might have been temporarily disappointed when MCS seemed to make LexBFS obsolete, but not for long! LexBFS has become useful in other contexts, including recognizing proper interval graphs, recognizing asteroidal triple-free graphs and several other algorithms. There are also results for LexBFS on the powers of chordal graphs and on distance hereditary graphs which have no analogous result for MCS. See Brandstädt, Le and Spinrad [1999, Chapter 5] which also surveys characterizations of many perfect graph families in terms of special kinds of vertex orderings.

Intersection Graphs on Trees

Let T be a tree and let $\{T_i\}$ be a collection of subtrees (connected subgraphs) of T. We may think of the host tree T either (1) as a continuous model of a tree embedded in the plane, thus generalizing the real line from the one-dimensional case, or (2) as a finite discrete model of a tree, namely, a connected graph of vertices and edges having no cycles, thus generalizing the path P_k from the one dimensional case.

The distinction between these two models becomes important when measuring the size of the intersection of two subtrees. For example, in the continuous model (1), we might take the size of the intersection to be the Euclidean distance of a longest common path of the two subtrees. In the discrete model (2), we might count the number of common vertices or common edges. We use the expressions "nonempty intersection" and "vertex intersection" to mean sharing a vertex or point of T, and "nontrivial intersection" and "edge intersection" to mean sharing an edge or otherwise measurable segment of T.

Using this terminology, Theorem 4.8 (page 92) stated the following.

Theorem 4.8. A graph is the *vertex intersection graph* of a set of subtrees of a tree if and only if it is a chordal graph.

In contrast to this, Golumbic and Jamison [1985a] observed that the family of *edge intersection graphs* of subtrees of a tree yield all possible graphs, proving the following:

Theorem 13.3. Every graph can be represented as the edge intersection graph of substars of a star.

Table 1. Graph classes involving trees

Type of Interaction	Objects	Host	Graph Class
vertex intersection	subtrees	tree	chordal graphs
vertex intersection	subtrees	star	split graphs
edge intersection	subtrees	star	all graphs
vertex intersection	paths	path	interval graphs
vertex intersection	paths	tree	path graphs or VPT graphs
edge intersection	paths	tree	EPT graphs
containment	intervals	line	permutation graphs
containment	paths	tree	? (open question)
containment	subtrees	star	comparability graphs

Proof. Let $G = (V, E)$ be any graph, and let $E = \{e_1, \ldots, e_m\}$. Consider the star T formed by a central node u and leaves $\bar{e}_1, \ldots, \bar{e}_m$. Define the substar corresponding to v_i to be the substar T_i of T induced by $\{u\} \cup \{\bar{e}_\ell \mid v_i \in e_\ell\}$. Clearly, $v_i v_j \in E$ if and only if T_i and T_j share an edge, namely edge ue_k of T where $e_k = v_i v_j$. $\blacksquare$

We will see below, in Theorem 13.6, that a graph is the *vertex intersection graph* of substars of a star if and only if it is a split graph.

Two different classes of intersection graphs also arise when considering simple paths (instead of subtrees) of an arbitrary host tree T. The *path graphs*, which we mentioned on page 94, are the subfamily of chordal graphs obtained as the "vertex intersection graphs of paths in a tree" and are also called *VPT graphs*. However, the graphs obtained as the "edge intersection graphs of paths in a tree", called *EPT graphs*, are not necessarily chordal. The class of EPT graphs are not perfect graphs, and the recognition problem for them is NP-complete, Golumbic and Jamison [1985a, 1985b]. See also Monma and Wei [1986] and Sysło [1985].

Table 1 summarizes the subtree graph classes we have discussed here and in Sections 13.5 and 13.6 below. A full treatment can be found in Golumbic and Trenk [2004, Chapter 11].

5. Comparability Graphs and the Dimension of Ordered Sets

Comparability Invariants

A graph can have many different transitive orientations, so there may be different partial orders with the same comparability graph. A property of partially

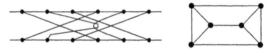

Fig.13.1. A function diagram and its intersection graph (which is isomorphic to $\overline{C}_6$).

ordered sets is called a *comparability invariant* if either all orders with a given comparability graph have that property, or none have that property. For example, we saw in Theorem 8.13 (page 187) and in Exercise 8.7 (page 194) that the properties of being a semiorder or an interval order are comparability invariants.

The dimension of a partial order is also a comparability invariant, that is, dim $P = \dim Q$ whenever P and Q have the same comparability graph G; hence, we can denote this common value by dim G. The proof we gave for Theorem 5.39 (page 139) is incomplete; a full correct proof can be found in Trotter [1992] or in Golumbic and Trenk [2004]. Those references may be consulted for further study on comparability invariant properties.

Function Diagrams: the Intersection Model for Cocomparability Graphs

Golumbic, Rotem and Urrutia [1983] have characterized the family of cocomparability graphs as the intersection graphs of function lines in a diagram which generalizes the matching diagrams (page 162) which represent permutation graphs. Their function diagrams are constructed as follows.

Let L_1 and L_2 be two horizontal lines. A continuous curve f connecting a point on L_1 with a point on L_2 is called a *function line* if, whenever two points (x, y) and (x', y') on f have the same horizontal value $y = y'$, the points must be equal, i.e., $x = x'$. A *function diagram* consists of L_1 and L_2 and a set of n function lines connecting points on L_1 and L_2. The function diagram in Figure 13.1 has six function lines. We note that a matching diagram is the special case in which the function lines are straight lines.

Consider the following special type of function diagram in which the curves are piecewise linear. Let $L_1, L_2, \ldots, L_{k+1}$ be horizontal lines each labeled from left to right by a permutation of the numbers $1, 2, \ldots, n$. For each i ($1 \leq i \leq n$) the curve f_i consists of the union of the k straight line segments which join i on L_t with i on L_{t+1} ($1 \leq t \leq k$). When $k = 1$, this is just a matching diagram; when $k \geq 2$, it is called the *concatenation* of k matching diagrams.

The following theorem is due to Golumbic, Rotem and Uruttia [1983], and a proof can also be found in Golumbic and Trenk [2004].

Theorem 13.4. The following are equivalent.
(i) G is the intersection graph of a function diagram.
(ii) G is a cocomparability.
(iii) G is the intersection graph of a concatenation of matching diagrams.

Moreover, if ℓ is the minimum value for which G is the intersection graph of a concatenation of ℓ matching diagrams, then dim $G = \ell + 1$.

Containment Graphs

The *containment graph* $G = (V, E)$ of a collection $\mathcal{F} = \{S_i\}$ of distinct subsets of a set **S** has vertex set $V = \{1, \ldots, n\}$ and edge set $E = \{ij \mid$ either $S_i \subset S_j$ or $S_j \subset S_i\}$. A graph with such a representation is called a *containment graph*. The class of containment graphs is equivalent to the class of comparability graphs. Moreover, Golumbic and Scheinerman [1989] observed that every comparability graph can be represented as the containment graph of a collection of subtrees (substars) of a star. Dushnik and Miller [1941] characterized the containment graphs of intervals on the line as precisely those having partial order dimension 2, and from Chapter 7, we recall that these are equivalent to the permutation graphs. Generalizing interval containment, Golumbic and Scheinerman [1989] also showed the following.

Theorem 13.5. A graph G is the containment graph of rectilinear boxes[4] in d-space if and only if $\dim(G) \leq 2d$.

Yannakakis [1982] has shown that the complexity of determining whether an order P has dimension $\leq k$, for any fixed $k \geq 3$ is NP-complete. This answers the open problem stated in the footnote on page 138. A proof can be found in Mahadav and Peled [1995, Chapter 7]. Therefore, as a corollary of Theorem 13.5, we conclude that *the recognition problem for the containment graphs of boxes in the plane is NP-complete*.

As early as the Banff Conference in 1984, we posed the problem, "Characterize the containment orders of circles in the plane and their comparability graphs", see Rival [1985, page 583]. Progress on this question can be found in Fishburn [1988], Scheinerman and Wierman [1988], and Scheinerman [1992]. Sphere orders are the generalization to higher dimension, and are also found in the literature. Characterizing the containment graphs of paths in a tree is still an open problem.

New Complexities for Comparability Graphs, Transitive Orientation and Permutation Graphs

New algorithms have been found for recognizing comparability graphs and permutation graphs, based on fast modular decomposition of graphs. Modular decomposition is the recursive version of the method we saw in Section 5.2. The

---

[4] Boxes with sides parallel to the axes.

first of these new algorithms were due to Spinrad [1985]. McConnell and Spinrad [1999] show how to find an orientation F of an arbitrary graph $G = (V, E)$ such that F is a transitive orientation (TRO) of G if and only if G is a comparability graph. This is very good if there is other information guaranteeing that G is a comparability graph. However, this alone does not recognize comparability graphs, since the algorithm simply produces an orientation which is *not* transitive when G is *not* a comparability graph. Hence, to complete it to a recognition algorithm, one must test F to determine if it is transitive.

The complexity of their method uses $O(n + e)$ time to produce a linear ordering L_1 of the vertices which is then applied to E to produce an orientation F_1. It then uses $O(n^\alpha)$ time to test whether F_1 is transitive, where $O(n^\alpha)$ is the complexity to perform transitive closure or $n \times n$ matrix multiplication (currently $n^{2.376}$). In a similar fashion, they can produce another linear ordering L_2 of the vertices which if applied to $\overline{E}$ will produce an orientation F_2 which will be transitive if and only if G is a cocomparability graph. So the complexity of recognizing cocomparabilty graphs is currently also $O(n^\alpha)$.

Interestingly, their method allows recognizing permutation graphs in $O(n + e)$ time, by first producing L_1, L_2 and F_1 and calculating the in-degrees and out-degrees of F_1 and F_2, but without actually producing F_2, since otherwise the complexity will hit $O(n^2)$. These enable us to follow the construction in the proof of our Theorem 7.1 (pages 158–159), yielding a permutation representation for G, if G is a permutation graph, or a contradiction among the degrees if G is not a permutation graph.

6. Split Graphs

Theorem 4.8 stated that a graph is the (vertex) intersection graph of a set of subtrees of a tree if and only if it is a triangulated graph. McMorris and Shier [1983] give an analogous version for split graphs, which we recall are characterized as being both triangulated and cotriangulated (Theorem 6.3, page 151). If the host tree T is a star $K_{1,n}$, then each induced subtree consists of either a substar containing the central node or just a single leaf node. It is easy to see that the graphs obtained in this restriction are precisely the class of split graphs, as observed in McMorris and Shier [1983].

Theorem 13.6. A graph G is the vertex intersection graph of distinct induced subtrees of a star $K_{1,n}$ if and only if G is a split graph.

Proof. Recall that a graph G is a split graph if its vertices can be partitioned into a clique $K = \{x_1, \ldots, x_k\}$ and a stable set $S = \{y_1, \ldots, y_\ell\}$. If G is a split graph, consider the star T formed by a central node u and leaves $\bar{x}_1, \ldots, \bar{x}_k, \bar{y}_1, \ldots, \bar{y}_\ell$, where the subtree corresponding to $y_i \in S$ is the single leaf $\bar{y}_i$ in T and the subtree corresponding to $x_i \in K$ consists of the substar of T induced by $\{u, \bar{x}_i\} \cup \{\bar{y}_j \mid y_j \in Adj(x_i)\}$. Clearly, this is an intersection representation for G. Conversely, if we are given a representation for G as the intersection graph of distinct induced substars of a star, then those substars containing the central node correspond to a clique in G and the remaining subtrees (the single leaves) correspond to a stable set of G.   ∎

7. Permutation Graphs and Applications from Circuit Design

Matching diagrams, like those we studied in Section 7.4, are used in circuit design for channel routing problems where a set of numbered pins on the upper side of the diagram must be connected (electrically) to a set of pins on the lower side. The area between the upper and lower horizontals is called the *channel*. When a pair of line segments connecting matched pins intersect, they must be placed on different silicon layers, similar to the altitudes for the aircraft in Application 7.1. Thus, the minimum number of layers needed to realize the diagram equals the clique number $\omega(G)$ of its permutation graph G, the value of which can be calculated in $O(n \log n)$ time. We will discuss briefly two similar graph problems originally motivated by circuit design. The books by Lengauer [1990] and Shrwani [1995] give a comprehensive treatment of other VLSI design and routing algorithms.

Cell Flipping in Matching Diagrams

Golumbic and Kaplan [1998] have considered the following generalization of the channel routing problem above, which is motivated by "standard cell" technology. The numbers on each side of the channel are partitioned into consecutive subsequences, or *cells*, each of which can be left unchanged or flipped (i.e., reversed). This takes place at a stage where the cell placement on horizontal rows has already been performed, and the only remaining degree of freedom is replacing some of the cells with their "mirror image" with respect to the vertical axis, i.e., cell flipping. The questions asked are:

MINFLIP: For what choice of flippings will the resulting clique number be minimized?

MAXFLIP: For what choice of flippings will the resulting clique number be maximized?

For example, let the upper sequence be partitioned $[3, 4, 7], [2, 6], [1, 5, 8]$ and let the lower sequence be $[6', 2', 5'], [4', 1', 7', 8', 3']$, where the brackets indicate cells. The clique number $\omega(G)$ is 4 with no flipping but is reduced to 3 if we flip $[2, 6]$, or is increased to 5 if we flip $[2, 6]$, $[6', 2', 5']$ and $[4', 1', 7', 8', 3']$.

The complexity of the MAXFLIP problem is $O(n^2)$ using a dynamic programming algorithm, whereas the MINFLIP problem is NP-complete (see Golumbic and Kaplan [1998]). When one side of the channel is fixed (no flipping on that side), the problem of finding a flipping for the other side which *maximizes* the clique number can be found in $O(n \log n)$ time. Verbin [2002] has shown the complexity of the one side flipping problem for *minimizing* the clique number to be NP-complete.

We note that the cell flipping problems could have been defined to minimize or maximize the stability number. It is clear, however, that these stability problems are computationally equivalent to our clique problems, since $\alpha(G) = \omega(\overline{G})$ and G is a permutation graph if and only if $\overline{G}$ is a permutation graph. For example, the 4 different one side flipping problems can be restated as follows: Given a permutation of the numbers $1, 2, \dots, n$ partitioned into cells, find flippings which minimize/maximize the longest increasing/decreasing subsequence.

Trapezoid Graphs

Dagan, Golumbic and Pinter [1988] extended the notion of a permutation diagram by replacing each matching segment i-i' by a trapezoid obtained from 4 points a_i, b_i, c_i, d_i where interval $I_i = [a_i, b_i]$ lies on the upper line of the diagram and interval $I_i' = [c_i, d_i]$ lies on the lower line of the diagram. The intersection graphs of these trapezoid diagrams are called *trapezoid graphs*. Trapezoid graphs are a subclass of cocomparability graphs, and their associated (trapezoid) orders are precisely those having interval dimension 2 (see Trotter [1992]).

Langley [1995] and Ma and Spinrad [1994] gave polynomial time algorithms for recognizing trapezoid graphs, and Felsner, Müller and Wernisch [1997] have given optimal maxclique and maxstable set algorithms for the class. Further investigation of the class can be found in Cheah and Corneil [1996]. As we will mention below, trapezoid graphs are equivalent to the bounded bitolerance graphs.

Cographs and Factoring Read-Once Functions

An important subfamily of permutation graphs are the *complement reducible graphs*, or *cographs*, which were investigated by Corneil, Lerchs and Burlingham [1981] and Corneil, Perl and Stewart [1985]. Cographs can be defined recursively as follows: (1) a single vertex is a cograph; (2) the disjoint union of cographs is a cograph; (3) the complement of a cograph is a cograph. Alternately, they can be defined by a restricted type of decomposition, as in our Section 5.2 (page 111) where G_R would be either a clique or a stable set. One can recognize whether a graph G is a cograph by repeatedly decomposing it in this way, and in linear time obtain a representation called its *cotree*. The cotree is useful in many algorithms for the class. The next theorem gives several characterizations of cographs.

Theorem 13.7. The following are equivalent for an undirected graph G.
(i) G is a cograph.
(ii) G is P_4-free.
(iii) For every subset X of vertices, either the induced subgraph G_X is disconnected or its complement $\overline{G}_X$ is disconnected.

Recognizing cographs and building the associated cotree has been recently used in an application involving multi-level logic synthesis. Golumbic and Mintz [1999] presented a factoring algorithm for general Boolean functions which is based on graph partitioning, and at the lower levels of the recursion, read-once functions are handled in a special manner. Read-once functions are very closely related to cographs, as we shall see.

A Boolean function F is called a *read-once function* if it has a factored form in which each variable appears exactly once. For example, the function $F_1 = aq + acp + ace$ is a read-once function since it can be factored into the formula $F_1 = a(q + c(p + e))$. The function $F_2 = ab + ac + bc$ is not a read-once function. Read-once functions have interesting special properties and account for a large percentage of functions which arise at the lower level real circuit applications. They have also gained recent interest in the field of computational learning theory.

Let $F = \alpha_1 + \cdots + \alpha_t$ be given in disjunctive normal form (sum-of-products), where each α_k is a product term, and let $V = \{v_1, \ldots, v_n\}$ be its set of variables. We consider the graph $\Gamma_F = (V, E)$ of F where $v_i v_j \in E$ whenever v_i and v_j appear together in some product term α_k. The function F is called *normal* if every clique of Γ_F is contained in one of the product terms. Our example F_2 above is not normal since Γ_{F_2} is a triangle, but all product terms are of size 2. The dual F^* of F is the function obtained from F by replacing products by sums and sums by products. By taking the disjunctive normal form of F^* one can construct the graph Γ_{F^*}. The next theorem is due to Gurvich [1991]:

Theorem 13.8. The following are equivalent:
(i) F is a read-once function.
(ii) Γ_F is P_4-free (i.e., a cograph) and F is a normal function.
(iii) $\overline{\Gamma}_F = \Gamma_{F^*}$.

Golumbic, Mintz and Rotics [2001] present a very fast method for recognizing and factoring read-once functions based on algorithms for cograph recognition and on checking normality.

8. Interval Graphs and Circular-arc Graphs

AT-Free Graphs

Three vertices v_1, v_2, v_3 in a graph $G = (V, E)$ form an *asteroidal triple* (AT) if, for all permutations i, j, k of $\{1, 2, 3\}$, there is a path from v_i to v_j which avoids using any vertex in the neighborhood $N(v_k) = v_k \cup Adj(v_k)$. An easy way to verify this for v_k is to delete $N(v_k)$ and test whether v_i and v_j remain in the same connected component of $G - N(v_k)$. It also follows from the definition that the three vertices of an asteroidal triple are pairwise nonadjacent. For example, the graph in Figure 8.15 (page 196) has 8 asteroidal triples[5].

A graph is called *asteroidal triple free* or *AT-free* if it contains no asteroidal triple. We saw in Theorem 8.4 (page 174) that interval graphs are characterized by being chordal and AT-free. Golumbic, Monma and Trotter [1984] proved the following result.

Theorem 13.9. Every cocomparability graph is an AT-free graph.

More recently, Corneil, Olariu and Stewart [1997, 1999] have given new algorithmic and structural results for AT-free graphs. Köhler [2000] also provides efficient recognition algorithms for the class.

Tolerance Graphs

Tolerance graphs were introduced by Golumbic and Monma [1982] to generalize some of the applications associated with interval graphs. Their motivation was the need to solve scheduling problems, more general than what we saw in Section 1.3, in which resources such as rooms, vehicles, support personnel, etc. may be needed on an exclusive basis, but where a measure of flexibility or tolerance would be allowed for sharing or relinquishing the resource when total exclusivity prevented a solution. The recent book by Golumbic and Trenk [2004] contains a thorough study of tolerance graphs and related topics.

An undirected graph $G = (V, E)$ is a *tolerance graph* if there exists a collection $\mathcal{I} = \{I_v\}_{v \in V}$ of closed intervals on the real line and an assignment of positive numbers $t = \{t_v\}_{v \in V}$ such that

$$vw \in E \Leftrightarrow |I_v \cap I_w| \geq \min\{t_v, t_w\}.$$

Here $|I_u|$ denotes the length of the interval I_u. The positive number t_v is called the *tolerance* of v, and the pair $\langle \mathcal{I}, t \rangle$ is called an *interval tolerance representation* of G. A tolerance graph is said to be *bounded* if it has a tolerance representation

---

[5] In this book, we incorrectly spelled these "astroidal" rather than asteroidal.

in which $t_v \leq |I_v|$ for all $v \in V$. Tolerance graphs generalize both interval graphs and permutation graphs, and in Golumbic and Monma [1982] it was shown that every bounded tolerance graph is a cocomparability graph. Golumbic, Monma and Trotter [1984] proved that tolerance graphs are perfect and are contained in the class of weakly chordal graphs.

Coloring bounded tolerance graphs in polynomial time is an immediate consequence of their being cocomparability graphs. Narasimhan and Manber [1992] use this fact (as a subroutine) to find the chromatic number of any (unbounded) tolerance graph in polynomial time, but not the coloring itself. Golumbic and Siani [2002] give an $O(qn + n \log n)$ algorithm for coloring a tolerance graph, given the tolerance representation with q vertices having unbounded tolerance (see Golumbic and Trenk [2004]). The complexity of recognizing tolerance graphs and bounded tolerance graphs remain open questions.

A variety of "variations on the theme of tolerance" in graphs have been defined and studied over the past years. By substituting a different "host" set instead of the real line and with a specified type for the subsets of that host instead of intervals, we obtain classes such as neighborhood subtree tolerance (NeST) graphs, tolerance graphs of paths on a tree or tolerance competition graphs. By changing the function *min* for a different binary function ϕ (for example, *max, sum, product*, etc.), we obtain a class that will be called ϕ-*tolerance* graphs. By replacing the measure of the length of an interval by some other measure μ of the intersection of the two subsets (for example, cardinality in the case of discrete sets, or number of branching nodes or maximum degree in the case of subtrees of trees), we could obtain yet other variations of tolerance graphs. When we restrict the tolerances to be 1 or ∞, we obtain the class of *interval probe graphs*. By allowing a separate leftside tolerance and rightside tolerance for each interval, various bitolerance graph models can be obtained. For example, Langley [1993] showed that *the bounded bitolerance graphs are equivalent to the class of trapezoid graphs*. Directed graph analogues to several of these models have also been defined and studied. For further study of tolerance graphs and related topics, we refer the reader to Golumbic and Trenk [2004].

Circular-Arc Graphs

In Section 8.6, we presented the early work on circular-arc graphs, the intersection graphs of arcs on a circle. The first polynomial time algorithm for recognizing circular-arc graphs was given by Tucker [1980] and had complexity $O(n^3)$. Over the years, more efficient algorithms were designed for the recognition problem first by Hsu [1995] in $O(ne)$ and Eshen and Spinrad [1993] in $O(n^2)$, and most recently by McConnell [2001, 2003] in $O(n + e)$ which is optimal. The coloring problem for circular-arc graphs was shown to be NP-complete by Garey, Johnson, Miller and Papadimitriou [1980]. The maximum stable set and the maximum clique problems are polynomial, see Golumbic and Hammer [1988], Hsu and Spinrad [1995], Hsu [1985], Apostolico and Hambrus [1987].

9. Superperfect Graphs

Superperfect Noncomparability Graphs

Progress has been made on the question asked in Section 9.4, "When does superperfect equal comparability?" Gallai [1967] published a complete list of the minimal noncomparability graphs, that is, the noncomparability graphs with the property that removing any vertex makes the remaining subgraph a comparability graph, see also Berge and Chvátal [1984, p. 78]. Using this list, Andreae [1985] determined all of the minimal noncomparability graphs which are superperfect. One of these is the graph in our Figure 1.18 on p.18 which we *incorrectly* placed in the position of non-superperfect in Figure 9.9 on p. 212. This graph *is* superperfect and a simple proof is also given in Kloks and Bodlander [1992, Theorem 3.2].

In Corollary 9.8, we saw that for split graphs, G is a comparability graph if and only if G is superperfect. This motivated our asking the question of whether or not this equivalence holds for triangulated graphs or for cotriangulated graphs. Using his forbidden subgraph characterization of superperfect noncomparability graphs, Andreae [1985] answered this question with "false" for triangulated graphs and "true" for cotriangulated graphs, showing the following.

Theorem 13.10.
(i) For cotriangulated graphs, G is a comparability graph if and only if G is superperfect.
(ii) For triangulated graphs, G is a comparability graph if and only if G is superperfect and contains no induced subgraph isomorphic to G_1, G_2 or G_3 of Figure 13.2.

As Andreae [1985] points out, this list of the minimal superperfect non-comparability graphs could also be used to answer our equivalence question for other hereditary classes of graphs. In a similar investigation for k-trees,

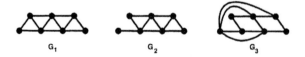

G_1 G_2 G_3

Fig.13.2. The minimal triangulated, superperfect noncomparability graphs.

Kloks and Bodlander [1992] gave a complete characterization, by means of forbidden subgraphs, of the superperfect 2-trees. The graph G_1 is a superperfect 2-tree. They also give an algorithm for testing superperfection in k-trees, whose complexity is linear for any fixed k, and can produce a complete forbidden subgraph characterization of superperfect k-trees.

10. Threshold Graphs

The undisputed authority on threshold graphs and related topics is the book by Mahadev and Peled [1995]. It is a masterpiece of clarity, and it presents a comprehensive coverage of Ferrers digraphs, threshold dimension, measures, weights, orders, enumeration, and a variety of generalizations, variations and specializations of threshold graphs. Any intention of updating or expanding upon Chapter 10 must therefore defer to their book.

New complexity results on the threshold dimension of a graph deserve mentioning. In Section 10.1, we defined the threshold dimension $\theta(G)$ of a graph G. Threshold graphs are those having threshold dimension at most 1, and they can be recognized in linear time. The complexity of determining $\theta(G)$ is NP-complete (page 223), and a stronger result is that determining whether $\theta(G) \leq k$, for any fixed $k \geq 3$, is NP-complete (Golumbic [1981], Yannakakis [1982]). So the remaining question, open for the next dozen years, was whether recognizing threshold dimension 2 could be done in polynomial time. A positive answer to this question follows from the next interesting result.

Two edges of G are said to *conflict* if their endpoints induce in G one of the forbidden subgraphs C_4, P_4 or $2K_2$ characterizing threshold graphs (see Theorem 10.7, page 227). The conflict graph G^* of G is defined by taking $V(G^*) = E(G)$ and by joining two vertices of G^* by and edge in $E(G^*)$ if and only if their corresponding edges in G conflict. For example, the conflict graph of a threshold graph is a stable set, and the conflict graph of C_4 is $2K_2$. We noted in Section 10.1, that we can take $\theta(G)$ to be the minimum number of threshold graphs needed to cover the edges of G. Thus, Chvátal and Hammer [1977] observed that such an edge covering of G leads to a valid coloring of G^*, that is, $\chi(G^*) \leq \theta(G)$ where each color represents a threshold graph in the edge covering. From this inequality, one can see that whenever G has threshold dimension 2, its conflict graph G^* must be a bipartite graph. The converse of this implication for $\theta(G) = 2$ was conjectured to hold by Ibaraki and Peled [1981], and it was finally proven by Raschle and Simon [1995], which we record as follows.

Theorem 13.11. A graph G has threshold dimension 2 if and only if its conflict graph G^* is a bipartite graph.

A proof of this theorem can be found in Mahadev and Peled [1995, Section 8.5]. Since constructing the conflict graph and testing whether it is 2-colorable can be done in polynomial time, we immediately obtain the following corollary.

Theorem 13.12. The problem of determining whether $\theta(G) \leq 2$ has polynomial time complexity.

Threshold Hypergraphs

In Section 10.4, we presented the definition of threshold hypergraph and gave an application in which they might be useful for the allocation of resources. Threshold hypergraphs are closely related to the class of threshold Boolean functions. We also posed a Research Problem (on page 233) giving three properties of r-regular hypergraphs, each being a generalization of threshold graphs to hypergraphs. Reiterman, et al. [1985] have shown that these three properties are different, and they have given a characterization of the most general of them (T_3) which are known as *shift stable* r-regular hypergraphs. Boros [1991] has generalized this further, giving a characterization of shift stable hypergraphs (not necessarily regular ones).

11. Circle Graphs

Circle graphs are the intersection graphs of chords of a circle. In Chapter 11, we called these "not so perfect graphs" rather as a joke, needing an excuse to include them in the book even though they are not perfect graphs. Circle graphs are an important and natural extension of permutation graphs, and their equivalence with overlap graphs raises their status even further. The recognition problem for circle graphs, which had been open, was solved independently by Bouchet [1987] and Gabor, Supowit, and Hsu [1989] who gave polynomial algorithms with complexity of $O(n^4)$ and $O(n^3)$, respectively. Subsequently, an $O(n^2)$ method was given by Spinrad [1994]. A further characterization of circle graphs appears in Bouchet [1994].

12. Chordal Bipartite Graphs

We defined the class of *chordal bipartite* graphs in Section 12.4 as those bipartite graphs which, for all $k > 4$, have no induced chordless cycle C_k. Thus, a chordal bipartite graph is not necessarily a chordal graph since the 4-cycle C_4

is chordal bipartite but not chordal[6]. Recalling the definition of weakly chordal graphs (Section 13.4 above), it is a simple exercise to show the following:

Theorem 13.13. An undirected graph G is chordal bipartite if and only if G is bipartite and weakly chordal.

Chordal bipartite graphs are also related to strongly chordal graphs and to totally balanced matrices. A 0/1 matrix is called *totally balanced* if it does not contain as a submatrix the (vertex-edge) incidence matrix of a cycle of length ≥ 3. A 0/1 matrix is called Γ-*free* if it does not contain a submatrix of the form $\binom{1\,1}{1\,0}$. Anstee and Farber [1984], Hoffman, Kolen and Sakarovitch [1985] and Lubiw [1987] proved the following:

Theorem 13.14. A 0/1 matrix is *totally balanced* if and only if its rows and columns can be permuted so that the resulting matrix is Γ-*free*.

Let $G = (V, E)$ be an undirected graph with $V = \{v_1, \ldots, v_n\}$. We define the *clique-vertex incidence graph* of G to be the bipartite graph $H(G)$ with the vertices of G on one side and the maximal cliques of G on the other side, such that a vertex v of G is adjacent to a clique K of G in $H(G)$ if and only if v is a member of K in G. In Section 8.2 (page 174), studying the characterizations of interval graphs, we defined a similar notion, the *clique-vertex incidence matrix* of G which we denote by $\mathbf{M} = \mathbf{M}(G)$.

Finally, we define $B(G) = (X, Y, E')$ to be the bipartite graph where $X = \{x_1, \ldots, x_n\}$, $Y = \{y_1, \ldots, y_n\}$ and $E' = \{x_i y_i \mid v_i \in V\} \cup \{x_i y_j \mid v_i v_j \in E\}$. The graph $B(G)$ is, in fact, the same as the *closed neighborhood-vertex incidence graph* of G.

The following theorem is due to Farber [1983].

Theorem 13.15. The following conditions are equivalent:
 (i) G is a strongly chordal graph.
 (ii) $H(G)$ is a chordal bipartite graph.
 (iii) $\mathbf{M}(G)$ is a totally balanced matrix.
 (iv) $B(G)$ is a chordal bipartite graph.

We mention another interesting equivalence which follows essentially from our proof of Theorem 12.1, and was stated explicitly later in Brandstädt [1991].

---

[6] When I introduced the term "chordal bipartite graph" in 1978, my thinking was that the word "chordal" in this context should be interpreted as an adjective permitting cycles of the smallest possible length (length 3 for graphs in general, or length 4 for bipartite graphs) but requiring that any larger cycle must have a chord. Thus, the meaning of "chordal" was context dependent, admittedly a somewhat confusing matter. Indeed, this criticism is voiced in Brandstädt, Le and Spinrad [1999, page 41]. Clearly, *a graph G is both a chordal graph and bipartite if and only if G is a forest of trees.*

Theorem 13.16. An undirected graph G is a chordal graph if and only if $B(G)$ is a perfect elimination bipartite graph.

Our Corollary 12.11 (page 263) related chordal bipartite graphs to perfect elimination bipartite graphs by adding the hereditary property to the latter. We now present two additional characterizations of chordal bipartite graphs.

Let $H = (X, Y, E)$ be a bipartite graph, and let $\sigma = [e_1, \ldots, e_m]$ be an ordering of the edges. Define $H_i = (X, Y, E_i)$ where $E_i = \{e_i, \ldots, e_m\}$, that is, H_i is the graph obtained by erasing the edges $e_1, \ldots, e_{i-1}$, but not their endpoints. We call σ a perfect *"edge-without-vertex" erasing* ordering if e_i is bisimplicial in H_i.

Denote by $\text{Split}_Y(H)$ the split graph obtained by completing Y into a clique.

In the following theorem, the equivalence (i) $\iff$ (ii) is due to Dalhaus [1991], and several researchers, according to Brandstädt, Le and Spinrad [1999], have observed (i) $\iff$ (iii).

Theorem 13.17. The following conditions are equivalent for a bipartite graph H:

(i) H is a chordal bipartite graph.

(ii) $\text{Split}_Y(H)$ is a strongly chordal graph.

(iii) H has an edge-without-vertex erasing order.

Further results on chordal bipartite graphs, relating them to vertex orderings, totally balanced matrices, matrices having a Γ-free ordering, and other classes, can be found throughout Brandstädt, Le and Spinrad [1999]. Minimum triangulations of chordal bipartite graphs are studied in Kloks [1994].

A Final Note on Terminology

Samuel Eilenberg, one of the leading mathematicians of the twentieth century, and my doctoral thesis advisor, objected strenuously against the use of mathematical terms such as "partial function". In his view, an adjective modifying a noun should specialize the mathematical concept represented by the noun and *not* generalize it. Since a partial function is not a function at all, but rather a mapping which is defined on only part of its domain, i.e., a partially defined function, he regarded such terms as an imprecise abuse of language. For him, semigroups were always monoids (with identity), and he would have disliked weakly chordal graphs.

Thus, quite correctly, a subset is a set, a recursive function is a function, an alligator purse is a purse, and a strongly chordal graph is a chordal graph, which in turn is a graph. The "is-a" hierarchy is a partial order (whoops!) is a partially ordered set (whoops again!) is simply an order.

The terminology we used in this book, and subsequently, has tried to follow this principle. For example, we prefer to use the terms interval tolerance graphs and neighborhood subtree tolerance graphs when referring to these classes, but never tolerance interval graphs since the concept of an interval graph is so firmly established in the literature. Similarly, we insist on using the term interval probe graphs. The use of chordal bipartite graph is consistent with this approach, but has caused confusion. It is never possible to get it right every time, but we hope that there will be enough tolerance on both sides of the interval to keep the conflict graph very sparse.

Bibliography

Allen, James F.
 [1983] Maintaining knowledge about temporal intervals, *Commun. ACM* **26**, 832–843.
Alon, Noga, and Scheinerman, Edward R.
 [1988] Degrees of freedom versus dimension for containment orders, *Order* **5**, 11–16.
Andreae, T.
 [1985] On superperfect non-comparability graphs, *J. Graph Theory* **9**, 523–532.
Anstee, Richaed P., and Farber, Martin
 [1984] Characterizations of totally balanced matrices, *J. of Algorithms* **5**, 215–230.
Apostolico, A., and Hambrus, S.E.
 [1987] Finding maximum cliques on circular-arc graphs, *Infor. Process. Lett.* **26**, 209–215.
Atkins, J.E. and Middendorf, M.
 [1996] On physical mapping and the consecutive ones property for sparce matrices, *Discrete Applied Math.* **71**, 23–40.
Avriel, M., Penn, M. and Shpirer, N.
 [2000] Container ship stowage problem: complexity and connection to the coloring of circle graphs, *Discrete Applied Math.* **103**, 271–279.
Berge, Claude, and Chvátal, V., eds.
 [1984] *"Perfect Graphs", Annals of Discrete Math., vol.* **21**, North-Holland, Amsterdam.
Berry, A., and Bordat, J. P.
 [2001] Asteroidal triples of moplexes, *Discrete Applied Math.* **111**, 219–229.
Bibelnieks, E., and Dearing, P. M.
 [1993] Neighborhood subtree tolerance graphs, *Discrete Applied Math.* **43**, 107–144.
Boros, Endre
 [1991] On shift stable hypergraphs, *Discrete Math.* **87**, 81–84.
Bouchet, A.
 [1987] Reducing prime graphs and recognizing circle graphs, *Combinatorica* **7**, 243–254.
 [1994] Circle graph obstructions, *J. Combin. Theory B* **60**, 8–22.
Brandstädt, A.
 [1991] Classes of bipartite graphs related to chordal graphs, *Discrete Applied Math.* **32**, 51–60.
 [1997] LexBFS and powers of chordal graphs, *Discrete Math.* **171**, 27–42.
Brandstädt, A., Le, V. B., and Spinrad, J. P.
 [1999] *"Graph Classes: A Survey"*, SIAM, Philadelphia.
Cheah, M., and Corneil, D. G.
 [1996] On the structure of trapeziod graphs, *Discrete Applied Math.* **66**, 109–133.
Chudnovsky, M., Cornuéjols, G., Liu, X., Seymour, P., and Vušković, K.
 [2002] Cleaning for Bergeness, submitted for publication.

Chudnovsky, M., Robertson, N., Seymour, P. and Thomas, R.
[2002] The Strong Perfect Graph Theorem, submitted for publication.
Chudnovsky, M., and Seymour, P.
[2002] Recognizing Berge graphs, submitted for publication.
Chvátal, V.
[1985] Star-cutsets and perfect graphs, *J. Combin. Theory B* **39**, 189–199.
Cook, W. J., Cunningham, W. H., Pulleyblank, W. R., and Schrijver, A.
[1998] *"Combinatorial Optimization"*, John Wiley & Sons, New York.
Corman, T. H., Leiserson, C. E., Rivest, R. L., and Stein, C.
[2001] *"Introduction to Algorithms"*, second edition, McGraw-Hill and MIT Press, Cambridge, Mass.
Corneil, D. G., Kim, H., Natarajan, S., Olariu, S., and Sprague, A.
[1995] Simple linear time recognition of unit interval graphs, *Infor. Process. Letters* **55**, 99–104.
Corneil, D. G., Lerchs, H., and Burlingham, L.
[1981] Complement reducible graphs, *Discrete Applied Math.* **3**, 163–174.
Corneil, D. G., Olariu, S., and Stewart, L.
[1997] Asteroidal triple-free graphs, *SIAM J. Discrete Math.* **10**, 399–430.
[1999] A linear algorithms for dominating pairs in asteroidal triple-free graphs, *SIAM J. Computing* **28**, 1284–1297.
Corneil, D. G., Perl, Y., and Stewart, L.
[1985] A linear recognition algorithm for cographs, *SIAM J. Computing* **14**, 926–934.
Cornuéjols, G., Liu, X., and Vušković, K.
[2002] A polynomial algorithm for recognizing Berge graphs, submitted for publication.
Dagan, Ido, Golumbic, Martin Charles, and Pinter, Ron.Y.
[1988] Trapezoid graphs and their coloring, *Discrete Applied Math.* **21**, 35–46.
Dahlhaus, E.
[1991] Chordale Graphen im besonderen Hinblick auf parallele Algorithmen, Habilitation thesis, Univ. Bonn.
Eschen, Elaine M., and Spinrad, J. P.
[1993] An $O(n^2)$ algorithm for circular-arc graph recognition, *Proc. 4th ACM-SIAM Symp. on Discrete Algorithms,* pp. 128–137.
Farber, M.
[1983] Characterizations of strongly chordal graphs, *Discrete Math.* **43**, 173–189.
Felsner, S., Müller, R., and Wernisch, L.
[1997] Trapezoid graphs and generalizations, geometry and algorithms, *Discrete Applied Math.* **74**, 13–32.
Fishburn, Peter C.
[1985] *"Interval Orders and Interval Graphs: A Study of Partially Ordered Sets"*, John Wiley & Sons, New York.
[1988] Interval orders and circle orders, *Order* **5**, 225–234.
Gabor, C., Supowit, K., and Hsu, W.
[1989] Recognizing circle graphs in polynomial time, *J. Assoc. for Comput. Mach.* **36**, 435–473.
Garey, M. R., Johnson, D. S., Miller, G. L., and Papadimitriou, C. H.
[1980] The complexity of coloring circular arcs and chords, *SIAM J. Alg. Discrete Meth.* **1**, 216–227.
Gattass, E. A. and Nemhauser, G. L.
[1981] An application of vertex packing to data analysis in the evaluation of pavement deterioration, *Operations Research Letters* **1**, 13–17.

Goldberg, A. V.
[1998] Recent developments in maximum flow algorithms, Technical Report 98-045, NEC Research Institute.

Goldberg, A. V., and Tarjan, R. E.
[1988] A new approach to the maximum flow problem, *J. Assoc. for Comput. Mach.* **35**, 921–940.

Goldberg, P. W., Golumbic, M. C., Kaplan, H., and Shamir, R.
[1995] Four strikes against physical mapping of DNA, *J. Comput. Biology* **3**, 139–152.

Golumbic, Martin Charles
[1984] Algorithmic aspects of perfect graphs, *Annals of Discrete Math.* **21**, 301–323.
[1988] Algorithmic aspects of intersection graphs and representation hypergraphs, *Graphs and Combinatorics* **4**, 307–321.
[1998] Reasoning about time, in *"Mathematical Aspects of Artificial Intelligence"*, F. Hoffman, ed., American Mathematical Society, *Proc. Symposia in Applied Math.*, vol 55, pp. 19–53.

Golumbic, Martin Charles, and Hammer, Peter L.
[1988] Stability in circular arc graphs, *J. of Algorithms* **9**, 314–320.

Golumbic, Martin Charles, and Jamison, Robert E.
[1985a] Edge and vertex intersection of paths in a tree, *Discrete Math.* **55**, 151–159.
[1985b] Edge intersection graphs of paths in a tree, *J. Combin. Theory B* **38**, 8–22.

Golumbic, Martin Charles, and Kaplan, Haim
[1998] Cell flipping in permutation diagrams, *Lecture Notes in Comput. Sci.* **1373**, Springer-Verlag, 577–586.

Golumbic, Martin Charles, Kaplan, Haim, and Shamir, Ron
[1994] On the complexity of DNA physical mapping, *Advances in Applied Math.* **15**, 251–261.
[1995] Graph sandwich problems, *J. of Algorithms* **19**, 449–473.

Golumbic, Martin Charles, and Laskar, Renu
[1993] Irredundancy in circular arc graphs, *Discrete Applied Math.* **44**, 79–89.

Golumbic, Martin Charles, and Mintz, Aviad
[1999] Factoring logic functions using graph partitioning, *Proc. IEEE/ACM Int. Conf. Computer Aided Design*, pp. 195–198.

Golumbic, Martin Charles, Mintz, Aviad, and Rotics, Udi
[2001] Factoring and recognition of read-once functions using cographs and normality, *Proc. 38th ACM Design Automation Conference*, pp. 109–114.

Golumbic, Martin Charles, and Monma, Clyde L.
[1982] A generalization of interval graphs with tolerances, *Proc. 13th Southeastern Conf. on Combinatorics, Graph Theory and Computing, Congressus Numerantium* **35**, Utilitas Math., Winnipeg, Canada, pp. 321–331.

Golumbic, Martin Charles, Monma, Clyde L., and Trotter, W. T.
[1984] Tolerance graphs, *Discrete Applied Math.* **9**, 157–170.

Golumbic, Martin Charles, Rotem, Doron, and Urrutia, Jorge
[1983] Comparability graphs and intersection graphs, *Discrete Math.* **43**, 37–46.

Golumbic, Martin Charles, and Scheinerman, Edward R.
[1989] Containment graphs, posets, and related classes of graphs, *Ann. N.Y. Acad. Sci.* **555**, 192–204.

Golumbic, Martin Charles, and Shamir, Ron
[1993] Complexity and algorithms for reasoning about time, *J. Assoc. for Comput. Mach.* **40**, 1108–1133.

Golumbic, Martin Charles, and Siani, Assaf
  [2002] Coloring algorithms for tolerance graphs: reasoning and scheduling with interval constraints, *Lecture Notes in Comput. Sci.* **2385**, Springer-Verlag, pp. 196–207.
Golumbic, Martin Charles, and Trenk, Ann N.
  [2004] *"Tolerance Graphs"*, Cambridge University Press.
Grötschel, M., Lovász , L., and Schrijver, A.
  [1981] The ellipsoid method and its consequences in combinatorial optimization, *Combinatorica* **1**, 169–197.
Hayward, Ryan B.
  [1985] Weakly triangulated graphs, *J. Combin. Theory B* **39**, 200–209.
Hayward, R. B., Hoàng, C. T., and Maffray, F.
  [1990] Optimizing weakly triangulated graphs, *Graphs and Combinatorics* **6**, 33–35. Erratum to *ibid*. **5**, 339–349.
Hertz, Alain
  [1995] Polynomially solvable cases for the maximum stable set problem, *Discrete Applied Math.* **60**, 195–210.
Hoffman, A. J., Kolen, A. W. J., and Sakarovitch, M.
  [1985] Totally-balanced and greedy matrices, *SIAM J. Alg. Discrete Meth.* **6**, 721–730.
Hougardy, S.
  [1998] Inclusions between classes of perfect graphs, Technical Report, Humbolt University, Berlin. Available at website http://www.informatik.hu-berlin.de/~hougardy/paper/classes.html (last checked April 2003).
Hsu, W. L.
  [1985] Maximum weight clique algorithms for circular-arc graphs and circle graphs, *SIAM J. Comput.* **14**, 224–231.
  [1995] $O(mn)$ algorithms for the recognition and isomorphism problems on circular-arc graphs, *SIAM J. Comput.* **24**, 411–439.
Hsu, W. L., and Spinrad, J. P.
  [1995] Independent sets in circular-arc graphs, *J. of Algorithms* **19**, 154–160.
Johnson, D. S. and McGeoch, C. C., eds.
  [1993] *"Network Flows and Matching: First DIMACS Implementation Challenge"*, American Mathematical Society, *DIMACS Series on Discrete Mathematics and Theoretical Computer Science*, vol. 12.
Johnson, D. S. and Trick, M. A., eds.
  [1996] *"Cliques, Coloring and Satisfiability: Second DIMACS Implementation Challenge"*, American Mathematical Society, *DIMACS Series on Discrete Mathematics and Theoretical Computer Science*, vol. 26.
King, V., Rao, S., and Tarjan, R. E.
  [1994] A faster deterministic maximum flow algorithm, *J. of Algorithms* **17**, 447–474.
Kloks, T.
  [1994] *"Treewidth. Computations and Approximations"*, *Lecture Notes in Comput. Sci.* **842**, Springer-Verlag.
Kloks, T., and Bodlaender, H.
  [1992] Testing superperfection of k-trees, *Lecture Notes in Comput. Sci.* **621**, Springer-Verlag, pp. 282–293.
Köhler, E.
  [2000] Recognizing graphs without asteroidal triples, *Lecture Notes in Comput. Sci.* **1928**, Springer-Verlag, pp. 255–266.

Langley, L.

[1995] Recognition of orders of interval dimension 2, *Discrete Applied Math.* **60**, 257–266.

Lengauer, T.

[1990] *"Combinatorial Algorithms for Integrated Circuit Layout"*, John Wiley, New York.

Lovász, L.

[1994] Stable sets and polynomials, *Discrete Math.* **124**, 137–153.

Lubiw, A.

[1987] Doubly lexical orderings of matrices, *SIAM J. Comput.* **16**, 854–879.

Ma, T. H., and Spinrad, J. P.

[1994] On the 2-chain subgraph cover and related problems, *J. of Algorithms* **17**, 251–268.

Mahadev, N. V. R., and Peled, U. N.

[1995] *"Threshold Graphs and Related Topics"*, North-Holland, Amsterdam.

McConnell, Ross M.

[2001] Linear time recognition of circular-arc graphs, *Proc. 42nd IEEE Symp. on Foundations of Computer Science,* pp. 386–394.

[2003] Linear time recognition of circular-arc graphs, *Algorithmica,* to appear.

McConnell, R. M., and Spinrad, J. P.

[1999] Modular decomposition and transitive orientation, *Discrete Math.* **201**, 189–241.

McKee, Terry A., and McMorris, Fred R.

[1999] *"Topics in Intersection Graph Theory"*, SIAM, Philadelphia.

Monma, C. L., and Wei, V. K.

[1986] Intersection graphs of paths in a tree, *J. Combin. Theory B* **41**, 141–181.

Narasimhan, G., and Manber, R.

[1992] Stability number and chromatic number of tolerance graphs, *Discrete Applied Math.* **36**, 47–56.

Natanson, A., Shamir, R., and Sharan, R.

[2001] Complexity classification of some edge modification problems, *Discrete Applied Math.* **113**, 109–128.

Pevzner, P.

[2000] *"Computational Molecular Biology"*, MIT Press, Cambridge, Mass.

Phillips, S., and Westbrook, J.

[1993] On-line load balancing and network flows, *Proc. 25th ACM Symp. on Theory of Computing,* pp. 402–411.

Prisner, E.

[1999] A journey through intersection graph county. Available at website http://www.math. uni-hamburg.de/spag/gd/mitarbeiter/prisner/Pris/Rahmen.html (last checked April 2003).

Ramirez-Alfonsin, J. L., and Reed, B., eds.

[2001] *"Perfect Graphs"*, Wiley, New York.

Reiterman, J., Rödl, V., Šiňajová, E., and Tůma, M.

[1985] Threshold hypergraphs, *Discrete Math.* **54**, 193–200.

Rival, I., ed.

[1985] *"Graphs and Order: The Role of Graphs in the Theory of Ordered Sets and Its Applications"*, Proc. NATO Advanced Institute on Graphs and Orders (Banff, Canada, May 18–31, 1984), D. Reidel Publishing, Dordrecht, Holland.

Scheinerman, E. R.

[1992] The many faces of circle orders, *Order* **9**, 343–348.

Scheinerman, E.R., and Wierman, J.

[1988] On circle containment orders, *Order* **4**, 315–318.

Shrwani, N.
  [1995] *"Algorithms for VLSI Physical Design Automation"*, Kluwer Academic Publishers, Boston.
Spinrad, J. P.
  [1985] On comparability and permutation graphs, *SIAM J. Comput.* **14**, 658–670.
  [1994] Recognition of circle graphs, *J. of Algorithms* **16**, 264–282.
Spinrad, J. P., Brandstädt, A., and Stewart, L.
  [1987] Bipartite permutation graphs, *Discrete Applied Math.* **18**, 279–292.
Spinrad, J. P., and Sritharan, R.
  [1995] Algorithms for weakly chordal graphs, *Discrete Applied Math.* **59**, 181–191.
Sysło, M. M.
  [1985] Triangulated edge intersection of paths in a tree, *Discrete Math.* **55**, 217–220.
Tarjan, R. E., and Yannakakis, M.
  [1984] Simple linear-time algorithms to test chordality of graphs, test acyclicity of hypergraphs and selectively reduce acyclic hypergraphs, *SIAM J. Comput.* **13**, 566–579.
Trotter, William T.
  [1992] *"Combinatorics and Partially Ordered Sets"*, Johns Hopkins, Baltimore.
Verbin, E.
  [2002] personal communication.
Waterman, Michael S.
  [1995] *"Introduction to Computational Biology"*, Chapman Hall, London.
Yannakakis, M.
  [1981] Computing the minimum fill-in is NP-complete, *SIAM J. Alg. Discrete Methods* **2**, 77–79.
  [1982] The complexity of the partial order dimension problem, *SIAM J. Alg. Discrete Methods* **3**, 351–358.

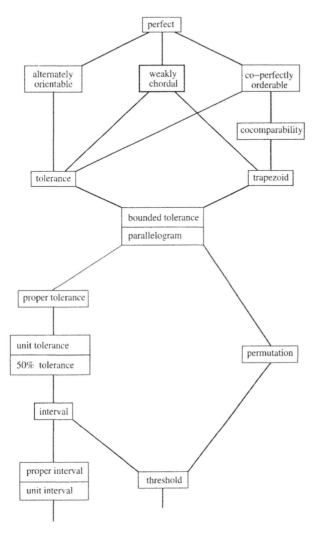

Fig.13.3. A complete hierarchy of classes of perfect graphs ordered by inclusion. Reprinted from Golumbic and Trenk [2004].

Index

Printed and bound by CPI Group (UK) Ltd, Croydon, CR0 4YY

03/10/2024

01040428-0009